APPLICATIONS OF HIGH-TC SUPERCONDUCTIVITY

Edited by **Adir Moysés Luiz**

Applications of High-Tc Superconductivity
Edited by Adir Moysés Luiz

Published by InTech
Janeza Trdine 9, 51000 Rijeka, Croatia

Publishing Process Manager Mirna Cvijic
Technical Editor Teodora Smiljanic
Cover Designer Jan Hyrat

Additional hard copies can be obtained from orders@intechweb.org

Applications of High-Tc Superconductivity, Edited by Adir Moysés Luiz
p. cm.
ISBN 978-953-307-308-8

For EU product safety concerns:
IN TECH d.o.o., Prolaz Marije Krucifikse Kozulić 3, 51000 Rijeka, Croatia,
info@intechopen.com or visit our website at intechopen.com.

Meet the editor

Adir Moysés Luiz is associate Professor at Instituto de Física at Federal University in Rio de Janeiro, (UFRJ) Brazil. He obtained doctor degree in Chemical Engineering (DSc.) at COPPE - "Universidade Federal do Rio de Janeiro", (UFRJ) in Brazil. The year of degree obtaining is 1982. The research areas which were always in the focus of his activities are: thermodynamics, solid state physics, materials science and superconductivity. His present interests are: superconductivity, superconducting materials, applications of superconductivity, superconducting devices, microscopic theories of superconductivity. Until now, Dr. Luiz has published 44 articles in scientific journals, 23 books and 62 articles in annals of events. He has also published 22 presentations of events (different congresses, conference or lectures), translated 19 books, and participated in 43 graduate boards. He has already been active as the Editor of 3 published books.

Contents

Preface XI

Chapter 1 **Overview of Possible Applications of High Tc Superconductors 1**
Adir Moysés Luiz

Chapter 2 **Some Contemporary and Prospective Applications of High Temperature Superconductors 15**
Z. Güven Özdemir, Ö. Aslan Çataltepe and Ü. Onbaşlı

Chapter 3 **Superconductivity Application in Power System 45**
Geun-Joon Lee

Chapter 4 **Current Distribution and Stability of a Hybrid Superconducting Conductors Made of LTS/HTS 75**
Yinshun Wang

Chapter 5 **Magnetic Relaxation - Methods for Stabilization of Magnetization and Levitation Force 97**
Boris Smolyak, Maksim Zakharov and German Ermakov

Chapter 6 **3-D Finite-Element Modelling of a Maglev System using Bulk High-Tc Superconductor and its Application 119**
Guang-Tong Ma, Jia-Su Wang, and Su-Yu Wang

Chapter 7 **Epitaxial Oxide Heterostructures for Ultimate High-Tc Quantum Interferometers 147**
Michael Faley

Chapter 8 **Thermophysical Properties of Bi-based High-Tc Superconductors 177**
Asghari Maqsood and M. Anis-ur-Rehman

Chapter 9 **Chemical Solution Deposition Based Oxide Buffers and YBCO Coated Conductors 193**
M. Parans Paranthaman

Chapter 10 **Superconducting Properties of Graphene Doped Magnesium Diboride 201**
Xun Xu, Wenxian Li, Xiaolin Wang and Shi-Xue Dou

Chapter 11 **Preparation of Existing and Novel Superconductors using a Spatial Composition Spread Approach 219**
Kevin C. Hewitt, Robert J. Sanderson and Mehran Saadat

Chapter 12 **Superhard Superconductive Composite Materials Obtained by High-Pressure-High-Temperature Sintering 237**
Sergei Buga, Gennadii Dubitsky, Nadezhda Serebryanaya, Vladimir Kulbachinskii and Vladimir Blank

Preface

The history of superconductivity is full of theoretical challenges and practical developments. Superconductivity was discovered in 1911 by Kamerlingh Onnes. About 75 years after this breakthrough, in 1986, it has been synthesized by Bednorz and Müller, an oxide superconductor with critical temperature (T_c) approximately equal to 35 K. This new breakthrough has given a tremendous impetus to this fascinating subject. Since this discovery, there are a great number of laboratories all over the world involved in research of superconductors with high T_c values, the so-called "high-T_c superconductors"(HTS). The discovery of a room temperature superconductor has been a long-standing dream of many scientists. The technological and practical applications of such discovery should be tremendous.

This book is a collection of works intended to study only practical applications of HTS materials. You can find here a great number of research on actual applications of HTS as well as possible future applications of HTS. Depending on the strength of the applied magnetic field, applications of HTS may be divided in two groups: large scale applications (large magnetic fields) and small scale applications (small magnetic fields).

In this book there are 12 chapters reporting fascinating studies about practical applications of HTS. In some chapters, you will also find many research on the synthesis of special materials that may be useful in practical applications of HTS.

The plan of this book is:

In chapters 1 and 2 are presented some interesting overviews about practical applications of HTS.

Chapter 3 contains a discussion concerning practical applications of superconductivity to electric power systems.

Chapter 4 is a discussion about current distribution and stability of a hybrid system containing a high-Tc superconductor and a low-Tc superconductor.

Chapter 5 presents the study concerning an important question of the stabilization of systems submitted to magnetic levitation forces.

Chapter 6 discusses a 3-D finite-element modeling of a maglev system.

Chapter 7 contains a research about quantum interferometers using high-Tc epitaxial oxide heterostructures.

Chapters 8, 9, 10, 11 and 12 are research about properties of high-Tc superconductors and experimental research about the synthesis of HTS materials with potential important applications.

The future of practical applications of HTS materials is very exciting. I hope that this book will be useful in the research activities of new radical solutions for practical applications of HTS materials and that it will encourage further experimental research of HTS materials with potential technological applications.

Adir Moysés Luiz
Instituto de Física, Universidade Federal do Rio de Janeiro
Brazil

1

Overview of Possible Applications of High Tc Superconductors

Adir Moysés Luiz
Instituto de Física, Universidade Federal do Rio de Janeiro
Brazil

1. Introduction

The history of high-T_c superconductors (HTS) begins in 1986 with the famous discovery of superconductors of the system Ba-La-Cu-O (Bednorz & Müller, 1986). Practical applications of superconductivity are steadily improving every year. However, the actual use of superconducting devices is limited by the fact that they must be cooled to low temperatures to become superconducting. For example, superconducting magnets used in most particle accelerators and in Magnetic Resonance Imaging (MRI) are cooled with liquid helium, that is, it is necessary to use cryostats that should produce and maintain temperatures of the order of 4 K. Helium is a very rare and expensive substance. On the other hand, because helium reserves are not great, the world's supply of helium can be wasted in a near future. Thus, because liquid nitrogen is not expensive and the reserves of nitrogen could not be wasted, it is important to use high-T_c superconductors cooled with liquid nitrogen. Superconductors with critical temperatures greater 77 K may be cooled with liquid nitrogen.

Copper oxide superconductors are the most important high-T_c superconductors (Cava, 2000). Up to the present time, after one hundred years of the first Kamerlingh Onnes discovery, the highest T_c is approximately equal to 135 K at 1 atm (Schilling & Cantoni, 1993), in superconductors of the Hg-Ba-Ca-Cu-O system. The discovery of a room temperature superconductor should trigger a great technological revolution. Nevertheless, in the meantime, waiting for this revolution, it is necessary to be prepared to apply existing technologies and develop new applications of HTS. The objective of this chapter is to give an overview of the most important applications of HTS. We shall discuss actual applications of HTS as well as possible applications of HTS in a near future.

Depending on the strength of the applied magnetic field, applications of HTS may be divided in two groups: large scale applications (large magnetic fields) and small scale applications (small magnetic fields).

Because HTS materials are brittle, the future of applications of HTS depends on the discovery of new radical solutions for this difficulty.

You will find in this chapter only discussions about practical applications of HTS. If you are interested in theoretical aspects of such applications, you may read a review book (Orlando & Delin, 1991).

The plan of this chapter is as follows:

The fabrication of HTS cables and coils are essential for all types of applications of HTS. Thus, in Section 2, we describe the state-of-the-art of the technology involved in the fabrication of cables, coils, electromagnets and magnets using HTS.
In Section 3 we study the most important projects involving large scale applications of HTS.
In Sections 4 and 5 we describe small scale applications of HTS. We claim that the most relevant small scale applications of HTS are applications of superconducting electronics, that is, the use of superconducting HTS devices in all types of electronic applications. Thus, in Section 5 we describe the researches involving applications of HTS in superconducting electronics.
In Section 6 some possible HTS applications in medicine are discussed.
Finally, in Section 7 concluding remarks are presented.

2. Uses of HTS in cables, coils, electromagnets and magnets

Because cables and coils are essential for all types of applications of HTS we begin the study of practical applications of HTS by this topic. It is well known that HTS are brittle materials. Thus, there is a technological difficulty to produce cables, tapes and coils using these materials. However the researches and developments in this area indicate that many solutions have been obtained and HTS equipments and devices will became commercially available in a near future.
It is well known that metals are appropriate to electric field screening. However, metals are not appropriate to magnetic field screening. One outstanding property of a superconductor is the capability of magnetic field screening. Thus, only superconductor coaxial cables and tapes can be used for the best electromagnetic screening. In a great number of small scale applications and in large scale applications of superconductivity it is very important to make electromagnetic screening. This is another possibility in HTS applications using cables and tapes with HTS materials. On the other hand, bulk HTS materials may also be used for this purpose.
The use of superconducting cables in high-voltage transmission lines is one of the most important applications of HTS materials. The performance of HTS cable depends on the quality of HTS tapes. HTS tapes for power transmission cables must be produced long enough to fulfill the required length of cable core to be installed. On the other side, it also must have sufficient critical current density and good mechanical characteristics.
Essential for the fabrication of coils, electromagnets and magnets is the development of new processes for the production of wires, cables and tapes using HTS. A study about the progress of researches for the production of wires, cables and tapes is available in chapters of a recent book (Polasek et al., 2009).

3. Possible large scale applications of HTS

Very high magnetic fields are involved in all possible large scale applications of superconductivity. Because HTS materials are type-II superconductors, it is crucial the use of HTS in the fabrication of coils, electromagnets and magnets.
The most important large scale applications of superconductivity are in: power transmission lines, energy storage devices, fault current limiters, fabrication of electric generators and motors, MAGLEV vehicles, in medicine (see Section 6) and applications in particle accelerators.

Now we discuss possible applications of HTS in the fabrication of electric generators and motors. The production of superconducting bearings is the crucial problem involved in the development of generators and motors. It is well known that a HTS material may levitate steadily above a magnet. The inverse position, that is, the levitation of magnets above superconductors is also stable (Davis et al., 1988). The stability of this levitation is due to the property of the magnetic flux quantization (see Section 4.1). Taking advantage of the capability of stable levitation of HTS materials it is possible to fabricate bearings for the development of generators and motors (Hull, 2000; Ma et al., 2003; Sotelo et al., 2009).

The development of an hydroelectric power generator has been successfully obtained (Fair et al., 2009).

In the next two sections we discuss the applications of HTS in energy storage devices, fault current limiters and applications in MAGLEV vehicles.

3.1 Fault current limiters and energy storage devices

It is well known that in electrical network, there are various faults produced by lightning, short circuits, etc. When these events occur, the current increases abruptly and there happens unexpected faults in the equipment, producing many damages, like fire and blackout. It is important to control these large currents for power system security. The objective of a Fault Current Limiter (FCL) is to limit very high currents in high speed when faults occur.

It seems that Superconducting Fault Current Limiters (SFCL) may provide the most promising solution of limiting the fault current in power systems. It is known that a superconductor has zero resistance when the current is lower than a certain critical current (I_c). If fault current exceeds I_c, superconductor becomes a normal conductor and this property may be used to design a SFCL.

An overview about the progress of the researches on high temperature superconductor fault current limiters is available in a review paper (Noe & Steurer, 2007).

Certainly energy storage devices are the most important equipments for energy conservation and ecological energy projects. The applications of solar energy, wind energy and other alternative energy sources, is limited by the fact that all these energies sources are intermittent. Thus, it is convenient to develop energy storage devices to storage these intermittent energies. HTS materials may be used in two important energy storage devices: in flywheels or in superconducting coils. The applications of HTS in flywheels is based on the use of HTS in superconducting bearings (see the end of the last section).

Because superconductors have zero resistance and considering the magnetic flux quantization rule, we conclude that the best method to storage energy is to maintain persistent currents in superconducting coils. Superconducting Magnetic Energy Storage (SMES) seems to be the best solution for energy storage projects.

A study about HTS energy storage devices is available in an article (Wolsky, 2002)

3.2 Applications of HTS in MAGLEV vehicles

The most relevant techniques for MAGnetic LEVitation (MAGLEV) vehicles are: (1) Electrodynamics Levitation (EDL), (2) Electromagnetic Levitation (EML), and (3) Superconductor Magnetic Levitation (SML).

EDL projects are based on Faraday-Lenz law: when a magnetic flux changes in he neighborhood of a conductor, a current is induced in the conductor. Superconductor magnets are maintained inside the train. There is an experimental project in Japan with two railway tracks between Osaka and Tokyo and the train based on this technique has reached a record speed of 582 km/h.

EML projects are based on the attractive force between an electromagnet and a ferromagnetic material. In this case it is not necessary to use superconductor magnets. It is well known that the levitation due to the force between an electromagnet and a ferromagnetic material is not stable and so it is necessary to use stabilization systems. There is a commercial train using this technique in China and a railway line with 30 km is used to transport people between Shanghai International Airport and Shanghai Lujiazui.

SML projects are based on the perfect diamagnetism of superconductors. It is well known that a HTS material may levitate steadily above a magnet. Conversely, the levitation of magnets above superconductors is also stable (Davis et al., 1988). Because HTS are type II superconductors, the magnetic flux exclusion (Meissner effect) is partial. Inside a type II superconductor there are Abrikosov vortices. A magnetic field may be maintained inside an Abrikosov vortice. Thus, the stability of this type of levitation is due to the property of the magnetic flux quantization (see Section 4.1). SML projects take advantage of this property. Thus, SML levitation is more stable than EDL and EML levitations.

Considering the above mentioned property, we conclude that a simple SML project for MGLEV vehicles is as follows. Permanent magnets may be used in the tracks and blocks of HTS materials may be used inside the train. The levitation and the motion of the vehicle is due to the magnetic repulsive force between the track and the train. There are some projects of application of HTS materials and permanent magnets in MAGLEV trains using this SML technique (David et al., 2006; Stephan et al., 2008; Sotelo et al., 2010).

4. Possible small scale applications of HTS

The most important small scale superconducting devices fall into two basic classes: (a) SQUID systems, which are designed to measure magnetic flux and other electromagnetic measurements, and (b) Josephson devices which take advantage of the electromagnetic characteristics of Josephson junctions to perform traditional electronic functions. We have divided the study of small scale applications in these two classes, but we emphasize that SQUIDs are fabricated using Josephson junctions as well. A collection of works about SQUIDs, Josephson junctions and other superconducting devices is available in a review book (Ruggiero & Rudman, 1990).

4.1 Magnetometers and other devices based on SQUIDs

It is well known that Superconducting QUantum Interference Devices (SQUIDs) are the most sensitive detectors of magnetic flux available. Basically, a SQUID is a flux-to-voltage transducer, providing an output voltage proportional to the magnetic flux.

SQUIDs combine two physical phenomena: flux quantization and tunneling (Josephson, 1962). Magnetic flux quantization is the most important macroscopic property of the superconducting state. Consider a closed loop in the bulk of a superconductor. It is known that quantum mechanics must be applied for the superconducting state. Applying the Bohr-Sommerfeld quantization rule to this loop we may write:

$$\oint \vec{p}.d\vec{l} = nh \tag{1}$$

where p is the linear momentum, dl is a line element, n is an integer and h is Planck's constant. The canonical momentum is given by:

$$p = mv + qA \tag{2}$$

where m is the mass, v is the velocity, q is the charge and A is the magnetic potential vector. Considering $v = 0$ in the bulk of the material, by equations (1) and (2) we have

$$\oint q\vec{A}.d\vec{l} = nh \tag{3}$$

Using the rotational theorem in equation (3) we find

$$\iint rot\ \vec{A}.\ d\vec{S} = nh / q \tag{4}$$

where dS is an area element. We know that $B = rot\ A$ and $q = 2e$ (Cooper pair). Thus, by equation (4) we have

$$\Phi = \iint \vec{B}.\ d\vec{S} = nh / 2e \tag{5}$$

Equation (5) is the flux quantization rule, that is, the magnetic flux Φ must be quantized in a superconducting loop according to the rule: $\Phi = n\ \Phi_0$, where Φ_0 is a quantum of magnetic flux:

$$\Phi_0 = nh/2e = 2{,}07 \times 10^{-15}\ \mathrm{Wb} \tag{6}$$

A SQUID is, in essence, a superconducting closed loop containing one or two Josephson junctions. Taking advantage of the flux quantization rule, it is possible to measure a very small magnetic flux of the order assigned in equation (6). On the other hand, because a SQUID is a flux-to-voltage transducer, providing an output voltage proportional to the magnetic flux, it is possible to measure quantities smaller than 10^{-15} Wb. By this reason, we conclude that SQUIDs are the most sensitive system for magnetic flux measurements. We conclude also that instruments based on SQUIDs are the most appropriate to be used in very high precision electric and magnetic measurements.

There are two kinds of SQUIDs: (a) dc SQUID and (b) rf SQUID. A dc SQUID consists of two Josephson junctions connected in parallel in a closed loop; it operates with a steady current bias (dc bias). The rf SQUID involves a single Josephson junction interrupting the current flow around the superconducting loop and it is operated with a radiofrequency bias.

Because it required only a Josephson junction, the rf SQUID was simpler to manufacture and became commercially available. However, in the mid of the 1970 decade, it was shown that dc SQUID is more sensitive than rf SQUID. Since then, there has been great developments of dc SQUIDs. By contrast, there has been little developments of rf SQUIDs in the last decades. Only low-Tc superconductors have been used in commercially available SQUIDs until 1988. However, in the last two decades HTS have been used in SQUIDs.

Because the tremendous sensitivity to magnetic flux, low-Tc and HTS SQUIDs remain the most practical ultra-sensitive magnetic field detectors. Thus, SQUID systems may be projected for a number of practical applications: submarine detection and relative motion magnetic field detectors, mineral surveying, medical diagnostics, and so on. On the other hand, with proper circuitry design, SQUID systems may be projected for a great number of

scientific instruments. A number of HTS SQUIDs have been projected in the last two decades. There is also advances in HTS thin-film SQUIDs (Koch et al., 1987).
There are many works about applications of HTS in SQUIDs. We list some of these works (Zimmerman et al., 1987; Golovashkin et al., 1989; Mankiewich et al., 1988).

4.2 Devices based on Josephson junctions

We study now the most important small scale superconducting devices based on Josephson junctions. For practical applications of Josephson effects there are two types of Josephson junctions: (a) Superconductor - Insulator - Superconductor (SIS) junction and (b) Superconductor - Normal - Superconductor (SNS) junction. SIS junctions are also known as tunneling junctions because it occurs tunneling of Cooper pairs from one superconductor to the other trough the insulator barrier. The tunneling of Cooper pairs was predicted by Josephson in 1962 (Josephson, 1962).
In the case of SNS junctions there is no insulator barrier, there are only two SN interfaces. Thus, it is easy to conclude that the current - voltage characteristic curve of a SIS junction should be completely different from the current - voltage characteristic curve of a SNS junction.
Interesting studies about Josephson effects and Josephson junctions may be found in review books (Barone & Paternò, 1982; Likharev, 1986).
A theoretical prediction of the current - voltage characteristic curve of a SNS junction has been successfully obtained (Kummel et al., 1990).
It is important to note that the current - voltage characteristic curve of a SNS junction exhibits a negative resistance region (Kummel et al., 1990). Taking advantage of this negative resistance region, two terminal devices based on SNS junctions may be projected for a great number of applications in superconducting electronics (Luiz & Nicolsky, 1991). In the next section we shall study such possible applications.

5. Applications of HTS in superconducting electronics

In Section 4 we have stressed that SQUIDs are fabricated using Josephson junctions. On the other hand, Josephson junctions are used directly in a great number of small scale applications of superconductivity. Thus, to study applications of HTS materials in superconducting electronics it is necessary to describe the properties and capabilities of Josephson junctions.
We claim that SNS junctions are more appropriate than SIS junctions for HTS small scale applications of superconductivity. This conclusion is based on the following comparison of 4 characteristics:

1. It is well known that in a SIS junction there is a very thin insulator between the two superconductors of the SIS junction. To occur tunneling, it is necessary that the thickness of the insulator layer should be of the order of the coherence length of the superconductor layer. The coherence length of a HTS is about 1000 times greater than the order of magnitude of the coherence length of a low-Tc metallic superconductor. For example, in a HTS material of the system Bi-Sr-Ca-Cu-O, the coherence length is approximately equal to 1 angstrom (10^{-10} cm) along the c-axis and approximately equal to 40 angstroms in the transverse direction (Davydov, 1990). Compare this value with the (isotropic) coherence length of a metallic superconductor which is of the order of 1000 to 10000 angstroms. It is known that it is not ease to make a SIS junction because the difficulties of fabrication of very thin layers of insulators. Thus, in the case of a SIS junction made with HTS this drawback is very enhanced.

In the case of a SNS junction there is no insulator barrier, no tunneling occurs in the SN interfaces, thus the above mentioned difficulties are not present in the fabrication of SNS junctions.

2. There is another important reason to use SNS junctions (instead of SIS junctions) in all possible applications of small scale applications of superconductivity using HTS. Generally a SIS junction is very small. To enhance the performance of a SIS junction it should be necessary to use arrays of a great number of SIS junctions. By the above mentioned reasons, to make arrays of SIS junctions is a very difficult task.
   However, because a SNS junction is a normal metal region between two superconductors, a SNS junction may have macroscopic dimensions. It is sufficient to make a constriction in a bulk superconductor to obtain a SNS junction. On the other hand, the so called microbridge may be actually realized with macroscopic dimensions. Consider a certain great current flowing in a HTS. Consider a constriction in this material. In the constriction, the current density increases. If the current density is greater than the critical current density of the HTS material considered, the constriction becomes normal and the system becomes a SNS junction. An important example of a SNS junction obtained with a HTS material (YBCO) is available (Alvarez et al., 1990).
3. It is well known that the current - voltage characteristic curve of a SIS junction exhibits hysteresis. However, it has been shown that in the current–voltage characteristic curve of a SNS junction there is no hysteresis (Kummel et al., 1990). Because in a great number of applications in superconducting electronics it is necessary to use devices without hysteresis, we conclude that, for those applications, SNS junctions are more appropriate than SIS junctions.
4. At last, we may compare the equivalent circuit of a SIS junction with the equivalent circuit of a SNS junction. Because there is an insulator barrier in a SIS junction, the equivalent capacitance of a SIS junction is greater than the equivalent capacitance of a SNS junction. Because in a great number of applications it is necessary to use low equivalent capacitances, it is obvious that, for those applications, SNS junctions are more appropriate than SIS junctions.
   In the past 50 years, the development of *semiconductor electronics* have produced a great technological revolution. With each generation of integrated circuits, the semiconductor devices became smaller, more complex and faster. However, the clock rate of semiconductor devices used in electronics has saturated around 5 GHz. The speed of the processors and all the devices of semiconductor electronics will soon reach a limit of this order of magnitude. One reason for this limit is not the switching speed of the transistors, but is due to power dissipation.

What is *superconducting electronics*? We may say that *superconducting electronics* is a new type of electronics based on superconducting devices.

There are two possible improvements in the traditional semiconductor electronics taking advantage of superconducting devices: (a) hybrid electronic systems, that is, systems containing semiconductors and superconductors, and (b) complete *superconducting electronics*, that is, electronic systems containing only superconducting devices, without semiconductor devices. A study about the state-of-the-art and future developments of superconducting electronics is available in a review article (Anders et al., 2010).

Until now, the most reasonable improvement in the performance of the traditional semiconductor electronics seems to be provided by hybrid electronic systems containing semiconductors and superconductors. We know that traditional semiconductor electronics has been the most reliable and modern technology in the past 50 years. However, the speed limit mentioned above is a fundamental difficulty in the further development of this technology. The prime reason for that limit is explained by Joule's law: $Q = RI^2$, where Q is the heat loss, R is the resistance and I is the current. The heat loss in the metallic interconnections can be avoided if superconducting interconnections could be used. In this case the speed of the processors and other devices should be increased.

In the above mentioned improvement in the traditional semiconductor electronics, we give an example of a solution involving an hybrid semiconductor-superconductor system.

Now we discuss the second possibility: a complete *superconducting electronics*, that is, electronic systems containing superconducting devices, without semiconductor devices. In the following sections we discuss this possibility.

5.1 Generators, amplifiers, mixers, detectors switches and thin-film filters using HTS materials

The most important electronic devices are generators, amplifiers, mixers, detectors and switches.

Superconducting devices based on SIS junctions and SNS junctions may be projected to substitute these and other semiconductor devices.

In this section and in the next 3 sections we discuss the possible use of superconducting devices in order to substitute semiconductor devices.

We have pointed out in the previous section that SNS junctions are more appropriate than SIS junctions in the prospective applications of Josephson junctions in superconducting electronics.

Combining a SNS junction with appropriate resonant circuits, it is possible to project many types of generators (Luiz & Nicolsky, 1990; Luiz & Nicolsky, 1991; Nicolsky & Luiz, 1992).

Taking advantage of the negative resistance region of SNS junctions, two-terminal devices based on SNS junctions may also be used to design electronic switches (Luiz, 1993; Luiz & Nicolsky, 1993).

On the other hand, using this same property of SNS junctions, it is possible to design mixers and detectors (Gorelov et al., 1997; Luiz et al., 1997; Luiz et al., 1999).

Signal amplification and harmonic generation may be obtained using SNS junctions with appropriate circuits (Luiz et al., 1998; Luiz et al., 1999).

Terahertz oscillations have also been obtained using HTS Josephson junctions (Güven et al., 2009; Minami et al., 2009; Machida & Tachiki, 2001).

In high frequency ranges up to 100 – 500 GHz the surface resistance of HTS like $YBa_2Cu_3O_7$ is so law that it becomes commercially interesting to build thin-film filters and resonators with quality factors of the order of 10^6.

Telecommunication applications of HTS are specially useful in the cellular phone market. For example, hundreds of superconducting filters have been installed in the USA in critical base stations for cellular phone communications (Anders et al., 2010).

5.2 Digital signal processing and analog signal processing

In the previous section we have pointed out that SNS junctions may be used for switching circuits and other superconducting electronic devices. The very high switching speeds that may be obtained using superconducting switching circuits suggests that wideband signal processing is an interesting possible application of HTS materials. A discussion about the possibility of applications of superconductor devices in digital signal processing is available in a review article (Van Duzer & Lee, 1990).

On the other hand, the use of other types of superconducting circuits is also possible in analog signal processing and in analog-to-digital converters. A discussion about the possibility of applications of superconductor devices in wideband analog signal processing is available in review articles (Clarke, 1988; Withers, 1990). Other reviews are available in a book (Van Duzer & Turner, 1998).

5.3 Three-terminal devices using HTS

In Section 5.1 we have described the possibility of applications of two-terminal superconducting devices based on SNS junctions for a number of applications o HTS in superconducting electronics. However, it is also feasible to use three-terminal superconducting devices in applications of HTS in superconducting electronics.

The most important three-terminal superconducting devices are superconducting transistors. A study about the characteristics and the performance of HTS transistors is available in a review article (Mannhart, 1996).

5.4 Digital computer, quantum computer and flux qubit

An exciting application of superconducting electronics should be provided by the possible applications of HTS in digital computers.

The most important components of a digital computer are memory units and arithmetic units.

The metallic interconnections of the traditional semiconductor digital computer should be substituted by superconducting interconnections.

The memory ferromagnetic units of the traditional digital computer should be substituted by superconducting memories containing superconducting loops. In Section 4.1 (equation 5), we have emphasized that in superconductors we must apply the flux quantization rule, that is, the magnetic flux Φ must be quantized in a superconducting loop according to the rule: $\Phi = n\ \Phi_0$, where Φ_0 is a quantum of magnetic flux. Thus, using appropriate circuits, the ferromagnetic memories of the traditional computers may be substituted by superconducting memories containing superconducting loops.

Arithmetic units are based on the action of transistors and other semiconductor devices. These arithmetic units may be substituted by superconducting devices described in Sections 5.1, 5.2 and 5.3. Therefore we conclude that digital superconducting computers will be feasible in a near future. The overall speed of a superconducting computer should be up to 1000 orders of magnitude greater than the speed of a traditional computer (Anders et al., 2010).

Superconducting digital technology is based on the Rapid Single Flux Quantum (RSFQ) logic. Another application of RSFQ is in superconducting **quantum bits (qubits)**. The *quantum computer* is based on qubit operations. Classical **bits** are used in traditional computers. However, in a quantum computer, the quantum bit (qubit) may carry two quantum states at the same time. Quantum mechanical phenomena such as quantum superposition, quantum entanglement and other quantum mechanical properties are the concepts involved in a *quantum computer*. Interesting discussions about flux qubits are found in the literature (Chiorescu et. al., 2003; Clarke & Wilhelm, 2008).

6. Possible applications of HTS in medicine

The ultimate objective of science and technology is human welfare. Thus, it is natural to ask how superconductivity may be applied in medicine. Medical applications of HTS involve small scale as well as large scale applications of superconductivity.

The most important large scale applications of superconductivity are Magnetic Resonance Spectroscopy (MRS) and Magnetic Resonance Imaging (MRI).

The most important small scale applications of superconductivity are those applications based on the properties of SQUIDs and Josephson junctions (Sections 4 and 5). We have pointed out that SQUIDs are the most sensitive devices for magnetic field measurements. It is well known that blood contains ions. Therefore, the circulation of blood produces small magnetic fields that can be detected using SQUIDs.

By measurement of the magnetic fields produced by blood circulation in the human body it is possible to make non-invasive diagnosis of diseases. The most important applications of HTS in medicine are (1) magnetoencephalography (MEG) for non-invasive tests of the brain activity and (2) magnetocardiography (MCG) for non-invasive tests of the heart activity. The magnetic activity of other regions of the human body may also be detected using SQUIDs.

A study about the state-of-the-art and future developments of applications of superconductivity in medicine is available in a review article (Anders et al., 2010).

7. Concluding remarks

In this chapter we have studied the most relevant questions about the possible applications of HTS. The history of superconductivity has not been smooth. Generally, very slow process was witnessed between breakthroughs. Practical applications of superconductivity follows a breakthrough with a time lag of about 30 years.

Practical applications of HTS are emerging steadily every year. However, as we have stressed in the Introduction, radical technological solutions should depend on the discovery of a HTS material with critical temperature in the neighborhood of room temperature. However, what happens to the basic science of HTS? As it has been noted in a recent book (Luiz, 2010), the microscopic mechanisms in HTS are unclear. However, nearly every year new theories are proposed and new HTS materials are synthesized.

Large scale applications of HTS have a bright future. Electric energy production and energy storage are the most important large scale applications of HTS. On the other hand, small scale applications of HTS have a bright future as well; these applications are more feasible than large scale applications of HTS. Because small scale applications of HTS generally involve small volumes, that is, very small Josephson junctions volumes, it is not necessary to use cryostats with liquid helium (or liquid nitrogen). In some very small systems, it is sufficient to use a thermodynamic cycle to maintain the temperature of the system at a value lower than the critical temperature of the material. On the other hand, experimental evidences show that HTS behave like a stack of superconductor-insulator-superconductor interfaces (SIS junctions) and superconductor-normal-superconductor interfaces (SNS junctions), that is, HTS materials may be considered as a network of intrinsic Josephson junctions (Kleiner & Müller, 1994; Machida & Tachiki M, 2001).

Finally, according to the study in Section 5, we claim that applications of HTS materials and SNS junctions should give new radical solutions for the use of superconducting devices in superconducting electronics.

The future of applications of HTS is very exciting. Because HTS materials are brittle, it is necessary to overcome this difficulty. Nevertheless, this drawback exists only in the fabrication of cables and coils (see Section 2). However, in a great number of HTS applications it is sufficient to use HTS cylindrical blocks. We hope that the researches described in this overview will be helpful for the discovery of new practical applications of HTS.

8. References

Alvarez, G.; Taylor, K. N. R. & Russell, J. G. (1990). Josephson behavior of variable thickness bridges in textured $YBa_2Cu_3O_7$. *Physica C*, 165, pp. 258-264

Anders, S.; Blamire, M.G.; Buchholz, F.-Im; Crété, D.-G.; Cristiano, R.; Febvre, P.; Fritzsch, L.; Herr, A.; Il'ichev, E.; Kohlmann, J.; Kunert, J.; Meyer, H.-G.; Niemeyer, J.; Ortelepp, T.; Rogalla, H.; Schurig, T.; Siegel, M.; Stolz, Tarte, E.; Brake, H.J.M.ter; Toepfer, H.; Villegier, J.-C.; Zagoskin, A.M. & Zorin, A.B. (2010). European roadmap on superconductive electronics - status and perspectives. *Physica C*, 470, 23-24, pp. 2079-2126

Barone, A. & Paternò, G. (1982). *Physics and Applications of the Josephson Effect*. John Wiley, New York

Bednorz, J. G. & Müller, K. A. (1986). Possible high T_c superconductivity in the Ba-La-Cu-O system. *Zeitschrift fur Physik B, Condensed Matter*, 64, 2, pp. 189-193

Cava, R. J. (2000). Oxide superconductors. *Journal American Ceramic Society*, 83, 1, pp. 5-28

Chiorescu, I.; Nakamura, Y.; Harmans, C. J. P. M. & Mooij, J. E. (2003). Coherent quantum dynamics of a superconducting flux qubit. *Science*, vol. 299, n. 5614, pp. 1869-1871

Clarke, J. (1988). Small-scale analog applications of high-transition-temperature superconductors. *Nature*, vol. 333, n. 5, pp. 29-35

Clarke, J. & Wilhelm, F. K. (2008). Superconducting quantum bits. *Nature*, vol. 453, n. 7198, pp. 1031-1042

Davis, L. C.; Logothetis, E. M. & Soltis, R. E. (1988). Stability of magnets levitated above superconductors. *J. Appl. Phys.*, Vol. 64, No. 8, pp. 4212–4218

David, E. G.; Stephan, R. M.; Andrade Jr, R. & Nicolsky, R. (2006). Feasibility study of an HTS-MAGLEV line at the Federal University of Rio de Janeiro. In *Proc. of MAGLEV 2006*, Dresden, Germany, pp. 749-752

Davydov, A. S. (1990). Theoretical investigation of high-temperature superconductivity. *Physics Reports (Review Section of Physics Letters)*, 190, 4-5, pp. 191-306.

Fair, R.; Lewis, C.; Eugene, J. & Ingles, M. (2009). Development of an hydroelectric power generator for the Hirschaid Power Station, Germany. In EUCAS - 2009, September 13-17, Dresden, Germany, p. 124

Golovashkin, A. I.; Gudkov, A. L.; Krasnosvobodtsev, S. I.; Kusmin, L. S.; Likharev, K. K.; Maslennikov, Yu. V.; Pashkin, Yu. A.; Petchen, E. V & Snigirev, O. V. (1989). Josephson effect and macroscopic quantum interference in high-Tc superconducting thin film weak links at T= 77 K. Report presented at 1988 Applied Superconductivity Conference. *IEEE Trans. Magn.*, 25, pp. 943-946

Gorelov, Y. A.; Luiz, A. M. & Nicolsky, R. (1997) Heterodyne Mixer Using a superconductor - normal metal - superconductor junction. *Physica C*, vol. 287, pp. 2491 - 2492

Güven Ö., Z.; Aslan, Ö. & Onbaşlı, Ü. (2009). Terahertz oscillations in mercury cuprate superconductors. *Pramana-Journal of Physics*, vol. 73, n. 4, pp. 755-763

Hull, J. R. (2000). Superconducting bearings. *Supercond. Sci. Technol.*, vol. 13, pp. R1–R15

Josephson, B. D. (1962). Possible new effects in superconductive tunneling. *Phys. Lett.*, 1, pp. 251-253

Kleiner, R. & Müller, P. (1994). Intrinsic Josephson effects in high-T_c superconductors, *Phys. Rev. B*, vol. 49, n. 2, pp. 1327-1341, ISSN: 1098-0121

Koch, R. H.; Umbach, C. P.; Clark, G. J.; Chaudhary, P. & Laibowitz, R. B. (1987). Quantum interference devices made from superconducting oxide thin films. *Appl. Phys. Lett.*, 51, pp. 200-202

Kummel, R.; Gunsenheimer, U. & Nicolsky, R. (1990). Andreev scattering of quasiparticle wave packets and current - voltage characteristics of superconducting metallic weak links. *Physical Review,* B42, pp. 3992-4009

Likharev, K. K. (1986). *Dynamics of Josephson Junctions and Circuits.* Gordon and Breach, New York

Luiz, A. M. (1993). Superconducting negative resistance switches. *Japanese J. of Applied Physics*, vol. 32, n. 11A, pp. 4971-4972

Luiz, A. M. (2010). A model to study microscopic mechanisms in high-Tc superconductors, *Superconductor*, Adir Moysés Luiz (Ed.), ISBN: 978-953-307-107-7, Sciyo, Available from the site: http://www.intechopen.com/articles/show/title/a-model-to-study-microscopic-mechanisms-in-high-tc-superconductors

Luiz, A. M.; Soares, V. & Nicolsky, R. (1999). Superconductor - Normal Metal - Superconductor junctions for signal amplification and harmonic multiplication. *IEEE Transactions on Magnetics,* vol. 35. pp. 4100-4102

Luiz, A. M.; Gorelov, Y. A. & Nicolsky, R. (1999). Simulation of conversion gain and reflectivity coefficients in heterodyne detector using a superconductor - normal metal - superconductor junction. *IEEE Transactions on Applied Superconductivity,* vol. 9. pp. 44-48

Luiz, A. M.; Soares, V. & Nicolsky, R. (1998). Simulations of signal amplification and oscillations using a SNS junction. *Journal de Physique IV France,* vol. 8, pp. 271 - 274

Luiz, A. M.; Pereira, L. A. A. & Nicolsky, R. (1997). Heterodyne detector using a SNS junction. *IEEE Transactions on Applied Superconductivity,* vol. 7, n. 2, pp. 3719 – 3721

Luiz, A. M. & Nicolsky, R. (1993). Negative Resistance Switch Using a SNS Junction. *IEEE Transactions on Applied Superconductivity,* vol. 3, pp. 2714-2715

Luiz, A. M. & Nicolsky, R. (1991). Microbridges and point contacts as negative differential resistance devices in the conventional generation of microwaves, up to the sub-millimeter range. *Physica C,* vol. C189, pp. 2589-2591

Luiz, A. M. & Nicolsky, R. (1991). Radiofrequency generation using a SNS microbridge. *IEEE Transactions on Magnetics*, vol. 27, n. 2, pp. 2712 - 2715

Luiz, A. M. & Nicolsky, R. (1991). Microwave generation using a SNS junction. *Japanese Journal of Applied Physics*, 30, pp. 1218-1219

Luiz, A. M. & Nicolsky, R. (1990). SNS microbridge as a superconducting harmonic generator. *Progress in High Temperature Superconductivity*, vol. 25. Editor: Roberto Nicolsky, World Scientific, Singapore, pp. 733-736

Ma, K. B.; Postrekhin, Y. V. & Chu, W. K. (2003). Superconductor and magnet levitation devices. *Rev. Sci. Instrum.*, vol. 74, n. 12, pp. 4989–5017

Machida M. & Tachiki M. (2001). Terahertz electromagnetic wave emission by using intrinsic Josephson junctions of high-Tc superconductors. *Current Appl. Phys.*, vol. 1, n. 4-5, 341-348, ISSN: 1567-1739. Editors: Orlando, T. P. & Delin, K. A. (1991). *Foundations of Applied Superconductivity*, Addison-Wesley Pub. Co., New York.

Mankiewich, P. M.; Schwartz, D. H.; Howard, R. E.; Jackel, L. D.; Straughn, B. L.; Burkhardt, E. G. & Dayem, A. H. (1988). Fabrication and characterization of a $YBa_2Cu_3O_7$/Au/ $YBa_2Cu_3O_7$ S-N-S microbridge. Fifth International Workshop on Future Electron Devices - High Temperature Superconducting Devices, June 2 - 4, pp. 157-160, Japan

Mannhart, J. (1996). High-Tc transistors. *Supercond. Science Technol.*, 9, pp. 49-67

Minami, H.; Kakeya, I. ; Yamaguchi, H.; Yamamoto, T. & Kadowaki, K. (2009). Characteristics of terahertz radiation emitted from the intrinsic Josephson junctions in high-*Tc* superconductor $Bi_2Sr_2CaCu_2O_{8+\delta}$. *Appl. Phys. Lett.*, vol. 95, n. 23, pp. 1- 3

Nicolsky, R. & Luiz, A. M. (1992). Superconducting metallic Josephson junctions as negative resistance devices in the conventional generation of microwaves. *Proceedings of the Asiatic-Pacific Microwave Conference*, Adelaide, Australia, pp. 15-18

Noe, M. & Steurer, M. (2007). High-temperature superconductor fault current limiters: concepts, applications, and development status. *Supercond. Science Technol.*, 20, pp. R15-R29

Polasek, A.; Serra, E. T. & Rizzo, F. C. (2009). On the melt processing of Bi-2223 high-Tc superconductor - challenges and perspectives. *In Superconducting Magnets and Superconductivity*. Editors: Tovar, H. & Fortier, J. Chapter 5, pp. 1-16, Nova Science Publishers, Inc., New York

Ruggiero, S. T. & Rudman, D. A. (Editors) (1990). *Superconducting Devices*, Academic Press, Inc., New York

Schilling, A. & Cantoni, M. (1993). Superconductivity above 130 K in the Hg-Ba-Ca-Cu-O system. *Nature*, 363, 6424, pp. 56-58

Stephan, R. M.; David, E. G. & de Hass, O. (2008). MAGLEV COBRA: An Urban transportation solution using HTS superconductors and permanent magnets. In *Proc. of MAGLEV 2008*, San Diego, California, pp. 1-4

Sotelo, G. G.; Dias, D. H. N.; Andrade Jr, R. & Stephan, R. M. (2010). Tests on a superconductor linear magnetic bearing of a full-scale maglev vehicle. *IEEE Trans. Appl. Supercond.*, doi: 10.1109/TASC.2010.2086034, pp. 1-5

Van Duzer, T. & Lee, G. (1990). Digital signal processing. In *Superconducting Devices*, Editors: Ruggiero, S. T. & Rudman, D. A., Academic Press, Inc., New York

Van Duzer, T. & Turner, G. (1998). *Principles of Superconductive Devices and Circuits.* Prentice Hall, New Jersey

Withers, R. S. (1990). Wideband analog signal processing. In *Superconducting Devices*, Editors: Ruggiero, S. T. & Rudman, D. A., Academic Press, Inc., New York

Wolsky, A. M. (2002). The status and prospects for flywheels and SMES that incorporate HTS. Physica C 372–376, pp. 1495–1499

Zimmerman, J. E.; Beall, J. A.; Cromar, M. W. & Ono, R. H. (1987). Operation of a Y-Ba-Cu-O RF SQUID at 81 K. *Appl. Phys. Lett.*, 51, pp. 617-618

2

Some Contemporary and Prospective Applications of High Temperature Superconductors

Z. Güven Özdemir[1], Ö. Aslan Çataltepe[2] and Ü. Onbaşlı[3]
[1]Yıldız Technical University, Physics Department, Davutpaşa Cad. Esenler, İstanbul, 34210
[2]Anatürkler Educational Consultancy & Trading Company, Cemil Topuzlu Cad. Üçem Konak No:73 D:1 Göztepe,Kadıköy
[3]Marmara University, Physics Department, Rıdvan Paşa Cad. 3. Sok. 4/12 Göztepe, Istanbul, Turkey

1. Introduction

High temperature superconductors (HTS) have a wide range of very sensitive and reliable advanced technological applications. In this chapter, some examples of contemporary and prospective usage of the superconductors such as in vivo living body measurements in medicine, terahertz equipments for security systems, quantum bit namely "qubit" applications in quantum computation and bolometers for some space investigations will be dealt with.

Especially in medicine, superconductors have been reliably utilized in Magnetic Resonance Imaging (MRI), Magnetic Resonance Spectroscopy (MRS), magnetoencephalography (MEG) and magnetocardiography (MCG) for both analysis of magnetic activity of different regions of the human body such as brain and heart's wave activities and very early diagnosis of several diseases. All equipments mentioned above contain Superconducting Quantum Interference Device (SQUID), which is based on the Josephson Effect. SQUID is a very sensitive magnetic detector to determine the change of the magnetic flux in material media of the order of 10^{-15} Tm^2, which coincides with the order of magnetic flux quanta, $\Phi_0 = 2.0678x10^{-15}\ Tm^2$. The sensitivity of SQUID is revealed easily by remembering the fact that the magnetic field of the Earth equals to $5x10^{-5}$ Tesla.

Besides these contemporary applications in medicine mentioned above, the Proton-MRS (P-MRS) measurements will be proposed for forensic science investigations as one of the prospective implementations of HTS's. Because of the fact that some in vivo investigations via P-MRS provide the facility to detect the minor changes of metabolites in human brain (Onbaşlı et al., 1999), the method, which enables to detect the mild head injuries, which cannot be recognized according to Glasgow Coma Scale (GCS) (Teasdale & Jennett, 1974) from the neurological point of view, will be presented in this chapter.

Moreover, one of the remarkable features of HTS is that some oxide layered HTS work as a terahertz wave source. From this point of view, HTS's are also utilized in security systems, in remote sensing and non-destructive diagnosis. As was previously determined that the

mercury based copper oxide layered HTS (mercury cuprates) act as a natural terahertz wave cavity at particular temperature interval (Özdemir et al., 2006; Güven Özdemir et al., 2009). Hence, some part of the chapter will be related to the investigation of the terahertz emission of the superconductor mentioned.
The last part of the chapter will be related to the facility of prospective application of the mercury cuprates as a quantum bit, "qubit" in particular flux qubit (Güven Özdemir, 2011), which is based on the occurrence of the two fluxoid states with equal energy but opposite circulating current at the same time in the system (Mooij et al., 1999). As is known, qubits are fundamentally considered as the main building block of both quantum computation, quantum communication etc. In recent years, quantum computers, which are based on qubit operations, have an increasing attention due to their both high speed and memory capacity. In the present superconducting qubit technology, some low temperature superconductors especially aluminum thin film superconductors have been extensively utilized. In this context, for a prospective usage of HTS, mercury cuprates as one of the oxide layered high temperature superconductors will be proposed as an intrinsic flux qubit in this chapter.

2. Clinical usage of Superconducting Quantum Interference Device (SQUID)

The superconducting magnets in ultra sensitive magnetic detectors (1.5 Tesla and above) namely SQUID magnetometer (Superconducting Quantum Interference Device) implement the reliable observation of the metabolites in the living organisms for the clinical applications.

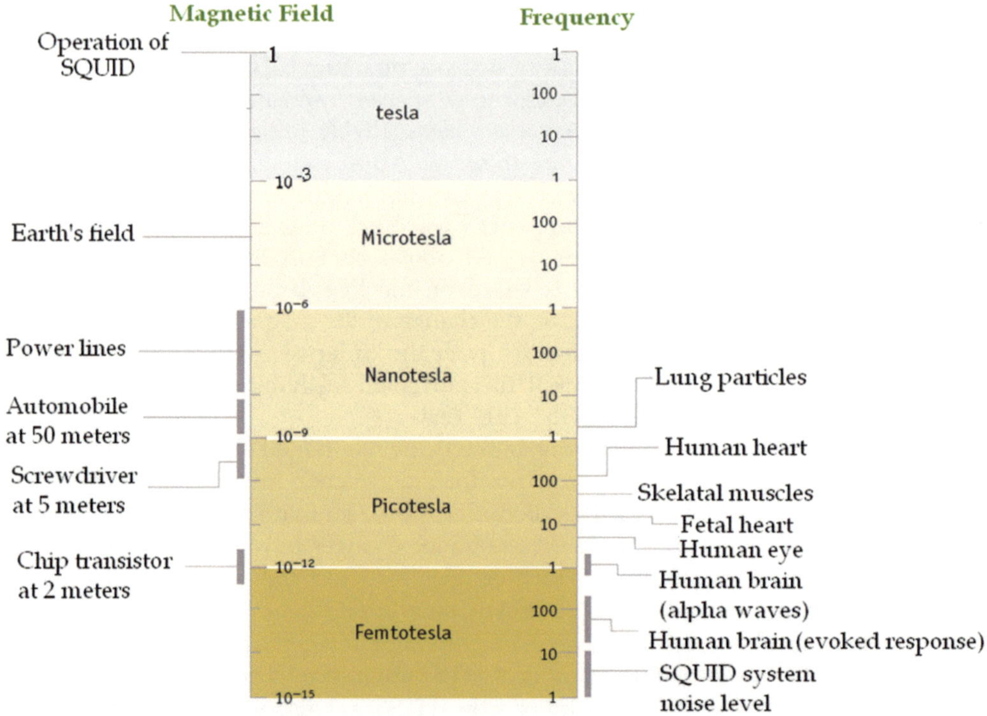

Fig. 1. Magnitudes of some biomagnetic fields (Fishbine, 2003).

As is known, the functions of human body are realized by the displacement of ions such as Na^+, K^+, Cl^- etc. The displacement of the ions corresponds to a current which produces a magnetic field. In Figure 1, the magnitudes of biomagnetic fields together with the other magnetic field sources are given.

According to Figure 1, especially biomagnetic fields produced by neuron cells' activities are very weak. They have magnetic field strengths of fT (femtoTesla i.e. 10^{-15}T). For comparison, the Earth's magnetic field is measured in micro Tesla and a magnetic resonance imaging system operates at several Tesla. The detection of such very small magnetic fields reliably is realized by the most sensitive magnetic field detector known as Superconducting Quantum Interference Device, namely SQUID.

A SQUID uses the properties of electron-pair wave coherence and Josephson Effect to detect very small magnetic fields. For example, to measure the magnetic field, that is produced by the electrical activity in brain, a special non-invasive imaging technique namely Magnetoencephalography (MEG), which contains SQUID sensors, are used (Fishbine, 2003).

In Figure 2, the representative illustration of detection of physical activities in human brain by MEG is given. The accuracy of measuring brain waves by MEG depends on the magnetic shielding of SQUID sensors from ambient magnetic fields such as the magnetic field of the Earth, power line's magnetic field and etc.

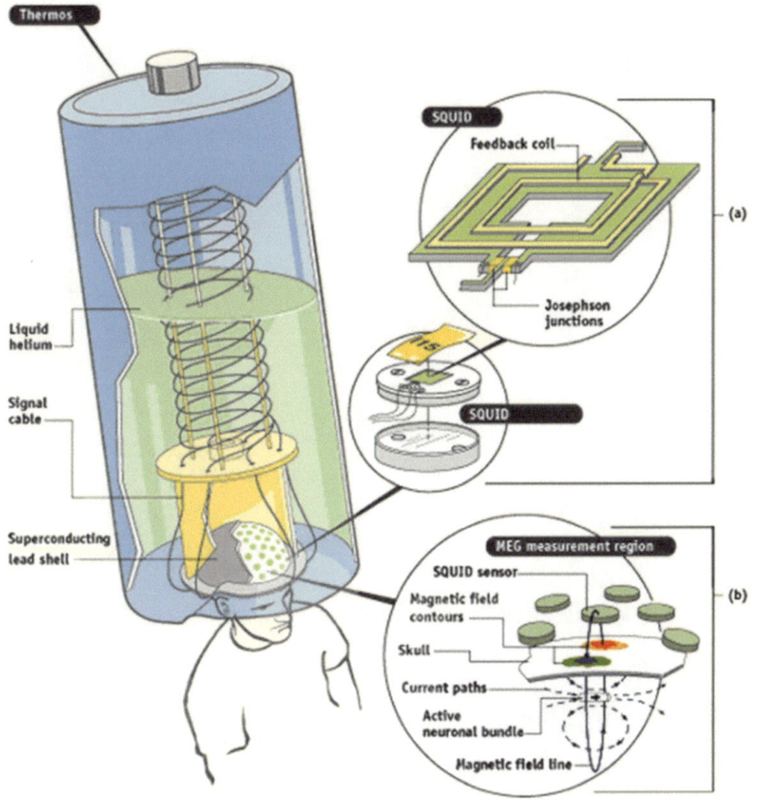

Fig. 2. Representative illustration of detection of physical activities in human brain by MEG (Fishbine, 2003).

As it is seen from Figure 2, SQUID sensors and the superconducting lead shell are cooled by immersion in liquid helium. In the superconducting state, lead shell expels ambient magnetic fields at all frequencies and hence it is ensured that SQUID sensor is only detect the magnetic field generated by brain waves. Each SQUID sensor contains a coil of superconducting wire that receives the brain fields and is magnetically coupled to the SQUID, which produces a voltage proportional to the magnetic field received by the coil. A computer program converts the SQUID data into maps of the currents flowing throughout the brain as a function of time (Fishbine, 2003).

2.1 Proton Magnetic Resonance Imaging (P-MRI) for medical diagnosis

Approximately 80% of human body is composed of water molecules and each water molecule consists of two hydrogen nuclei i.e. protons. In P-MRI measurements, the nuclear spins of hydrogen nuclei are aligned in one direction by applying strong magnetic fields of 1.5-3T that are generated by superconducting magnets. Afterwards, the polarized spins in one direction are excited by properly tuned radio frequency radiation. When the influence of short pulse of radio waves is removed, they drift back to their initial position, thereby

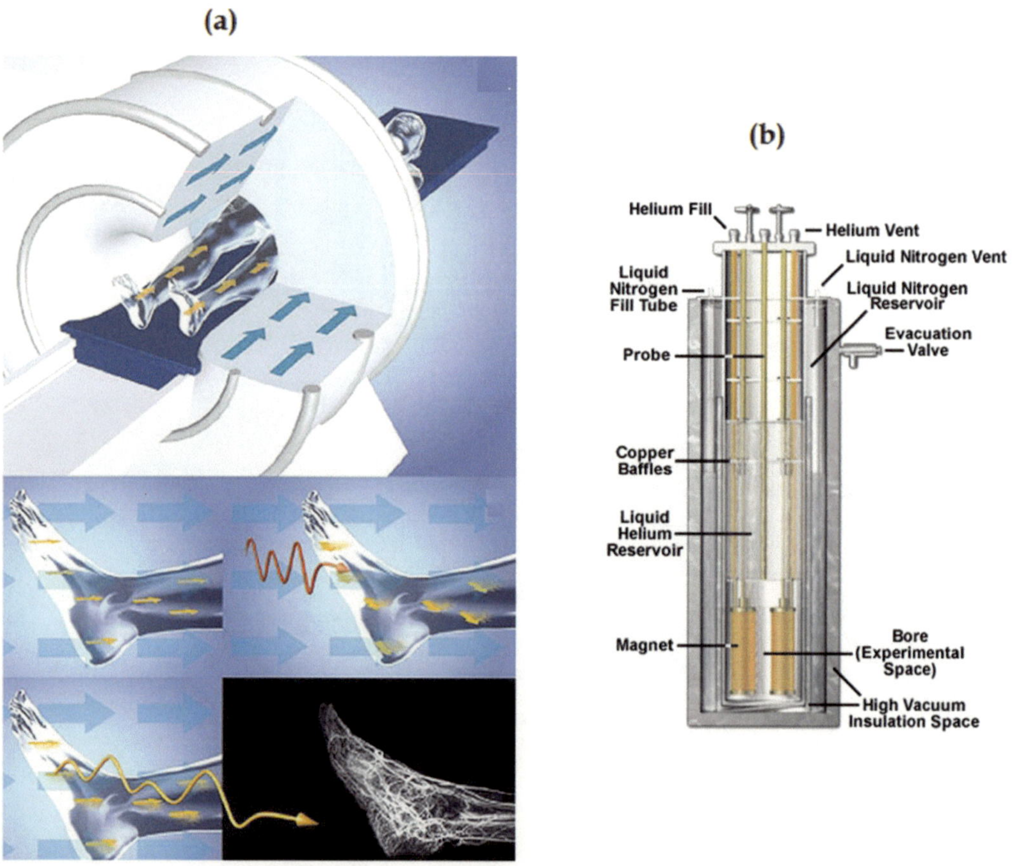

Fig. 3. (a) The working principle of P-MRI (Bayer, 2010). (b) The schema of superconducting magnets (National High Magnetic Field Laboratory, FSU, 2010).

emitting electromagnetic signals that can be used to reconstruct an image of the inside of the body. The protons in different tissues of the body (e.g. fat, muscle and etc.) realign at different speeds, so that the different structures of the body can be revealed (Georgia State University, 2010; Wikipedia, 2010; Bayer, 2010). The main steps of P-MRI are given in Figure 3(a).

The MRI technique has been extensively used for especially imaging the brain, heart, muscles and joints, for early diagnosis of cancer cells.

Superconducting magnets have a crucial role in P-MRI measurements. A superconducting magnet is an electromagnet made from coils of superconducting wire. Superconducting magnets can produce stronger and homogeneous magnetic fields than iron-core magnets. The most remarkable feature of the superconducting magnets is their capability of supporting very high current density with a vanishingly small resistance.

In conventional MRI devices, low temperature type II superconductors such as NbTi alloy are used to make coil windings for superconducting magnets. The illustration of typical superconducting magnet is shown in Figure 3(b).

Although, superconducting magnets are relatively more expensive to be build than ordinary iron-core magnets and require a constant supply of liquid helium closed cycle system, superconducting magnets have great advantages to detect very weak signals come from different sections of brain. If an ordinary magnet is used in MRS, only a giant water peak is observed, since approximately 80% of the tissue investigated consists of water molecule. Usage of superconducting magnet, which generates strong magnetic fields of 1,5-3T, enables both to suppress the signal comes from water molecules and distinguish the very weak signals, which have slightly different frequencies and that come from different metabolites secreted at the vicinity of the related tissue (Salibi & Brown, 1998).

2.2 Proton Magnetic Resonance Spectroscopy (P-MRS)

P-MRS generally produces spectra via Fourier transformation process of the tissue investigated instead of creating an image of the tissue. In vivo P-MRS, an appropriate radio frequency is applied the tissue and the signal comes from the tissue is measured and the

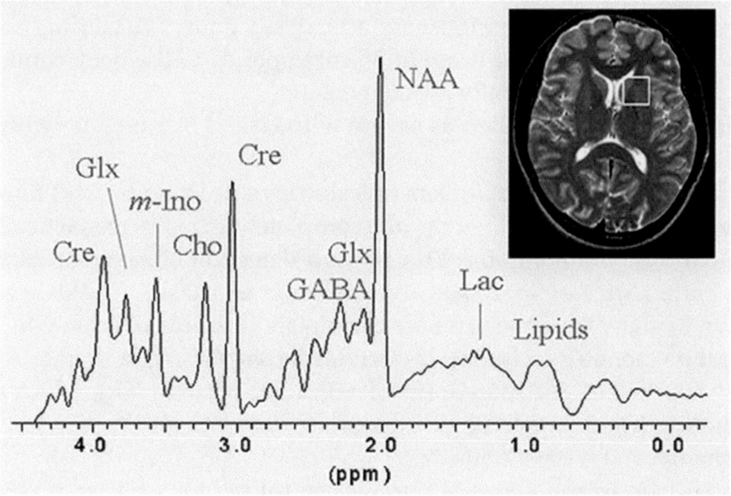

Fig. 4. P-MRS results of healthy human brain by means of metabolites. (Blamire, 2010)

Fourier transformation technique is used. In vivo P-MRS use position, signal intensity and line width along with spectral patterns to display chemical information about the related metabolites. A "ppm i.e. parts per million" scale describes the position of the peaks or resonances on the x-axis. Each peak in the spectrum arises from different brain metabolites. The heights of the peaks in the vertical axis refer to the concentrations of metabolites in arbitrary intensity scale.

In Figure 4, the proton magnetic resonance spectroscopic curve of a healthy human brain tissue with N-acetylaspartate (NAA), creatine (Cr), cholin (Cho) and etc. metabolite peaks is shown. Some of the metabolite peaks appear at the specific ppm value in the x-axis of the spectroscopic data. For example NAA peak appears at 2.0 ppm and NAA is accepted as the neural marker. While the primary resonance of Cr lies at 3.03 ppm, Cho resonates at 3.2 ppm (Danielsen &Ross, 1999; Blamire, 2010; Frahm et al., 1989).

2.3 *In vivo* experimental detection of Mild Traumatic Brain Injury (MTBI) by means of Proton Magnetic Resonance Imaging (P-MRI) and Proton Magnetic Resonance Spectroscopy (P-MRS)

In this chapter, we will focus on the detection of mild traumatic brain injuries (MTBI) by means of both Proton Magnetic Resonance Imaging (P-MRI) and Proton Magnetic Resonance Spectroscopy (P-MRS) techniques based on our experimental investigations. Although, the physical principle of both neuro-imaging techniques are the same, the information comes from these methods are completely different. Whereas the magnetic interaction between applied magnetic field and hydrogen nuclei is considered for P-MRI, the detection of chemical variation at the vicinity of hydrogen nuclei is realized in P-MRS. The main goal of this section is to determine mild traumatic brain injury (MTBI) at which no significant pathology is seen by imaging studies and that is considered as healthy based upon the Glasgow Coma Score (GCS).

As is known, a neurological GCS, which was published in 1974 (Teasdale & Jennett, 1974), aims to give a reliable information about the status of the central nervous system in three types of tests: eye, verbal and motor responses of the patient. The GCS is given in Table 1.

In GCS, the sum of the value that is related to the three types of tests (eye, verbal and motor) is used to assess the level of consciousness after head injury. According to the values in Table 1, while "the lowest possible score of 3" corresponds to the deep coma or death, "the highest score of 15" indicates the fully awake person.

Generally brain injuries are classified as severe with GCS $\leq$ 8, moderate with CGS:9-12 and minor GCS $\geq$ 13.

In MTBI, often no lesion or a few numbers of lesions are detected by MRI findings so that in many cases neuoro-imaging findings do not completely explain the clinical symptoms. In this context, P-MRS is commonly used for in vivo detection of variation of the quantity of metabolites in brain (Ariza et al., 2004; Govindaraju et al., 2004). P-MRS is a very sensitive and noninvasive in vivo technique to assess the metabolic status of brain which can quantify selected cerebral metabolites including N-acetylaspartate (NAA), a marker of neuronal and axonal viability (Bachelard & Badar-Goffer, 1993), total creatine (Cr), which reflects energy status, total choline (Cho) a marker of membrane metabolism, lactate (Lac), which is an indicator of ischemia and mobile lipids.

Although, the MRI findings indicate a normal central nervous status in post MTBI, it has been determined by P-MRS that concentrations of the cerebral metabolites mentioned above

Score	Infant (<1 year old)	Child (1-4 years old)	Adult
RESPONSES of EYES			
4	Open	Open	Open
3	To voice	To voice	To voice
2	To pain	To pain	To pain
1	No response	No response	No response
VERBAL RESPONSES			
5	Coos, babbles	Oriented, speaks, interacts, social	Oriented and alert
4	Irritable cry, consolable	Confused speech, disoriented, consolable	Disoriented
3	Cries persistently to pain	In appropriate words, inconsolable	Nonsensical speech
2	Moans to pain	Incomprehensible, agitated	Moans, unintelligible
1	No response	No response	No response
MOTOR RESPONSES			
6	Normal, spontaneous movement	Normal, spontaneous movement	Follow commands
5	Withdraws to touch	Localizes pain	Localizes pain
4	Withdraws to pain	Withdraws to pain	Withdraws to pain
3	Decorticate flexion	Decorticate flexion	Decorticate flexion
2	Decerebrate extension	Decerebrate extension	Decerebrate extension
1	No response	No response	No response

Table 1. Glasgow Coma Score (Teasdale & Jennett, 1974).

change. Several studies with P-MRS have indicated that it is possible to detect neural injury i.e. loss of neuron by comparing these concentrations and the related ratios of pre and post MTBI (Luyten &Den Hollander, 1986; Cecil et al., 1998; Friedman et al., 1998; Friedman et al., 1999; Onbaşlı et al., 1999; Garnett et al., 2000; Brooks et al., 2000; Holshouser, 2000; Garnett et al., 2001; Rao et al., 2006). Whereas death of neurons manifests itself as the deficiency of NAA concentration, the increased Cho concentration is related to the cell membrane breakdown (Brooks et al., 2001). These changes both occur in occipital, parietal and frontal lobes and the splenium of the corpus callosum of brain regions (Ross et al., 1998). The related regions of brain are seen in Figure 5.

In MTBI, the metabolic abnormality is relatively small and due to this reason no lesion in brain tissue is observed in P-MRI results and the related GCS is higher than 14. Moreover, as is known from medical literature, these patients generally exhibit prolonged neurological deficits. In this context, P-MRS is considered as the most sensitive probe for detection of minor changes in brain metabolites. Determination of MTBI has a significant role for both

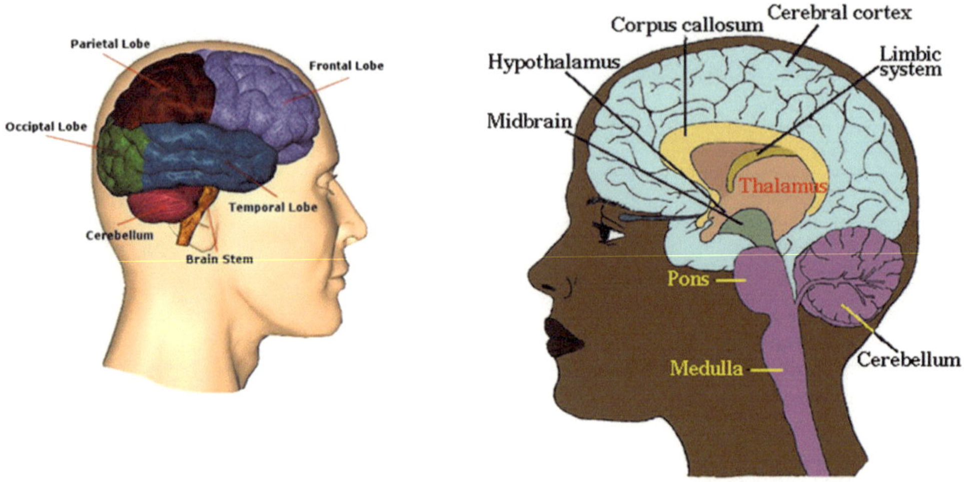

Fig. 5. The brief anatomy of human brain (Royal Adelaide Hospital web site, 2010; Weber State University web page, 2010).

medical diagnosis about the neuron loss percentage and that for forensic science investigations.

In this context, the MTBI has been investigated via four volunteers experimental subjects[1] (mean age: 25 years old; age range: 23-27 years old; mean weight: 75 kg; sex: male; profession: sportsman) by detection of NAA, Cho and Cr metabolites with P-MRS for pre and post trauma for both corpus callosum splenium and frontal white lobe of brain.

To obtain MTBI, sportsmen performed boxing for five minutes as a practice by using boxing gloves and head guard for protection. All experimental subjects were fully awake according to GCS of "15" for pre and post MTBI. No lesion such as hematoma or ecchymose was viewed in P-MRI experiments in pre and post MTBI. For comparison, MRI photographs of sagittal section for pre and post MTBI of one of the experimental subjects are given in Figure 6 (Eruygun, 1998). The results have been confirmed by professional neurologist that no difference between the images given in Figure 6, which were taken from corpus callosum splenium and frontal white lobe of brain, has been detected before and after minor trauma.

After taking the MRI for post trauma, P-MRS experiments were realized within one and half hour after trauma. Proton single voxel spectroscopy[2] experiments have been performed for corpus callosum splenium and frontal white lobe of brain. The data obtained by pre and post P-MRS experiments are listed in Table 2 for the sections of brain mentioned above.

---

[1] All volunteers have been informed about the investigation process and they all accepted the experiment.

[2] The single voxel techniques generate a cubic or rectangular shaped volume element i.e. voxel with maximum $8cm^3$ volume for a region of the sample with P-MRS. Two pulse sequences are used in P-MRS: The STEAM (STimulated Echo Acquisition Mode) and PRESS (Point RESolved Spectroscopy). Both STEAM and PRESS sequences generate cubic or rectangular shaped voxel by acquisition of three orthogonal slice selective 90° pulses and 90° pulse followed by two 180° pulses, respectively. In the STEAM technique, the signal contamination outside the tissue observed is minimum.

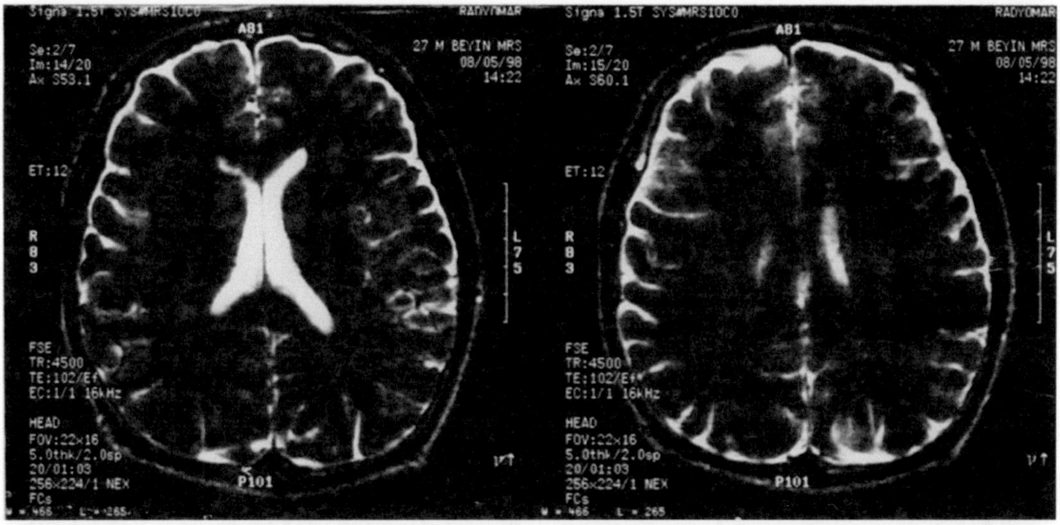

(a) Pre-trauma for sagittal section

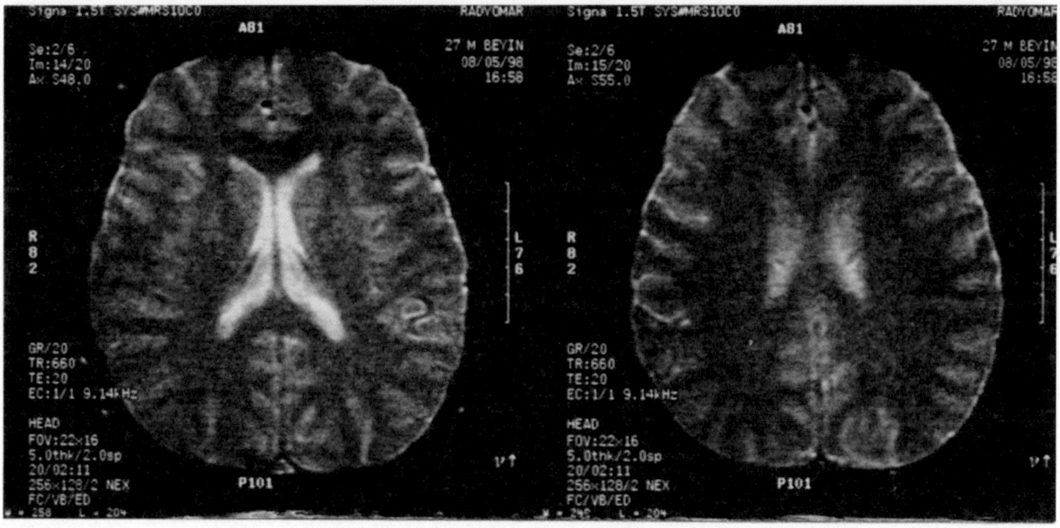

(b) Post trauma for sagittal section

Fig. 6. MRI images of experimental subject 1 for (a) pre trauma and (b) post trauma from sagittal section. MR imaging of these regions were performed by General Electric Signa 1.5T MRI device.

As is seen from Table 2, the neuron marker, NAA, decreases after the minor brain trauma, whereas the replenishing metabolites of Cholin, increases for every lesions of the all volunteers' brain.

The pre and post trauma P-MRS photographs of one of the experimental subject for the both corpus callosum splenium and the white frontal lobe of brain are given in Figure 7 and 8, respectively. The heights of peaks of the metabolites for pre and post trauma are marked on the left and the right of the photographs, respectively.

As is seen from Figure 7 and 8 no pathologic peak has been observed.

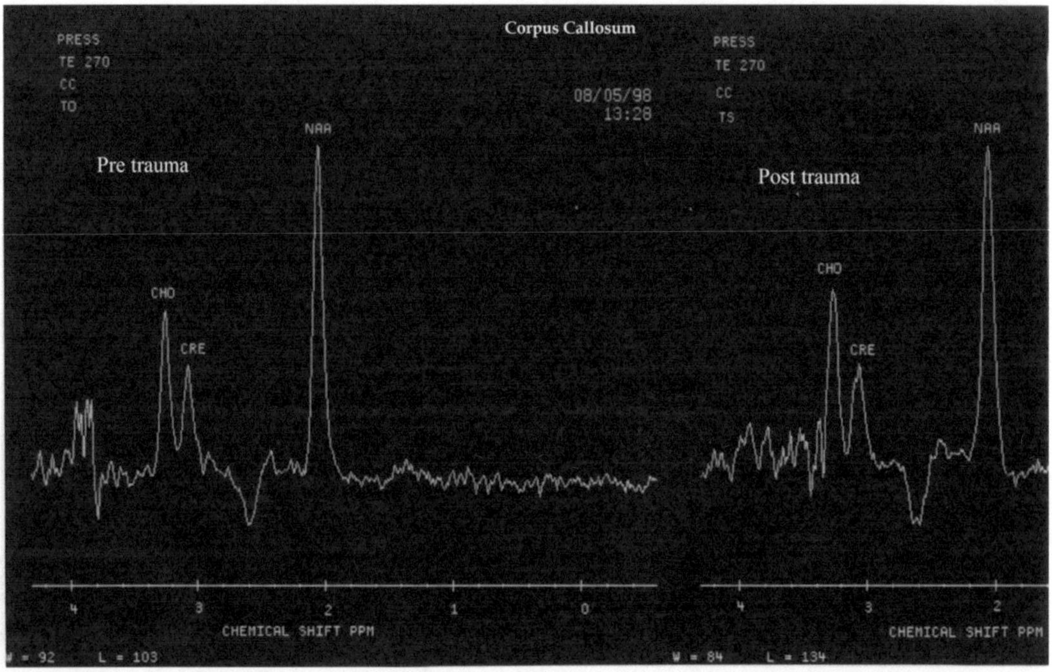

Fig. 7. Brain P-MRS images one of the experimental subject for corpus callosum splenium.

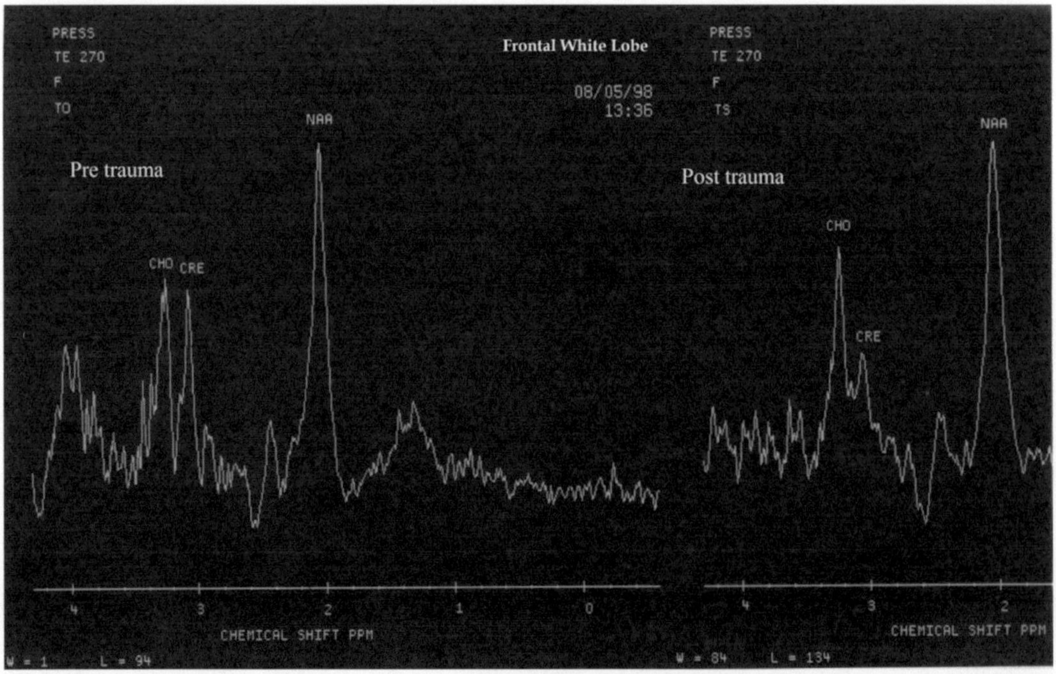

Fig. 8. Brain P-MRS images one of the experimental subject for white frontal lobe.

Brain Sections	Corpus Callosum Splenium Pre trauma			Corpus Callosum Splenium Post trauma			White Frontal Lobe Pre trauma			White Frontal Lobe Post trauma		
Metabolites	NAA	Cr	Cho	NAA	Cr	Cho	NAA	Cr	Cho	NAA	Cr	Cho
Experimental Subject 1	126	46	57	118	34	63	123	42	53	120	47	60
Experimental Subject 2	120	52	47	120	49	73	123	54	52	120	53	84
Experimental Subject 3	128	43	63	122	38	70	129	60	66	123	44	81
Experimental Subject 4	118	42	57	114	40	75	123	46	55	116	61	77

Table 2. By means of chemical shift of metabolites in MRS results for pre and post MTBI (Eruygun, 1998; Onbaşlı et al, 1999).

2.4 Conclusion

The observation of the change in the amplitude of brain metabolites at the vicinity of hydrogen nuclei via P-MRS has been only used for clinical diagnosis for several years. In addition to clinical researches, a new method for determining of closed and mild brain injuries, which are caused due to psychological pressure, beating or any kind of abusing, has been introduced for forensic science investigation for the first time. Moreover, in order to make clear beyond a reasonable doubt for the detection of such mild brain trauma, the P-MRS has been suggested as the most reliable scientific tool for forensic science investigations as well as medical diagnosis.

3. High temperature superconductors as a new terahertz wave sources

Terahertz waves have various advanced technological applications including medical diagnosis, security, biomedical imaging, atmospheric researches, drug and food inspection, gas tracing etc. (Tonouchi, 2007). Terahertz waves also known as T-waves, exist a frequency region between microwaves and far infrared of the electromagnetic spectrum. As is seen from Figure 9, terahertz waves occupy a region from 300 GHz to 10 THz that can provide imaging and sensing technologies not available through conventional technologies such as X-rays.

In recent years, T-waves are extensively utilized for non-destructive security devices since many materials and living tissues are semi-transparent at terahertz wavelengths and also have distinct THz absorption spectra namely "finger prints". Unlike X-rays, THz radiation posses a little or no health threat since that T-ray photons are both not strong enough to ionize atom or molecules and not able to break the chains of chemical bonds[3]. Moreover, T-waves, especially in 1-10 THz region, have great advantages for biomedical imaging such as tumor recognition, detection of dental cavities etc. due to its non-invasiveness and low Rayleigh scattering property factor compared with optical waves (Lu et. al., 2010). Furthermore, DNA molecules and many proteins have collective vibration and rotation

---

[3] THz waves have low photon energies (1 THz = 4.1 meV), one million times weaker than X-rays, and do not cause harmful photoionization in biological tissues.

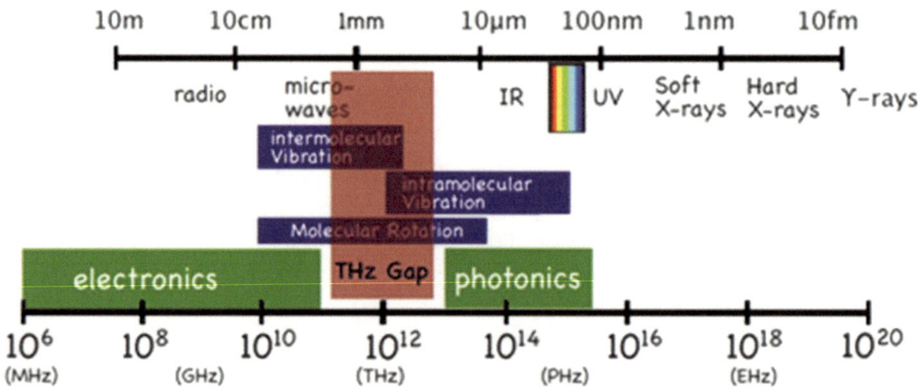

Fig. 9. Terahertz waves in the electromagnetic spectrum (Zomega Terahertz Corp, 2010). The terahertz gap lies between electronics and photonics.

modes that coincide with THz region. Due to this reason, T-waves have also a crucial importance for molecular imaging (Zuhang et.al, 1990; Young et al., 1990; Feng et.al, 1991; Nossal & Lecar, 1991).

There are many ongoing researches for developing terahertz wave sources. One of the most popularly used terahertz wave sources is optoelectronic terahertz wave generators. In this process, near-infrared laser light illuminates a metal-semiconductor-metal structure that results in a photocurrent which is utilized as terahertz wave. Ultrafast fiber lasers and diode lasers & photo-mixers are used for "pulsed" and "continuous wave" sources, respectively. In recent years, the semi-conducting light sources have extended the limit of emission of T-rays to 1.6 THz for quantum cascade lasers and 1.5 THz for unitravelling-carrier photo diodes (Ito et. al. 2005; Walther et. al, 2006). On the other hand, the frequency of alternating currents in semiconductors is limited by finite electron velocities, while that of solid state lasers are limited by thermal energies which prevent the small electronic transition required for lasing at THz wavelengths.

Among the sources mentioned above, superconductors, especially the high temperature superconductors that display "Intrinsic Josephson Effect" (IJE), may be the excellent promising candidate for terahertz source due to their extremely low noise factor and wide frequency coverage (Emuidzinas & Richards, 2004; Güven Özdemir et al, 2009, Moody, 2009). Very recently, continuous and monochromatic terahertz wave emitter with the frequency of 0.63THz has been achieved for the intrinsic Josephson junctions in high temperature $Bi_2Sr_2CaCu_2O_{8+\delta}$ superconductor by applying d.c. voltage in the order of milivolt (Minami et. al. 2009). The voltage due to the fluxon flow mechanism in the system excites the Josephson plasma with terahertz frequency. The frequency of 0.63THz can be considered as the beginning of the filling the THz gap, that is defined between 300 GHz to 10 THz, by high temperature superconducting terahertz sources. The long term goal of the scientists working on that area is to cover the whole THz gap by superconducting coherent terahertz wave sources.

In this chapter, the copper oxide layered mercury based cuprate family superconductors, $HgBa_2Ca_2Cu_3O_{8+x}$, which have the highest Meissner transition temperature of 140K ever obtained at normal atmospheric pressure and display IJE, is proposed as an intrinsic terahertz wave source, since there is no need to apply any bias voltage to the system.

3.1 Mercury cuprates as an intrinsic terahertz wave sources

As is known that the copper oxide layered high temperature superconductors such as $Bi_2Sr_2CaCu_2O_8$, $Tl_2Ba_2Ca_2Cu_3O_{10}$, $HgBa_2Ca_2Cu_3O_{8+x}$ etc. have a common structure in which superconducting copper oxide layers are separated by a thin insulating layer. Copper oxide layers are electromagnetically coupled together by Josephson tunneling process. According to the experimental evidences, cuprates such as $Bi_2Sr_2CaCu_2O_8$, $Tl_2Ba_2Ca_2Cu_3O_{10}$ and $YBa_2Cu_3O_{7-x}$ behave like a stack of superconductor-insulator-superconductor structure i.e. intrinsic Josephson junctions (IJJ) (Kleiner & Müller, 1994; Özdemir et. al, 2006; Güven Özdemir et. al., 2009)

As is known, the main Josephson plasma excitation modes in weakly Josephson coupled layered superconductors are longitudinal (along the c-axis) and transversal (in the ab-plane) plasma modes. Whereas the transverse-mode plasma oscillations can be converted into electromagnetic waves at the boundary of the junctions (Machida & Tachiki, 2001), the longitudinal plasma propagation modes in an array of IJJ do not lead the electromagnetic wave radiation, since the electromagnetic wave has only transverse mode (Bae &Lee, 2006).

Mercury based superconductors have been considered as an array of IJJ. Since, the Interlayer Theory[4] is valid at low temperatures for the optimally oxygen doped mercury cuprates (Özdemir et. al, 2006), all copper oxide layers along the c-axis of the system are in the resonance with the Josephson plasma frequency that coincides to the THz gap (Güven Özdemir et. al, 2007). This type of Josephson plasma resonance mode cannot be attributed to the fluxon flowing mechanism, since the working magnetic field interval never exceeds the lower critical magnetic field of the system. Hence, the magnetic flux is totally expelled from the superconductor. The phenomenon is displayed as a perfect diamagnetic response on SQUID data. Ultimately, the terahertz plasma oscillations determined in the system provide the bulk mercury cuprate superconductor as an intrinsic coherent terahertz wave source without application of any external voltage.

The Josephson plasma frequency values for the mercury cuprates had already been calculated via magnetic critical current density J_c, which was deduced by magnetization versus magnetic field curves obtained by SQUID[5] (Özdemir et. al, 2006; Güven Özdemir et. al., 2009). The Josephson plasma frequency, f_P is calculated via Josephson penetration depth, λ_J .

$$f_P = \frac{c}{2\pi\lambda_J} \tag{1}$$

where λ_J describes the penetration depth of the magnetic field induced by the supercurrent flow in the superconductor. The Josephson penetration depth is defined as

---

[4] According to Interlayer theory, electron pairing in the superconducting state makes the transport process along the c-axis to be coherent via Josephson (Lawrence-Doniach like) coupling between the superconducting copper oxide layers. Moreover, in order to provide the coherent transport mechanism along the c-axis, the Josephson coupling energy must be equal to the superconducting condensation energy (Anderson, 1997; Anderson 1998).

[5] The critical current density at the vicinity of lower critical magnetic field had already been calculated by Bean critical state model.

$$\lambda_J = \sqrt{\frac{c\phi_0}{8\pi^2 J_c d}} \tag{2}$$

where, c is the speed of light, J_c is the magnetic critical current density, ϕ_0 is the magnetic flux quantum, and d is the average distance between the copper oxide layers (Ferrel & Prange, 1963; Ketterson & Song, 1999; Fossheim & Sudbo, 2004).

We have calculated the intrinsic Josephson plasma frequencies for the optimally and over oxygen doped Hg-1223 superconductors. The related data have been listed in Table 3.

The optimally oxygen doped Hg-1223 superconductors		The over oxygen doped Hg-1223 superconductors	
Temperature (K)	Plasma Frequency, f_p (Hz)	Temperature (K)	Plasma Frequency, f_p (Hz)
4.2	8.303x10^13	5	3.295x10^13
27	3.363x10^13	17	2.175x10^13
77	8.303x10^12	25	1.981x10^13
		77	1.866x10^12
		90	1.537x10^12

Table 3. The Josephson plasma frequencies for the optimally and over oxygen doped Hg-1223 cuprate superconductors (Özdemir et al., 2006; Güven Özdemir et al., 2007; Güven Özdemir, 2007).

In Figure 10, the Josephson plasma frequency versus temperature curves and the related fitting functions for both the optimally and over oxygen doped $HgBa_2Ca_2Cu_3O_{8+x}$ superconductors are shown.

The temperature dependence functions of the Josephson plasma frequency for the optimally and over oxygen doped Hg-1223 superconductors have been determined by Origin Lab 8.0 graphic program. According to the fitting functions given in Eqs. (3) and (4), an extra doping of oxygen atoms to the optimally doped system changes the temperature dependence of f_P and also lowers the plasma frequencies.

$$f_P(T) = 8.303\times10^{12} + \frac{\left(8.303\times10^{13} - 8.303\times10^{12}\right)}{1+\left(\dfrac{T}{21.65914}\right)^3} \tag{3}$$

for the optimally oxygen doped Hg-1223 superconductor

$$f_P(T) = 1.537\times10^{12} + \left(3.29503\times10^{13} - 1.537\times10^{12}\right)\left[\frac{0.5}{1+10^{-(11-T)0.05941}} + \frac{1-0.5}{1+10^{-(15-T)0.05882}}\right] \tag{4}$$

for the over oxygen doped Hg-1223 superconductor

3.2 Brief conclusions for mercury cuprate intrinsic terahertz wave sources

Both the optimally and over oxygen doped mercury cuprate samples exhibit intrinsic terahertz Josephson plasma frequencies at relatively high temperatures. The optimally oxygen doped system starts to emit a coherent terahertz waves at the vicinity of liquid nitrogen temperature of 77K, whereas for the over oxygen doped samples, this effect starts

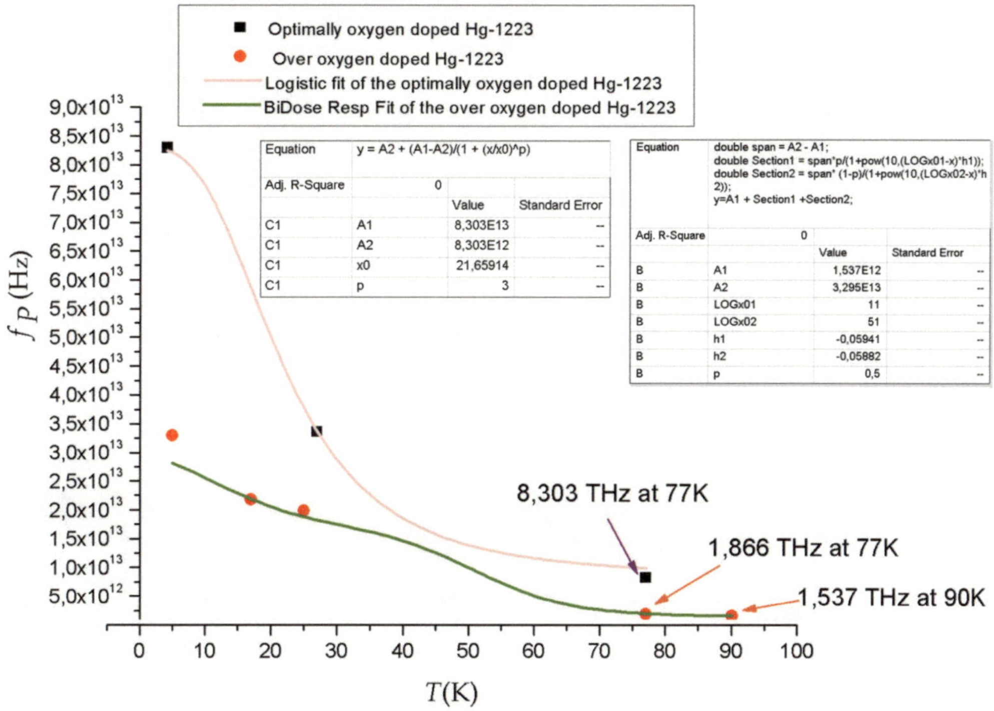

Fig. 10. The intrinsic Josephson plasma resonance frequency versus temperature curves and fitting functions for both the optimally and over oxygen doped mercury cuprates.

above 77K. So it has been concluded that the excess oxygen only affects the starting temperature of emission of coherent terahertz wave in the superconducting system.

4. Intrinsic Quantum Bit "Qubit" operations with mercury cuprate high temperature superconductors

4.1 Introduction to quantum computers and qubit

In recent years, quantum computers have an increasing attention due to their both high speed and memory capacity. As is known that quantum computers are completely different from the classical computers which are based on the standard semiconductor transistor technology. While classical bit is used in the classical computers, the quantum bit namely "qubit", which can carry two quantum states at the same time, is used in the quantum computers. Quantum computers are operated by some quantum mechanical phenomena such as quantum superposition, quantum entanglement and quantum teleportation.

A quantum computer maintains a sequence of qubits. A single qubit is represented by $|0\rangle$, $|1\rangle$ or crucially any quantum superposition of these states. The quantum superposition of these orthogonal states is defined by

$$|\psi\rangle = c_1|0\rangle + c_2|1\rangle \tag{5}$$

The squares of the complex coefficients c_1^2 and c_2^2 represent the probabilities for finding the particle in the corresponding states. Pair of qubits can be in any quantum superposition of 4

(=2^2) states and three qubits in any superposition of 8 (=2^3) states etc. While, for the classical computer one of these states has the probability of "1", for the quantum computer, the sum of the probabilities of these states equals to "1". In this point of view, quantum superposition allows a particle to be in two or more quantum states at the same time. So that the quantum computation is a parallel computation in which all 2^M basis vectors are acted upon at the same time. This parallelism allows a quantum computer to work on a million computations at once, while the desktop PC works on one. A 30-qubit quantum computer would equal the processing power of a conventional computer that could run at 10 **teraflops** (trillions of floating-point operations per second). Today's typical desktop computers run at speeds measured in gigaflops (billions of floating-point operations per second) (Deutchs, 1997). In Figure 11, the difference between classical and quantum computers is illustrated representatively in the context of computation process.

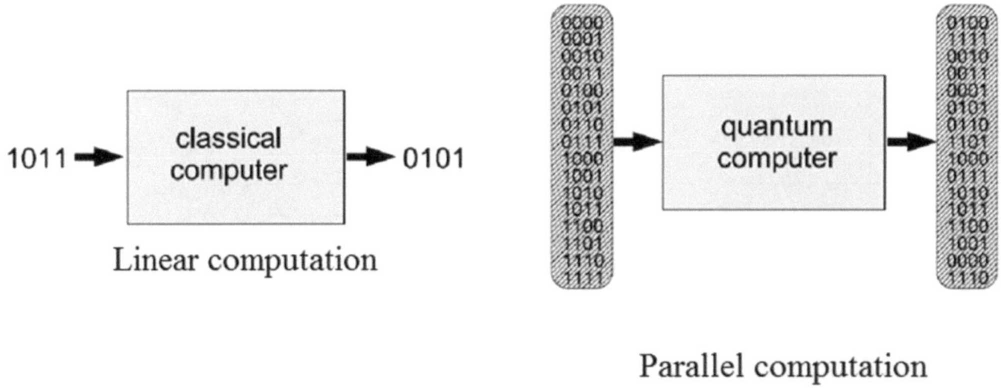

Fig. 11. The main difference between classical and quantum computers by means of computation process (Optical Lattices & Quantum Information Web Site, 2011).

Quantum computers also use a special quantum mechanical phenomenon called as "Quantum Entanglement". In the quantum entanglement, it is possible to link together two quantum particles such as photons or atoms in a special way that makes them effectively two parts of the same entity. Then you can separate them as far as you like, a change in one part is instantly reflected in the other and collectively they constitute a single quantum state (Clegg, 2006). Two entangled particles often must have opposite values for a property, for example, opposite spin. For instance, two photons can be entangled such that if one of them is horizontally polarized, the other is a vertically polarized. It is not important how far they are located, the change in one also reflected in the other. So that quantum entanglement allows particles to have a much closer relationship than is possible in classical physics (Dumé, 2004).

In quantum teleportation, complete information about the quantum state of a particle is instantaneously transferred by the sender to a receiver. This is a great advantage for quantum computing.

In the quantum computers, data is stored by using atoms, photons or fabricated microstructures. In recent years, low temperature superconductors such as Nb and Al have been widely used for qubit technology. Superconducting qubits have an increasing attention due to their collective coherent behavior. As it is well known that the superconducting system can be considered as a condensed state like superfluids so that the all electron pairs are described by the single quantum state with the quantum wave function, Ψ, which is directly related to the phase difference, φ (Annett, 2004; Clarke & Wilhelm, 2008). In this point of view, this quantum mechanically coherent superconducting system is considered as the most viable for the qubit applications. Furthermore, superconductors provide the general requirement of the quantum circuits such as low dissipation and low noise. The zero resistance phenomenon of the superconducting state provides low dissipation and operating them at low temperatures offers a low noise.

Moreover, the formation of an energy gap between the electron pairs energy states and the free electron energy states has a crucial role in the superconducting qubit technology, since a significant amount of energy is needed for escaping the electron from this collective coherent state. So that it is very difficult to destroy the coherence of the physical superconducting qubit system. The mentioned collective behavior of superconductors that yields to a macroscopic quantum wave function, Ψ is connected to two crucial effects: "Flux quantization" and "Josephson effect" (Mooij, 2010).

Flux quantization is a fundamental quantum phenomenon in which the magnetic field is quantized in the unit of $\Phi_0 = \frac{h}{2e} = 2.068x10^{-15} Tm^2$, flux quantum. The flux quantization also occurs in Type II superconductors between lower (H_{c1}) and upper (H_{c2}) critical magnetic fields since magnetic field begins to penetrate above the lower critical magnetic field of H_{c1} through the superconductor in discrete (quantized) units while the system is still a superconductor.

In the Josephson effect, electron pairs can quantum mechanically tunnel the thin insulating layer due to the phase difference, φ between the adjacent superconducting layers. The supercurrent (I_s) across the Josephson junction, which consists of two superconducting layers separated by thin insulating layer, is directly related to the gauge invariant phase difference, φ.

$$I_s = I_{\max} \sin\varphi \tag{6}$$

where I_{max} represents the maximum current through the Josephson junction. In this situation, no voltage is applied to the junction. If an appropriate dc voltage is applied to the junction, the supercurrent oscillates with a characteristic angular frequency ω (Josephson, 1962).

$$\begin{gathered} \frac{d\varphi}{dt} = \omega = \frac{2eV}{\hbar} \rightarrow \omega = \frac{2\pi V}{\Phi_0} \\ I_s = I_{\max} \sin\left(\varphi + \frac{2eV}{\hbar} t\right) \end{gathered} \tag{7}$$

So that any change in the Josephson current results in a finite voltage across the junction. Hence the Josephson junction behaves as a nonlinear inductor.

Since both the phase coherence and long range order are the essence of the Josephson effect, they both play key roles in the qubit technology.

In the present superconducting qubit technology, some low temperature superconducting tunnel junctions have been utilized and their coherence times are around several microseconds while the operating time of qubit is in the order of nano seconds. The coherence times need some improvement. On the other hand, some high temperature superconductors such as Bi-

family superconductors have been tested for qubit operations but they did not give good results due to their high decoherence that results the loss of quantum information.
If one could fabricate a qubit with high temperature superconductors, it would have great advantages such as multiply connected and coupled millions of qubits in the thickness of 1mm. Since some of the high temperature superconductors consist of intrinsic Josephson junction array, there will no need to fabricate a Josephson junction one by one. Moreover, they will operate at significantly high temperatures such at 100K and above so the system will work with very low cost. On the other hand most the copper oxide layered high temperature superconductors such as Bi-family, Y-family superconductors are considered as two-dimensional superconductors due to their high anisotropy. Among other high temperature copper oxide layered superconductors, mercury cuprate family superconductors, $HgBa_2Ca_2Cu_3O_{8+x}$ have remarkable features for the superconducting qubit technology particularly, flux qubit. Due to this reason, in the following section the working principle of the flux qubit will be reviewed. Afterwards, the essential features of the mercury cuprates such as intrinsic Josephson junction structure, occurrence of the Paramagnetic Meissner effect, the electromagnetic wave cavity behavior, occurrence of the spatial resonance and etc. have been discussed in the context of bulk flux qubit. The last section is devoted to determine the required conditions for operating the bulk mercury cuprate superconductors that work as a flux qubit.

4.2 The working principles of flux qubits

Superconducting qubits are classified by comparing the Josephson coupling energy and the charging energy. Josephson coupling energy is defined by

$$E_J = \frac{\Phi_0 I_{max}}{2\pi} \tag{8}$$

The Josephson coupling energy characterizes the coupling strength between the adjacent superconducting layers. The charging energy is related to the occurrence of the electric field due to the motion of electron pairs in the junction that described as

$$E_C = \frac{(2e)^2}{2C} \tag{9}$$

where C is the capacitance of the junction.
Charging energy is important for small Josephson junctions. In the flux qubit, Josephson coupling energy is significantly larger than the charging energy ($E_J >> E_C$). The phase of the superconducting wave function is more important than the charge. Different value of the total phase change is connected with the different circulating current.
As is known that flux qubit consists of multiple connected Josephson junctions, typically three Josephson junctions (Fig. 12). If the zero magnetic flux is trapped in the qubit loop, the lowest energy is obtained at the zero phase change with zero current. If the magnetic flux quantum is trapped in the qubit loop, the lowest energy is obtained at the phase change of 2π. If half of a magnetic flux quantum is trapped in the qubit loop, the lowest energy is obtained at the phase change of π and the two fluxoid states have equal energies with opposite circulating currents. This is the basis of the working principle of the flux qubit (Mooij et al., 1999).

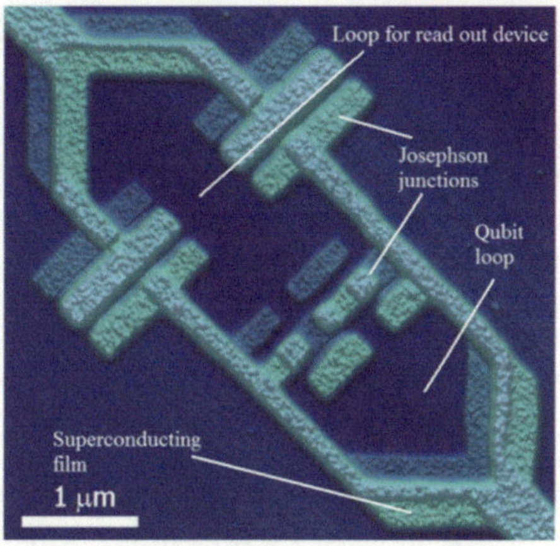

Fig. 12. The configuration of flux qubit consist of three Josephson junctions (Hans Mooij's research group at Delft University of Technology Physics, 2005)

According to Mooij et al., if ($f\Phi_0$) magnetic flux is applied to the qubit loop, where f is slightly smaller than 0.5, the system has two stable magnetic-flux states namely $|0\rangle$ and $|1\rangle$ quantum states. As is shown in Fig. 13, one magnetic flux state corresponds to a current, which is the order of several microamperes, of flowing clock wise, the other magnetic flux state corresponds to the same amount of current flowing anti-clock wise (Chiorescu et al., 2003).

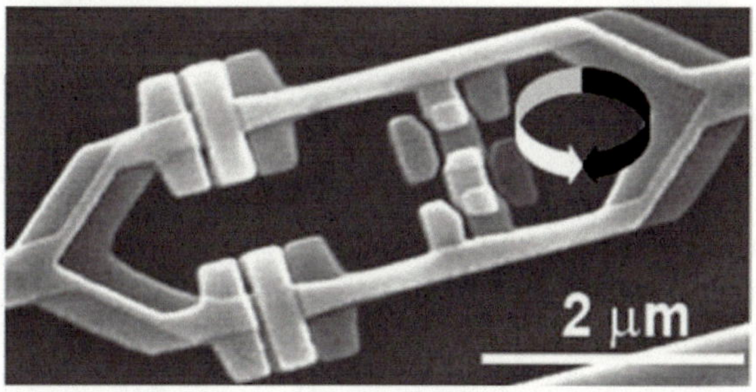

Fig. 13. The Scanning Electron Microscopy (SEM) photography of the micrometer sized superconducting flux qubit. Arrows indicate the clock wise and anti-clockwise currents.

Moreover, the quantum superposition of two states ($|0\rangle$ and $|1\rangle$) is also manipulated by resonant microwave pulses and applying strong microwaves to the system induces hundreds of coherent oscillations. This phenomenon is known as "Rabi oscillations" and it is the basis of quantum gate operations (van der Wal et al., 2000).

One of major problems of superconducting qubits is decoherence which causes the loss of information. Since the phase of quantum wave function dominates the effect of charge in the flux qubit, flux qubit circuits are directly affected by the external flux and its noise that results to cause decoherence (Wellstood et al., 1987; Mooij et al., 1999; Friedman et al., 2000). As is known that quantum information processing is limited by the coherence times. The increasing the coherence time makes possible to carry out a real effective quantum computer in future. From this respect, mercury cuprates have a great potential for flux qubit technology due to their long coherence times as will be expressed in the next section.

4.3 The general properties of mercury based copper oxide layered ceramic superconductors in the context of flux qubit

Besides the fact that the $HgBa_2Ca_2Cu_3O_{8+x}$ cuprate superconductors have the highest Meissner critical transition temperature of 140K at normal atmospheric pressure (Onbaşlı et al, 2009), they have remarkable features for the superconducting flux qubit technology. The crucial advantages of the mercury cuprates have been listed below.

a. **Intrinsic Josephson junction array**: Mercury cuprate family superconductors consist of typical superconducting copper oxide layers which are separated by thin insulating layers. Due to that fact the system is considered as an intrinsic Josephson junction network (Kleiner & Müller, 1994; Özdemir et al., 2006). This property of mercury cuprates removes the problem of the fabrication of the Josephson junctions separately. As is known, in order to build a real quantum computer one needs many coupled qubits. According to the relevant qubit technology, the connected qubit circuits are designed in a special way that allow to 10^5 operations (Mooij, 2010). However, utilizing the bulk mercury cuprate superconductor may increase the number of operation, since the connections between intrinsic Josephson junctions are naturally realized.

b. **The confirmation of interlayer theory and occurrence of spatial resonance**: The interlayer theory, which was proposed by P.W. Anderson for explaining the mechanism of superconductivity in the copper oxide layered high temperature superconductors, has been confirmed for the mercury cuprate family superconductors (Özdemir et al., 2006). According to the interlayer theory of high temperature oxide superconductors, the interlayer coupling correlates electromagnetic coupling along the c-axis with superconducting condensation energy of the superconductor (Anderson, 1997; Anderson 1998). In other words, the Josephson coupling energy equals to superconducting condensation energy in the mercury cuprates at around liquid helium temperature. In this point of view, all superconducting copper oxide layers along the c-axis are in the resonance. Hence, the system behaves like a three dimensional electromagnetic wave cavity. Also it has been determined that the mercury cuprate family superconductors behave like an electromagnetic wave cavity with the frequency of microwave, terahertz and infrared depending on the temperature dealt with (Özdemir et al., 2006; Güven Özdemir et al. 2007; Güven Özdemir et al., 2009).

In this context, the intrinsic Josephson junctions are connected via electromagnetic coupling in the bulk mercury cuprate so that the intrinsic Josephson junctions are in the lossless and perfect communication which has a crucial role in the qubit interactions in quantum computation. Moreover, the spatial microwave electromagnetic wave cavity also is utilized for the manipulation of the quantum states intrinsically.

c. **d-wave symmetric order parameter:** As is known, mercury cuprate superconductors have $d_{x^2-y^2}$-wave symmetric order parameter (Panagapoulos et al., 1996; Panagapoulos & Xiang, 1998; Onbaşlı et al, 2009). According to Taffuri et. al, the qubit proposals basically utilized the fact that the Josephson junctions with a π-shift in phase can be produced by a d-wave order parameter symmetry. This may lead to intrinsically double degenerated system, i.e. systems based on Josephson junctions with an energy-phase relation with two minima (Tafuri et al., 2004). This condition is intrinsically occurs in the mercury cuprate family due to the d-wave symmetry of its order parameter.

d. **Paramagnetic Meissner Effect (PME):** Some superconductors acquire a net positive magnetic moment when they are cooled in weak magnetic fields such as in the order of 1 Gauss. This phenomenon is known as paramagnetic Meissner effect (PME).

Paramagnetic Meissner effect has been observed on both very cleanly prepared some high temperature superconductors and some low temperature superconductors (Braunisch et al., 1992; Braunisch et al., 1993; Schliepe et al., 1993; Khomskii, 1994; Riedling et al., 1994; Thompson et al., 1995; Onbaşlı et al., 1996; Magnusson et al., 1998; Patanjali et al., 1998; Nielson et al, 2000).

Paramagnetic Meissner effect has been observed on both d.c. and a.c. magnetic moment versus temperature data of the mercury cuprate superconductors (Onbaşlı et al, 1996; Onbaşlı et al, 2009). As is seen from Fig. 14, the temperature of T_{PME}, at which the maximum paramagnetic signal is observed on the imaginary component of magnetic moment, is known as PME temperature.

One of the main theoretical explanations of the PME is that the π-junctions between weakly coupled superconducting grains cause spontaneous orbital currents in arbitrary direction. An application of a very weak magnetic field aligns these orbital currents in the opposite direction to diamagnetic Meissner current and hence the system gains a net positive magnetic moment (Braunisch et al., 1992).

The origin of the PME is based on the weakly coupled π-junctions in which the phase difference is π. On the other hand, phase difference is associated with supercurrent of the system. From this respect, the mercury cuprates intrinsically provide the phase change of π which has a key role in the flux qubit as it was mentioned in the previous section.

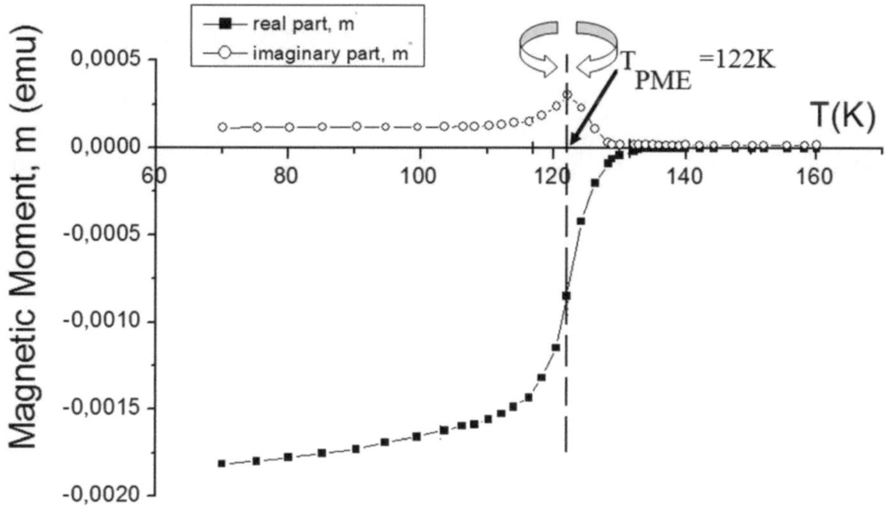

Fig. 14. The a.c. magnetic moment versus temperature data of the optimally oxygen doped mercury cuprates. The data has been taken from the MPMS-5S model quantum design SQUID magnetometer by applying 1 Gauss a.c. magnetic field. The clock wise and anti-clock wise orbital currents both exist at 122K.

As is seen from Fig. 14, for temperatures lower than T_{PME}, the imaginary component of the magnetic moment increases and the orbital current is circulating in one direction clock wise or anti-clock wise. For the temperatures higher than T_{PME}, the imaginary component of the magnetic moment decreases and the orbital current is circulating in the opposite direction to the previous state. So that at T_{PME} temperature, the clock wise and anti-clock wise currents exist. In this point of view, it has been proposed that the mercury cuprate system can be utilized as an intrinsic bulk flux qubit.

4.4 Concluding remarks on bulk flux qubit character of mercury based copper oxide layered superconductors

As it has been stated in the previous section, the general requirements of the flux qubit are fulfilled by the bulk mercury cuprate superconductors which have been summarized in the following items:

- There is no need to fabricate single Josephson junctions one by one since mercury cuprates intrinsically behaves as a Josephson junction network. Moreover, occurrence of the spatial resonance in the system also forms perfect (lossless) communication between the intrinsic Josephson junctions. In this respect, utilizing mercury cuprates for qubits may increase the present speed of quantum computations.
- There is no need to apply external ($\Phi_0/2$) magnetic flux to the qubit loop to achieve the opposite circulating currents at the same time. As is known that, in order to apply external ($\Phi_0/2$) magnetic flux to the qubit loop, rather complicated, high sensitive and expensive techniques have been used. On the other hand, the existence of opposite circulating orbital currents (clock wise and anti-clock wise) at the same time has been achieved spontaneously by the weakly coupled π -junctions in the mercury cuprates at the T_{PME}.
- In the standard qubit technology, strong microwave pulses have been utilized for manipulating the quantum superposition of these opposite circulating fluxoid states and obtaining the coherent oscillations for quantum gate operations. In this point of view, for qubits produced by the mercury cuprates, the intrinsic microwave cavity behaviour also provides continuous coherent oscillations for the lossless communicated intrinsic qubits in the bulk mercury cuprates.
- One of the main aims of qubit investigations is to fabricate a quantum computer one day. This aim will come true only by obtaining many connected qubits with long coherence times. In this respect, this work may give an insight to obtain a huge number of coupled (lossless communicated) qubits.
- Another important element is that, the opposite circulating orbital currents preserve their state as long as it operates at the temperature of T_{PME}. A remarkable point that the T_{PME} temperature (122K) is approximately 20K below the critical transition temperature of 140K. In this point of view, the required working temperature is very high relative to present low temperature superconducting qubits. In the present superconducting qubit technology, superconducting Aluminum thin films have been extensively used and its critical transition temperature is just 1.2K. So that working with mercury cuprates would lower the cost for technological applications.
- Morever, to fabricate the single intrinsic flux qubit with mercury cuprates is possible by referring to the Scanning Electron Microscopy (SEM) data of the optimally oxygen doped mercury cuprates. The mentioned intrinsic layered structure is shown in Fig.15. The primitive cell of the mercury cuprate contains two intrinsic Josephson junctions in the thickness of approximately 1.5 nm (Aslan et al., 2009). In this context, by using an appropriate technology, it is possible to extract three intrinsic Josephson junctions of about 2.25nm to fabricate single flux qubit with mercury cuprates.

Fig. 15. SEM photography of the optimally oxygen doped $HgBa_2Ca_2Cu_3O_{8+x}$ superconductors. The experiment has been performed JSM-5910 LV Scanning Electron Microscopy.

5. Bolometer applications of high temperature superconductors

Cosmology experiments show that the Universe consists of 73% Dark Energy, 23% Dark Matter and only 4% ordinary matter. The acceleration of the Universe occurs by unknown forces due to the increasing dominance of a mysterious dark energy. In order to resolve the nature of the dark energy and matter, a new generation of telescopes, which are designed to measure the polarization in the cosmic microwave background, is needed. For this kind of telescopes, the advanced detectors called as bolometers are required (Kuzmin, 2006).

The bolometer is a thermal detector, which employs an electrical resistance thermometer to measure the temperature of a radiation absorber. In the bolometer, the higher the energy is absorbed, the higher the temperature will be. The variation of the temperature can be measured via an attached thermometer. Today, in the most bolometers semiconductor or superconductor absorptive elements are used instead of metals. These devices can be operated at cryogenic temperatures, enabling significantly greater sensitivity.

One of the promising devices made of high temperature superconducting materials are edge transition bolometers. Upon the discovery of high temperature superconductors, many studies have been focused on the application of these materials in different types of bolometers for the microwave to infrared wavelength regime. The superconducting bolometers consist of patterned thin or thick superconducting film. Their operation is based on their sharp drop in the resistance, *R* at their transition temperature, T_c. The detector is kept at a temperature close to the middle of the superconducting transition, where the dR/dT is maximum. The edge transition superconductive bolometers have been investigated in various studies (Skchez et al., 1997; Fardmanesh, 2004; Cámara Mayorgaa et al., 2006).

Photo-mixing devices needed for hot electron bolometers, which have been verified with a superconductor-insulator-superconductor (SIS) mixer, have tremendous potential for various applications such as radio astronomy, terahertz imaging, high-resolution spectroscopy, medicine, security, and defense (Cámara Mayorgaa et al., 2006). Moreover, it has been reported that the design, fabrication and performance of a high temperature $GdBa_2Cu_3O_{7-x}$, superconductor bolometer positioned on a thick silicon nitride membrane. The technological feasibility of this high-Tc superconductor transition edge bolometer investigated could satisfy the requirements of a Fabry-Perot (FP) based satellite instrument designed for remote sensing of atmospheric hydroxyl ion (Skchez et al., 1997). Furthermore, the SQUID readout has been already developed for bolometers such as Cold Electron Bolometer (CEB) (Kuzmin, 2006). In addition to these works, it has been reported that a superconductor-insulator-metal bolometer with microwave readout is suitable for large format arrays (Schmidt et al., 2005).

In this study, the mercury based copper oxide superconductors have been proposed as a sensitive and reliable microwave bolometers to be used for the cosmic researches due to the special effect occurs at the paramagnetic Meissner temperature (T_{PME}) which coincides to the space temperature (Aslan, 2007; Aslan et al.,2009). As was explained in the previous section, the PME is intrinsic property which is observed at the vicinity of TPME=122K for the mercury based cuprate superconductors. At this temperature, clock wise and anti-clock wise orbital currents cancel the noise factor in the system. Moreover, at this temperature, the system emits microwaves intrinsically. Since, the temperature of T_{PME} approximately equals to the space temperature, the system reliably works at the space as an intrinsic microwave bolometers for the investigations of dark energy and dark matter qualitatively. Moreover, the temperature of T_{PME} can be modified by oxygen doping rates which enable us to obtain wider range of bolometric measurements. Furthermore, an alternative method for differential resistivity measurements, observation of change in orbital current has been suggested to detect of dark energy very precisely via PME.

6. References

Anderson, P.W. (1997). *The Theory of Superconductivity in the High-Tc Cuprates*, Princeton University Press, ISBN: 0-691-04365-5, Princeton, New Jersey.

Anderson, P.W. (1998). C-Axis electrodynamics as evidence for the interlayer theory of high temperature superconductivity. *Science*, Vol. 279, No. 5354, (20 February 1998) 1196-1198, ISSN:0036-8075 (Print)

Annett, J.F (2004), *Superconductivity, Superfluids and Condensates*, Oxford University Press, ISBN: 0-19-850755-0 (Hbk), Great Britain.

Ariza, M.; Junqué, C.; Mataro, M., Poca, M.A.; Bargallo, N.; Olondo, M. & Sahuquillo, J. (2004) Neuropsychological Correlates of Basal Ganglia and Medical Temporal Lobe NAA/Cho Reductions in Traumatic Brain Injury. *Arch Neurol* Vol.61, No.4 (April 2004), pp 541-544 ISSN 0375-8540 (Online).

Aslan, Ö. (2007). Investigation of the symmetries and the breakages in relativistic and nonrelativistic regions in high temperature superconductors. PhD Thesis Marmara University Institute of Pure and Applied Sciences, Turkey.

Aslan, Ö.; Güven Özdemir, Z.; Keskin, S.S. & Onbaşlı, Ü. (2009). The chaotic points and XRD analysis of Hg-based superconductors". *Journal of Physics: Conference Series*, Vol 153, Number. 1/ 012002 1-9, ISSN:1742-6596 (online), ISSN:1742-6588 (print)

Bachelard, H. & Badar-Goffer, R. (1993). NMR Spectroscopy in Neurochemistry. Review, *J. Neurochem* Vol.61, No.2, pp. 412-429, ISSN 0022-3042.

Bae M.H. & Lee, H.J. (2006). Progress in THz generation using Josephson fluxon dynamics in intrinsic junctions. *IEICE Trans. Electron*. Vol. E89-C, No. 2, 106-112, ISSN:0916-8516 (Print).

Bayer Schering Pharma Web site, (December 2010), Available from: http://www.diagnostic-imaging.bayerscheringpharma.de/scripts/pages/en/public/modalities/magnetic_resonance_imaging_mri/index.php

Blamire, A.M. (2010) NewCastle University, NewCastle Mangetic Resonance Center Web page, Avaliable from: http://www.ncl.ac.uk/magres/research/brain/

Braunish, W.; Knauf, N.; Kataev, V.; Neuhausen, S.; Grutz, A.; Kock, A.; Roden, B.; Khomskii, D. & Wohlleben, D. (1992). Paramagnetic Meissner effect in Bi high-temperature superconductors. *Physical Review Letters*, Vol. 68, Issue 12 1908-1911, ISSN:1079-7114 (Online), ISSN:0031-9007 (Print).

Braunish, W.; Knauf, N.; Bauer, G.; Kock, A.; Becker, A.; Freitag, B.; Grutz, A.; Kataev, V.; Neuhausen, S.; Roden, B.; Khomskii, D. & Wohlleben, D. (1993). Paramagnetic Meissner effect in high-temperature superconductors. *Physical Review B*, Vol. 48, Issue 6 4030-4042, ISSN:1550-235X (Online), ISSN:1098-0121 (print),

Brooks, W.M; Stidley, C.A.; Petropoulos, H.; Jung, R.E.; Weers, D.C.; Friedman, S.D.; Matthew A. Barlow, M.A. Sibbitt Jr., W.L. & Yeo, R.A. (2000). Metabolic and Cognitive Response To Human Traumatic Brain Injury: A Quantitative Proton Magnetic Resonance Study. *J Neurotrauma* Vol.17, No.8 (August 2000), pp. 629–640, ISSN 1557-9042 (Online).

Brooks, W.M.; Friedman, S.; Gasparovic, C. (2001). Magnetic Resonance Spectroscopy In TBI. *J. Head Trauma Rehabil*, Vol.16, No.2, pp. 149-164, ISSN 1550-509X (Online).

Cámara Mayorgaa, I. ; Muñoz Pradas, P.; Michael, E. A. ; Mikulics, M.; Schmitz, A. ; van der Wal, P.; Kaseman, C.; Güsten, R.; Jacobs, K.; Marso, M.; Lüth, H. & Kordoš, P. (2006). Terahertz photonic mixers as local oscillators for hot electron bolometer and superconductor-insulator-superconductor astronomical receivers, *Journal Of Applied Physics*, Vol. 100, No.4, 043116/1-4, ISSN:0021-8979.

Cecil, K.M.; Hills E.C.; Sandel, M.E.; Smith D.H.; McIntosh, T.K. ; Mannon, L.J.; Sinson, G.P.;. Bagley, L.J.; Grossman, R.I. & Lenkinski, R.E. (1998). Proton Magnetic Resonance Spectroscopy For Detection Of Axonal İnjury In The Splenium Of The Corpus Callosum Of Brain-Injured Patients. *J* Neurosurg, Vol. 88, No.5, (May 1998), pp. 795–801, ISSN 1933-0693 (Online).

Chiorescu, I.; Nakamura, Y.; . Harmans, C.J.P.M & Mooij, J.E. (2003). Coherent quantum dynamics of a superconducting flux qubit, *Science*, Vol. 299, No. 5614, 1869-1871, ISSN: 0036-8075 (Print).

Clarke, J. & Wilhelm, F.K. (2008). Superconducting quantum bits. *Nature* Vol. 453, No. 7198, (19 June 2008), 1031-1042, ISSN:0028-0836.

Clegg, B. (2006). *The God Effect: Quantum Entanglement, Science's Strangest Phenomenon*, St. Martin's Griffin, ISBN-10: 0-312-34341-8, USA.

Danielsen, E.R & Ross, B. (1999). *Magnetic Resonance Spectroscopy Diagnosis of Neurological Diseases*, Marcel Dekker Inc, ISBN 0-8247-0238-7, USA.

Deutchs, D. (1997). *The Fabric of Reality: The Science of Parallel Universes--And Its Implications*, Penguin Group, ISBN-13: 978-0-140-14690-5, England.

Dumé, B. (30 June 2004). Entanglement breaks new record. Avaliable from: http://physicsworld.com/cws/article/news/19793

Emuidzinas J. & Richards P.L. (2004). Superconducting Detectors and Mixers for Millimeter and Submillimeter Astrophysics. *Proc. IEEE* Vol. 92, pp.1597–1616, ISSN 0018-9219.

Eruygun, T.O. (1998), Usage of Advanced Superconducting Devices for Forensic Science, Master Thesis Supervised by Ü. Onbaşlı, İstanbul University, Forensic Science Institution.

Fardmanesh, M. (2004). Response analysis and modeling of high temperature superconductors edge transition bolometers. In: *High Temperature Superconductivity 2, Engineering materials,* A.V.Narlikar (Ed.), 477-536, Springer-Verlag, ISBN 3-540-40639-5, Germany.

Feng, Y.; Zhuang, W. & Prohofsky, E.W. (1991). Calculation of Temperature Dependence of Interbase Breathing Motion of a Guanine-Cytosine DNA Double Helix With Adenine-Thymine Insert. *Phys. Rev. A* Vol. 43, No. 2 pp. 1049–1053, ISSN 1094-1622 (Online).

Ferrell R. & Prange R. (1963). Self field of Josephson tunneling of superconducting electron pairs. *Physical Review Letters*, Vol. 10, No. 11, 479-481, ISSN:1079-7114 (Online).

Fishbine, B. (Spring 2003). SQUID Magnetometry, In: *Los Alamos National Laboratory Web Site* (December 2010), Avaliable from: <http://www.lanl.gov/quarterly/q_spring03/meg_helmet_measurements.shtml>

Fossheim, K. & Sudbo, A. (2004) *Superconductivity: Physics and Applications,* John Wiley & Sons, Ltd, ISBN-10: 0-470-84452-3, Great Britain.

Frahm, J.; Bruhn, H.; Gyngell, M.L.; Merboldt, K.D.; Hänicke, W. & Sauter, R. (1989). Localized High-Resolution Proton NMR Spectroscopy Using Stimulated Echoes: Initial Applications To Human Brain *In Vivo. Magn. Reson. Med. Vol. 9, No.1, (January 1989), pp.79-93, ISSN 1522-2594 (Online).*

Friedman, J.R.; Patel, V.; Chen, W.; Tolpygo, S.K. & Lukens, J.E. (2000). Quantum superposition of distinct macroscopic states. *Nature* Vol. 406, No. 6791, 43-46, ISSN: 0028-0836.

Friedman, S.D.; Brooks, W.M.; Jung , R.E.; Chiulli, S.J.; Sloan, J.H.; Montoya, B.T.; Hart, B.L. & Yeo, R.A. (1999). Quantitative Proton MRS Predicts Outcome After Traumatic Brain İnjury. *Neurology* Vol.52, (April 1999), pp. 1384–1391, ISSN 1526-632X (Online).

Friedman, S.D.; Brooks, W.M.; Jung, R.E.; Hart, B.L. & Yeo, R.A. (1998). Proton MR Spectroscopic Findings Correspond To Neuropsychological Function In Traumatic Brain İnjury. *AJNR Am J Neuroradiol,* Vol. 19 No.10, (November/December 1998), pp. 1879–1885, ISSN 1936-959X (Online).

Garnett, M.R.; Blamire, A.M.; Corkill, R.G.; Cadoux-Hudson, T.A.; Rajagopalan, B. & Styles, P. (2000). Early Proton Magnetic Resonance Spectroscopy In Normal-Appearing Brain Correlates With Outcome In Patients Following Traumatic Brain İnjury. *Brain* Vol.123, No.10, (October 2010), pp.2046–2054, ISSN 1460-2156 (Online).

Garnett, M.R.; Corkill, R.G.; Blamire, A.M.; Rajagopalan, B.; Manners, D.N.; Young, J.D.; Styles, P. & Cadoux-Hudson, T.A.D. (2001). Altered Cellular Metabolism Following

Traumatic Brain Injury: A Magnetic Resonance Spectroscopy Study. *J Neurotrauma* Vol.18, No.3, (March 2001), pp. 231-240 ISSN 1557-9042 (Online).

Georgia State University (n.d.) Web site (December 2010), Avaliable from: <http://hyperphysics.phy-astr.gsu.edu/hbase/nuclear/mri.html>

Govindaraju, V.; Gauger, G.E.; Manley, G.T.; Ebel, A.; Meeker, M. & Maudsley, A.A. (2004). Volumetric Proton Spectroscopic Imaging of Mild Traumatic Brain Injury. *AJNR Am J Neuroradiol* Vol.25, (May 2004), pp. 730-737, ISSN 1936-959X (Online).

Güven Özdemir Z.; Onbaşlı, Ü. & Aslan, Ö. (2007). Calculation of microwave plasma oscillation in high temperature superconductors. In: *The Seventh International Conference on Vibration Problems ICOVP 2005 Springer Proceedings in Physics*, E. İnan & E. Kırış (Eds.), 377-382, Springer, ISBN: 978-1-4020-5400-6, Dordrecht, The Netherlands.

Güven Özdemir, Z. (2007). Determination of Electrical and Magnetic Properties of Mercury Based CuO_2 Layered Superconductors. PhD Thesis Marmara University Institute of Pure and Applied Sciences, Turkey.

Güven Özdemir, Z.; Aslan, Ö. & Onbaşlı, Ü. (2009). Terahertz oscillations in mercury cuprate superconductors. *Pramana-Journal of Physics* Vol.73, No.4 pp. 755-763, ISSN 03044289.

Güven Özdemir, Z. (2011). Chaotic Point Works as Qubit in High Temperature Superconductors, *Journal of Applied Functional Analysis (JAFA)* Vol. 6 No 2, pp 165-172, ISSN 1559-1948 (Print) 1559-1956 (Online).

Hans Mooij's research group at Delft University of Technology, (2005). Avaliable from: http://www.physorg.com/news4996.html

Holshouser, B.A.; Ashwal, S.; Shu, S.; Hinshaw, D.B. Jr. (2000). Proton MR Spectroscopy In Children With Acute Brain Injury: Comparison Of Short And Long Echo Time Acquisitions. *J Magn. Reson Imaging* Vol.11, No. No.1, pp. 9–19, ISSN 1522-2586 (Online).

Ito, H..; Nakajima, F.; Furuta, T. & Ishibashi, T. (2005). Continuous THz-wave generation using antenna-integrated uni-travelling-carrier photodiodes. *Semicond. Sci. Technol* Vol. 20 No. 7, S191-S198, ISSN: 0268-1242 (Print).

Josephson, B.D. (1962). Possible new effects in superconducting tunneling. *Physics Letters*, Vol. 1 No. 7, (1 July 1962) 251-253, ISSN: 0375-9601.

Ketterson, J. B. & Song S. N. (1999) *Superconductivity*, Cambridge University Press, ISBN:0-521-56295-3, United Kingdom.

Khomskii, D. I. (1994). Wohlleben effect (Paramagnetic Meissner effect) in high-temperature superconductors, *Journal of Low Temp. Physics*, Vol. 95, No. 1-2 205-223, ISSN: 0022-2291.

Kleiner, R. & Müller, P. (1994). Intrinsic Josephson effects in high-T_c superconductors, *Phys. Rev. B*, Vol. 49,No. 2, 1327-1341, ISSN: 1098-0121 (Print).

Kuzmin, L. (2006) Optimal Cold-Electron Bolometer with a Superconductor-Insulator-Normal Tunnel Junction and an Andreev Contact, 17th International Symposium on Space Terahertz Technology TH3-5, 183-186.

Lu, J.Y; Chang, H.H.; Chen, L.J.; Tien, M.C. & Sun, C.K. (2010). Terahertz-Wave Molecular Imaging Based on a Compact High-Efficiency Photonic Transmitter.

http://gra103.aca.ntu.edu.tw/gdoc/93/D91941004a.pdf

Luyten, P.R. & Den Hollander, J.A. (1986). Observation of Metabolites In The Human Brain By MR Spectroscopy. *Radiology* Vol.161, No.3, (December 1986), pp. 795-798, ISSN 1527-1315.

Machida M. & Tachiki M. (2001). Terahertz electromagnetic wave emission by using intrinsic Josephson junctions of high-Tc superconductors. *Current Appl. Phys.* Vol. 1, No.4-5, 341-348, ISSN: 1567-1739.

Magnusson, J. ; Papadopoulou, E.; Svedlindh, P. & Nordblad, P. (1998). Ac susceptibility of a paramagnetic Meissner effect sample, *Physica C*, Vol. 297, No:3-4, (10 March 1998) 317-325, ISSN:0921-9601.

Minami, H.; Kakeya, I. ; Yamaguchi, H.; Yamamoto, T. & Kadowaki, K. (2009). Characteristics of terahertz radiation emitted from the intrinsic Josephson junctions in high-*Tc* superconductor $Bi_2Sr_2CaCu_2O_{8+\delta}$. *Appl. Phys. Lett.* Vol. 95, No. 23 232511/1- 3, ISSN:0003-6951 (Print).

Moody, N.A. (2009). New Waves: Solid State THz Source, Los Alamos National Laboratory, Avaliable from:
http://www.lanl.gov/science/ldrd/LDRD-Day-2009/posters/Moody.pdf

Mooij, J.E. ; Orlando, T.P.; Levitov, L.; Tian, L.; . van der Wal, C. H. & Lloyd S. (1999). Josephson persistent current qubit , *Science* Vol .285, no 5430, 1036-1039, ISSN: 0036-8075 (Print).

Mooij, H. (2010). Superconducting qubits: quantum mechanics by fabrication Avaliable from: ftp://ftp.cordis.europa.eu/pub/ist/docs/fet/qip2-eu-29.pdf

National High Magnetic Field Laboratory, Magnet Lab web site, (2010), Avaliable from:
http://www.magnet.fsu.edu/education/tutorials/magnetacademy/magnets/page5.html

Nielsen, A. P.; Cawthorne, A.B.; Barbara, P.; Wellstood, F.C.; Lobb, C.J.; Newrock, R.S. & Forrester, M.G. (2000). Paramagnetic Meissner effect in multiply-connected superconductors, *Physical Review B*, Vol. 62, No.21 14380-14383, ISSN: 1550-235X (Online).

Nossal, R. & Lecar, N. (1991). *Molecular & Cell Biophysics*, Perseus Books, ISBN-10: 0201195607, Cambridge, MA.

Onbaşlı, Ü. ; Wang, Y.T.; Naziripour, A.; Tello, R.; Kiehl, W. & Hermann, A.M. (1996). Transport properties of high Tc mercury cuprates. *Physica Status Solidi B* Vol. 194, 371-382, ISSN: 0370-1972

Onbaşlı Ü.; Eruygun, T.O. & Dinçer, A. (1999), Sensitive Usage of Proton MRS in Forensic Medicine, *4th European Conference on Applied Superconductivity*, Sitges, Spain.

Onbaşlı , Ü. ; Güven Özdemir, Z. & Aslan, Ö. (2009). Symmetry breakings and topological solitons in mercury based d-wave superconductors. *Chaos, Solitons & Fractals* Vol. 42, No. 4, (30 November 2009) 1980-1989, ISSN:0960-0779 (Online).

Optical Lattices and Quantum Information web site (2011). Avaliable from:
http://olaqui.df.unipi.it/beginners.html

Özdemir, Z.G.; Aslan Ö. & Onbaşlı Ü. (2006). Determination of c-axis Electrodynamics Parameters of Mercury Cuprates. *Journal of Physics and Chemistry of Solids*, Vol.67 No.1-3 (January-March 2006) pp. 453-456 ISSN 0022-3697.

Panagopoulos, C.; Cooper, J.R.; Peacock, G.B.; Gameson, L.; Edwards, P.P.; Schmidbauer, W. & Hodby, J.W. (1996). Anisotropic magnetic penetration depth of grain-aligned $HgBa_2Ca_2Cu_3O_{8+\delta}$. *Phys. Rev B*, Vol. 53, No. 6, R2999-R3002, ISSN: 1098-0121 (Print).

Panagopoulos, C. & Xiang, T. (1998). Relationship between the superconducting energy gap and the critical temperature in high-Tc superconductors. *Physical Review Letters*, Vol. 81, 2336–2339, ISSN: 1079-7114 Online), ISSN: 0031-9007 (Print).

Patanjali, P. V. ; Seshu Bai, V.; Kadam, R.M. & Sastry, M.D.(1998). Anomalous microwave absorption in GdBCO powder: π–junctions and the paramagnetic Meissner effect. *Physica C*, Vol. 296, No. 3-4, (20 February 1998) 188-194, ISSN:0921-4534.

Rao, V.; Spiro, J.; Degoankar, M.; Horská, A.; Rosenberg, P.B.; Yousem, D.M.; Barker, P.B. & Phil, D. (2006), Lesion Location in Depression Post Traumatic Brain Injury Using Magnetic Resonance Spectroscopy: Preliminary Results From a Pilot Study. *Eur. J. Psychiat.* Vol.20, No.2, pp. 65-73, ISSN 0213-6163.

Riedling, S. ; Brauchle, G. ; Lucht, R.; Röhberg, K.; Löhneysen, H.V.; Claus, H.; Erb, A.& Müler-Vogt, G.(1994). Observation of the Wohlleben effect in $YBa_2Cu_3O_{7-\delta}$ single crystals. *Physical Review B*, Vol. 49, No.18 13283-13286, ISSN: 1550-235X (Online).

Ross, B.D.; Ernst, T.; Kreis, R.; Haseler, L.J.; Bayer, S.; Danielsen, E.; Blüml, S.; Shonk, T.; Mandigo, J.C.; Caton, W.; Clark, C.; Jensen, S.W.; Lehman, N.L.; Arcinue, E.; Pudenz R. & Shelden, C.H. (1998). 1H MRS In Acute Traumatic Brain Injury. *J. Magn Reson Imaging*, Vol.8, No.4 (July/August 1998), pp. 829–840, ISSN 1522-2586 (Online).

Royal Adelaide Hosptital web site, (2010), Avaliable from: http://www.rah.sa.gov.au/birs/bi_brain.php

Salibi, N.M. & Brown, M.A. (1998) *Clinical MR Spectroscopy: First Principles*, Wiley-Liss Inc., ISBN-10: 0-471-18280-X, USA.

Schliepe, B.; Stindtmann, M.; Nikolic, I. & Baberschke, K. (1993). Positive field-cooled susceptibility in high-Tc superconductors. *Physical Review B*, Vol. 47, No. 13 8331-8334, ISSN: 1550-235X (Online).

Schmidt, D. R.; Clark, A. M.; Duncan, W. D.; Irwin, K. D.; Miller, N.; Ullom, J. N. & Lehnert, K. W. (2005). A superconductor-insulator-normal metal bolometer with microwave readout suitable for large-format arrays, *Appl. Phys. Lett.* Vol. 86, No. 5, 053505/1-3, ISSN:0003-6951 (Print).

Skchez, S.; Elwenspoek, M.; Gui, C. ; de Nivelle, M.J.M.E. ; de Vries, R.; de Korte, P.A.J.; Bruijn, M.P.; Wijnbergen, J.J.; Michalke, W.; Steinbeie, E.; Heidenblut, T. & Schwierzi, B. (1997). A high-Tc superconductor bolometer on a silicon nitride membran, In: *Micro Electro Mechanical Systems, Proceedings, IEEE MEMS '97*, 506-511, Print ISBN: 0-7803-3744-1

Tafuri, F.; Kirtley, J.R.; Lombardi, F.; Bauch, T.; Il'ichev, E.; Miletto Granozio, F.; .Stornaiuolo, D & Scotti di Uccio, U. (2004). Flavours of Intrinsic d-Wave Induced Effects In $YBa_2Cu_3O_{7-\delta}$ Grain Boundary Josephson Junctions, In Institute of Physics Conference Series Number 181: Proceedings of the Sixth European Conference on Applied Superconductivity, A Andreone, G.P. Pepe, R. Cristiano &G. Masullo (Eds.), 273-282, IOP Publishing, United Kingdom.

Teasdale, G. & Jennett, B. (1974). Assesment of Coma and Impaired Consciousness: A Practical Scale" *Lancet*, Vol.2, No.7872, (13 July 1974), pp. 81-84, ISSN 0140-6736.

Thompson, D.J.; Minhaj, M.S.M.; Wenger, L.E. & Chen, J.T. (1995). Observation of paramagnetic Meissner effect in niobium disks. *Physical Review Letters* Vol. 75, No.3 529-532, ISSN: 1079-7114 (Online).

Tonouchi, M. (2007). Cutting-Edge Terahertz Technology, *Nature Photonics* Vol. 1 , pp.97-105, ISSN 1749-4885.

Van der Wal, C. H. ; Haar, A.C.J.T.; Wilhelm, F.K.; Schouten, R.N.; Harmans, C.J.P.M.; Orlando, T.P. ; Lloyd, S. & Mooij, J.E. (2000). Quantum superposition of macroscopic persistent-current states, *Science*, Vol. 290, No. 5492, 773-777, ISSN: 0036-8075 (Print).

Young, L.; Prabhu, V.V. & Prohofsky, E.W. (1990). Prediction of Modes With Dominant Base Roll and Propeller Twist in *B*-DNA Poly(dA)-Poly(dT), *Phys. Rev. A* Vol. 41, No. 12, pp. 7020–7023, ISSN 1094-1622 (Online).

Walther, C.; Scalari, G..; Faist, J. ; Beere, H. & Ritchie, D. (2006). Low frequency terahertz quantum cascade laser operating from 1.6 to 1.8 THz. *Appl. Phys. Lett.* Vol. 89, No. 23 231121/1- 3, ISSN:0003-6951 (Print).

Weber State University web site, (2010), Avaliable from: http://departments.weber.edu/chfam/2570/Neuro-3.jpg

Wellstood, F.C.; Urbina C. U. & Clarke, J. (1987), Low-frequency noise in dc superconducting quantum interference devices below 1 K. *Appl. Phys. Lett.*, Vol. 50, No. 12, 772/1-3, ISSN:0003-6951 (Print).

Wikipedia, (December 2010), Avaliable from: <http://en.wikipedia.org/wiki/Magnetic_resonance_imaging>

Zhuang W.; Feng, Y. & Prohofsky, E.W. (1990). Self-Consistent Calculation of Localized DNA Vibrational Properties At A Double-Helix–Single-Strand Junction With Anharmonic Potential. *Phys. Rev. A* Vol. 41, No. 12, pp.7033–7042, ISSN 1094-1622 (Online).

Zomega Terahertz Corp. Web site, (2010), Avaliable from: http://www.zomega-terahertz.com/what-is-thz.html

3

Superconductivity Application in Power System

Geun-Joon Lee
Chungbuk Provincial College
Republic of Korea

1. Introduction

Electric power system is one of the most important infra-structure of modern digital society. This energy, which is easy to control, to be converted any type of energy, and clean, is becoming the standard how the society is developed well and the demand of electricity is increasing rapidly over the world.

However, in most highly developed electrical power system, there are several difficulties related from generation to distribution. Usually, power generation is located remote area from the load center, long transmission and distribution lines have to be constructed and maintained to meet required reliability, power quality and economic point of views. Reliable, cheap, efficient conductor is required to support desirable electric power systems.

Most of conductors used in modern power system facilities, for example, generator, transformer, transmission line, cable, motor etc., are copper or aluminum. They have resistance R which restricts the capability of thermal rating of electric facilities with the ohmic loss. If there is a conductor with no loss, we can make efficient electrical facilities. Superconductor, which is zero resistance, is a promising solution to make innovation on electric facilities.

This chapter introduces various power system facilities based on superconductor application. First of all, superconducting cable is most applicable solution to solve transmission congestion problem in high power density area such as metropolitan cities with its high density transmission capability. Recently developed superconducting cable in distribution class can deliver about 5 times more power than conventional XLPE cable at same dimension. DC superconducting cable is also in developing stage to eliminate AC loss in superconductor, and will be applied to HVDC transmission system. Section 2 introduces superconducting cable in power system.

Second promising one is Superconducting Fault Current Limiter (SFCL).With the development of power system, short circuit fault currents are increasing much more than conventional power system which is the components of present system. For example, a lot of circuit breakers have to be replaced higher level break capacity in case of source impedance is reduced by increased power system generation and/or reinforced transmission and distribution system.

SFCL can limit fault current fast, within 1/2 cycle, using quench effect of superconductor in case of current exceeds specified fault current. Also, it can supply a solution on power system voltage sag problem. Section 3 introduces various type of SFCL and their application.

Other promising applications in power system are Superconducting Synchronous Condenser (DSC : SuperVar) and Superconducting motor. SuperVar is a good solution as

reactive power compensator which can be applied to increase power transmission capability on voltage stability limited system. Also, it can support industry sector which require high voltage quality service. Section 4 introduces SuperVar and superconducting motor with their application.
There are a lot of superconductor application field in power system. However, the basic discussion has to be start with the study whether the power system requirements can have better solution from superconducting electric facilities. In this discussion, we will present to supply some examples how to consider superconducting facilities on modern electric power system. Lastly, we will discuss how to apply superconducting facilities to electric power system.

2. Superconducting cable

Traditionally, the main stream of power delivery system are composed by ACSR(Aluminum Cable Streel Reinforaced) in overhead line and XLPE(Cross Link Poly Etheline) underground cable. In modern highly industrialized society, which requires much higher capacity in transmission and distribution line with the increase of electricity consumption due to energy transition to electricity and population convergence into metropolitan area. However, it is almost impossible to build new power delivery system in metropolitan area in environmental point of view.

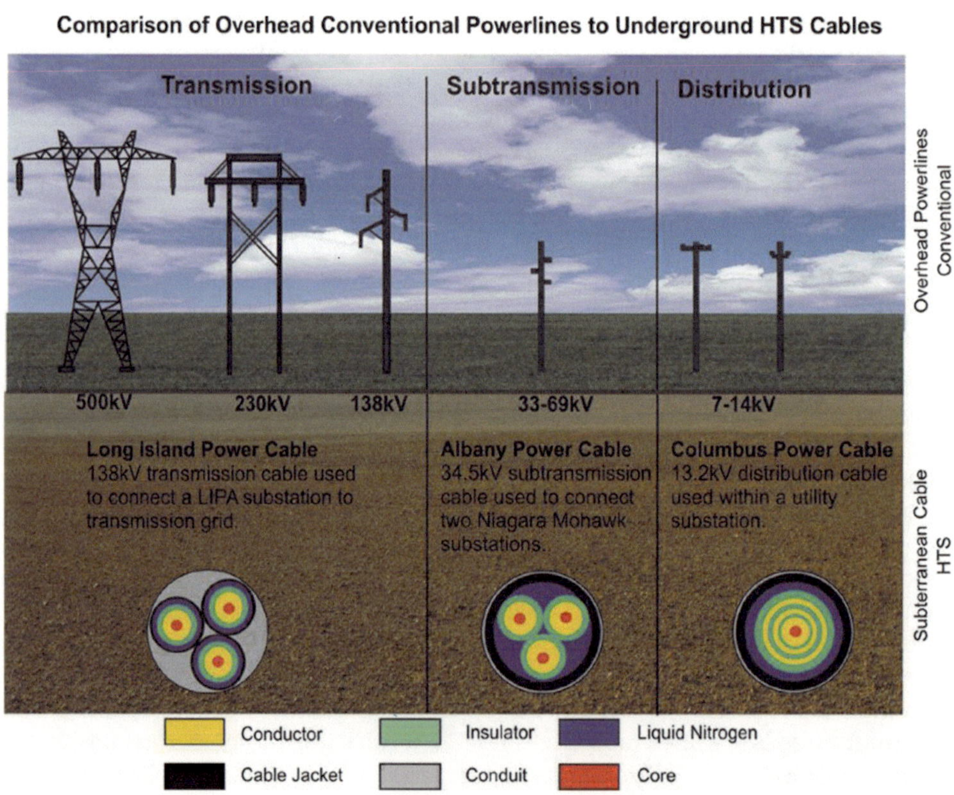

Fig. 1. Comparision of overhead power lines to HTS cable (http://www.DoE.gov)

Since superconducting phenomenon was developed by Kamerling Onnes in 1911, research and development on superconducting materical has been progressed actively over the world. After McFee suggessed superconducting cable at first in 1961, R&D on low temperature superconducting(LTS) cable using Helium cooling system had been studied during 1970's and 1980's.

In 1986, high temperature superconducting(HTS) material which use liquid Nitrogen(LN) instead of Helium had developed by Bednorz and Muller, research on HTS cable has been progressed continuously, and is in industrial application stage at present[1 ~ 3]. Several leading countries, including USA, China, Japan, Europe and Korea already experienced HTS(High Temperature Superconducting) cable test operation[ ], and finding good applicable places in engineering point of view.

HTS superconducting cable, which has zero resistance and low inductance, can increase power transfer capacity about 3~5 times more than conventional XLPE cable with the same size of underground right of way, and can reduce power transmission loss and construction cost. By DoE, USA, three level of HTS cable is compared to substitute the overhead lines. Below figure shows the relative power increase compare HTS cable to XLPE cable.

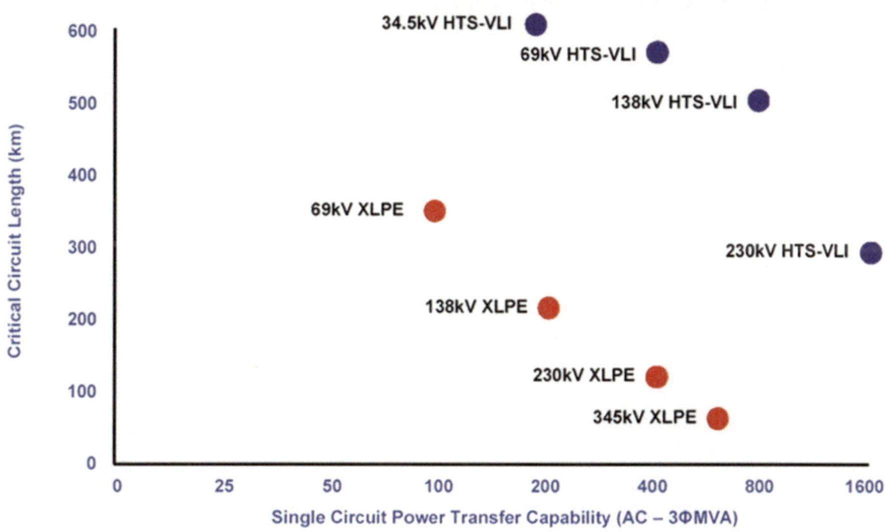

Fig. 2. Comparision of conventional cable to HTS cable

2.1 Type classification

Superconducting cables are classified various point of view. By the electrical source, it is classified AC and DC. Also, by the superconductor material, it is classified HTS(High Temperature Superconductor) which is non-metal, Oxide compound substances such as BSCCO seires and LTS(Low Temperature Superconductor) which is mainly metal seires, such as NbTi.

LTS is cooled by liquid Helium because it has superconducting property nearly absolute temperature(-273.16℃). It is very hard to get near absolute temperature with normal materals and cooling system. Also, Liquid Helium is too expensive to normal use. LTS is easy to make conductor with its ductility, but operation in near zero absolute temperature is very difficult to be utilized in industrial field, such as power transmission system.

However, HTS is cooled by liquid Nitrogen[LN2] as it has superconducting property about 70[K], temperature gradient between HTS and normal room temperature are much more reduced than LTS case, it makes easier to design cooling system for HTS cable. HTS conductors are more difficult to manufacture and handle as its plasticity is worse than LTS, however it is recognized as cost effective measure compare to LTS as power cable application. At present, LTS conductors are used for special application such as MRI(magnetic resonance imaging) system. Therefore, our discussion on power cable will focus on HTS cable, later.
HTS cable for power transmission is developed two types of design. The one is WD(Warm Dielectric Design), the other is CD(Cold dielectric coaxial Design).
Fig. 3 (a) shows the cross section of WD HTS cable. LN2 flows in the tube type former which sustains HTS cable on its outer circle. HTS conductors are surrounded by cryostate which insulates heat transfer. The dielectric is located outer of the cryostate. Therefore the dielectric does not to be cooled with LN2(Warm Dielectric). Because WD type HTS cable can not only preserve conventional cable dimension and use proved dielectric materials, but also limited HTS conductors are used(omit HTS shield), it is cost effective and efficient in design of cooling system. However, omitting shield layer produces magnetic interaction between phase to phase and limit power transfer capacity.
However, in fig 3 (b) which is the cross section of CD HTS cable, LN2 flows the outer and inner duct of cable and it cools not only HTS conductor but also dielectric material. Another important difference between CD and WD is that CD has return HTS screening conductors which shields outer magnetics and make low inductance.

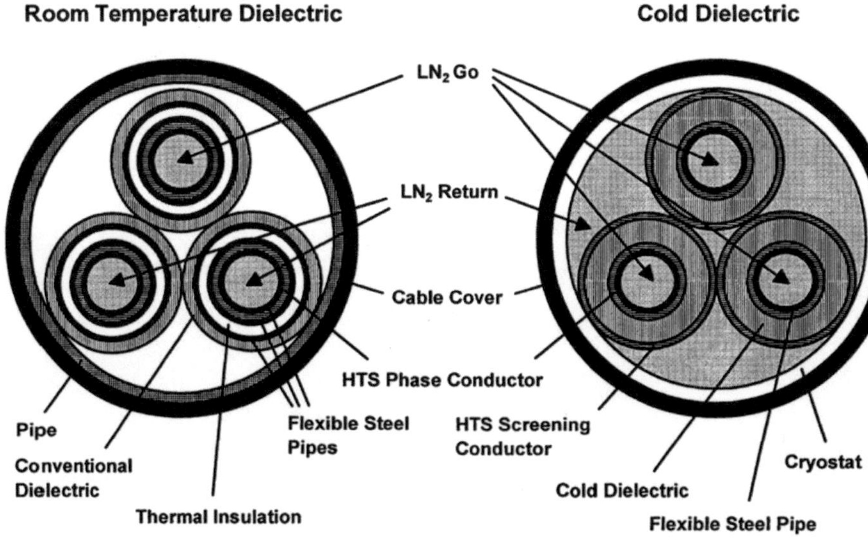

(a) WD (b) CD

Fig. 3. WD and CD HTS cable

2.2 HTS cable system

General conceptual diagram of HTS cable system is shown as below. The main components of HTS cable system are HTS cable, cooling facility, terminal and monitoring system.

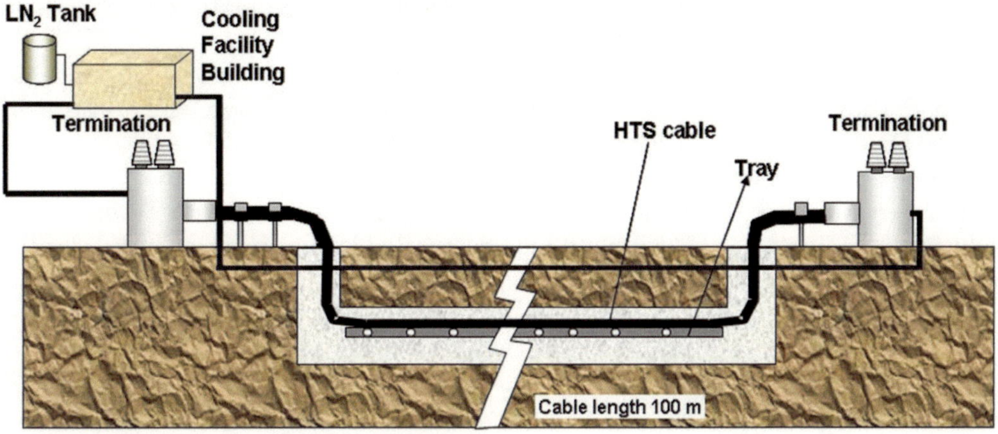

Fig. 4. General conceptual diagram of HTS cable system

2.2.1 HTS cable

Three kinds of HTS cable in outward appeareance are developed. Fig 5 shows single core cable, co-axial core cable, tri-axial cable.

(a) Single core cable

(b) tri-axial cable

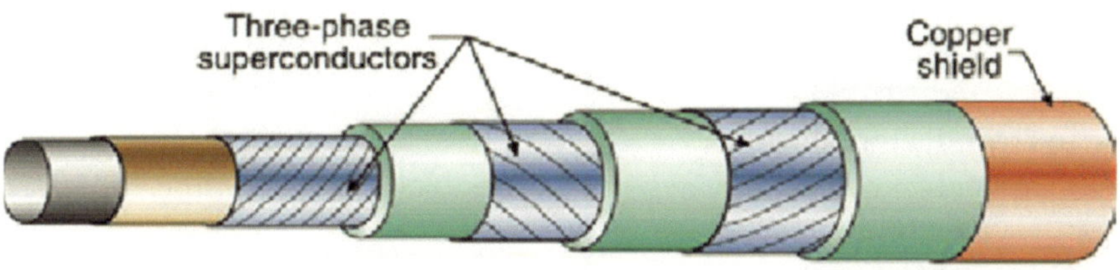

(c) Co-axial cable

Fig. 5. HTS cable type classified by core

Usually, single core type is for transmission, tri-axial type is for subtransmission and co-axial type is distribution.

The performance of HTS cable depends on the quality of HTS tape. HTS tape for power cable has to be produced long enough to fulfill the required length of cable core to be installed, also have sufficient critical current density and uniform current and good mechnical characteristics.

Recently, the improvement of critical current and length in Bismuth series high temperature superconducting wire make possible to realize HTS power cable application in real field. BSCCO-2223, the recently developed HTS conductor which has almost 110[K] critical temperature, is mainly applied to make HTS cable.

Fig. 6 shows CD type HTS cable cross section. It is composed with Former(copper), conductor(HTS), Electrical Insulation(PPLP), electrical shielding(HTS), stainless sheath for thermal insulation and cladding material.

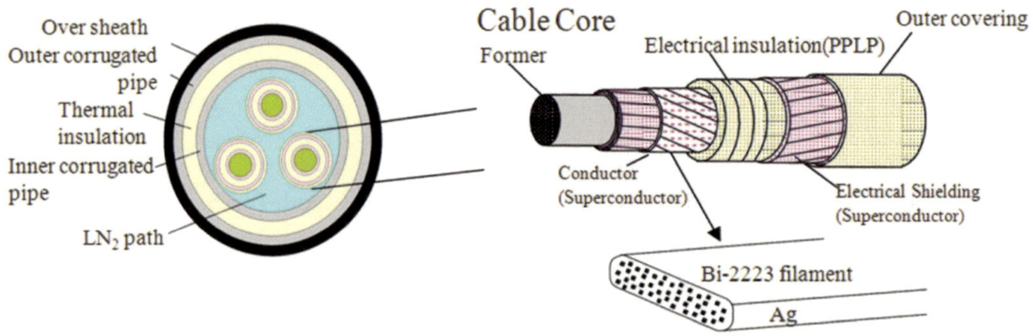

Fig. 6. Cross section of HTS cable(CD type, for distribution system)

Table 2 shows one of HTS cable specification for 22.9kV distribution line. It is designed for replace present distribution cable system without changing underground right of way.

Item	Specification
Former	Stranded copper
Conductor	Bi-2223, 2 layer
Shield layer	Bi-2223, 1 layer
Electrical insulation	PPLP, 4.5 mm
Cable core diameter	35 mm
Superconductor/shield	Bi-2223 tape
Thermal insulation	Double corrugated pipe, MLI, Vacuum
Oversheath	PE
Cable outer diameter	130 mm

Table 1. Example of HTS cable specifications (CD type, for distribution system)

2.2.2 Cooling facility

Cooling facility is another important component of HTS cable system to maintain superconductivity with sufficient low temperature at various operating conditions. In fig.7, LN2 flows LN2 line, superconducting cable, refrigerator and pump. Cryostat prevents heat transfer from cable inner and outer.

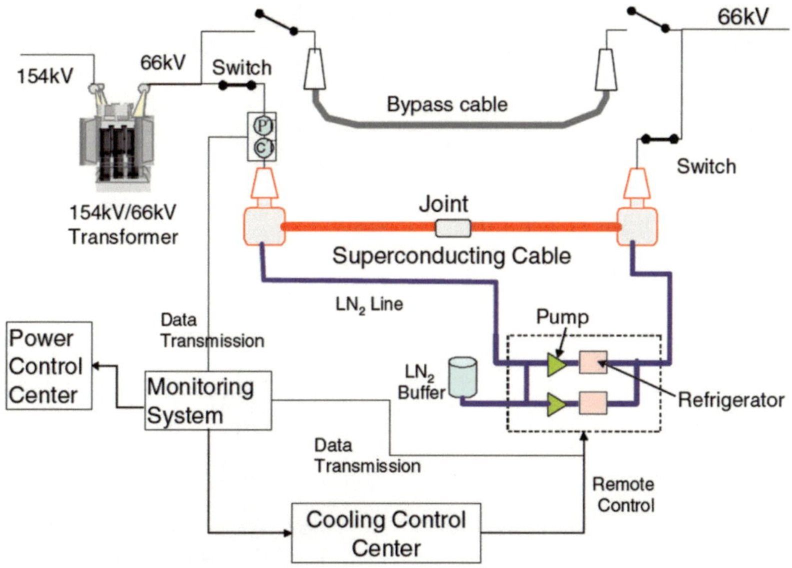

Fig. 7. HTS cable system at Albany project

2.2.3 Termination

Termination locates both ends of HTS cable. It connects HTS cable and normal temperature power line. Because of large difference of temperature between HTS cable and outer weather, termination has to sustain temperature difference and pump out heat from joint resistance.

2.2.4 Monitoring system

Monitoring system checks electrical and thermal status of HTS cable system. Electrical variables are currents and voltages. Thermal variables are temperatures of every components, such as cable inlet, outlet, refrigerator inlet and outlet etc.

2.3 Characteristics of HTS cable

2.3.1 Electrical characteristics

Brief comparison of electric characteristics among power delivery systems are suggested in table 2.

WD type can transfer about 2 times power than conventional cable at same power loss, however, CD type can transfer about 4.5 times power. Below table shows brief comparison between WD and CD type.

	conventional	HTS(WD)	HTS(CD)
Pipe outer diameter(mm)	200	200	200
Voltage(KV)	115	115	115
Power rating(MVA)	220	500	1000
power loss(W/MVA)	300	300	200

Table 2. Comparison of ratings between WD and CD HTS power cable

The capacity of WD HTS cable is about 2.5[kA] per phase at 132/150~400[kV] transmission voltage and 500~2000[MVA] per system[2]. CD type has better current capacity than WD type, 8[kA]/phase. Also, DC HTS cable can transfer 15[kA] and more at same design.

Power delivery system	Cable dimension			Electrical constants(Z_1 / Z_0)		
	Inside Radius [mm]	Outside Radius [mm]	Shield Radius [mm]	Resistance [Ω/km]	Inductance [mH/km]	Capacitance [nF/km]
Conventional XLPE	2	25	40	0.03/0.15	0.36/1.40	257/175
HTS WD type	12.7	14	29	0.0001/ 0.12	0.39/1.47	217/175
HTS CD type(VLI)	12.7	14	29	0.0001/ 0.03	0.06/0.10	200/140

Table 3. Comparision of electrical constants between WD and CD HTS power Cable

Table 3 introduces the electrical constants of HTS cable. We can find that CD type cable has only 1/6 positive sequence inductance over WD and XLPE cable which acts as impedance in AC system. This tells us CD type HTS cable shows excellent power transfer capability at steady state.

However, it has quench property if the conductor temperature rise over critical temperature, the resistivity increase dramatically. See Fig.8.

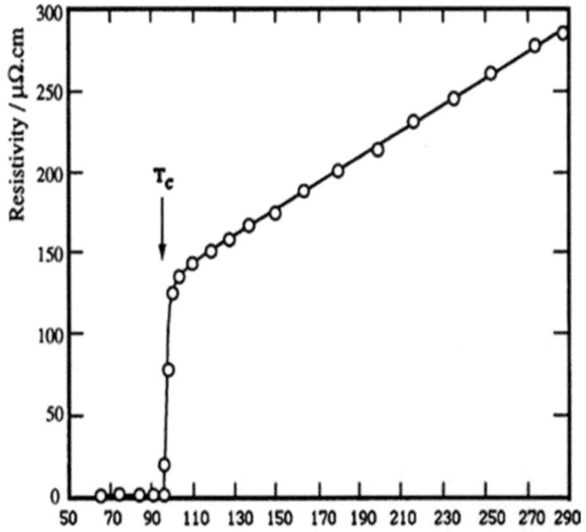

Fig. 8. Temperature and Resistivity of HTS conductor

2.3.2 Thermal characteristics

To sustain superconductivity of HTS cable in normal operation, it is very important to keep the temperature of cable system within permissible range. Depend on above figure, if temperature rise over about 97[K], quench happens.

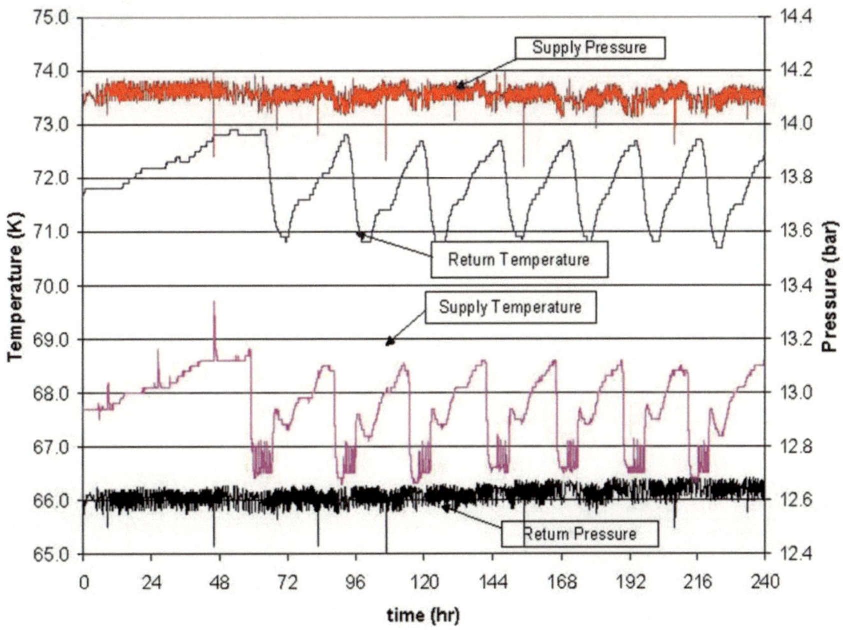

Fig. 9. Inlet and outlet temperature of HTS cable

Above figure shows the temperatures of inlet and outlet of HTS cable during load cycling operation. At both terminal, temperatures are below 73[K] and there are about 24 degrees temperature margin.

2.3.3 Operational characteristics of HTS cable system in sample system

In this section, a sample of distribution level HTS cable operation status shall be introduced to understand each electrical components response to steady and transient state. HTS cable may be operated at unbalanced 3 phase currents, harmonics, various fault condition. Well designed HTS system has to survive expected abnormal state.

2.3.3.1 Sample system

22.9kV, 50MVA distribution CD type HTS cable applied sample system is introduced in Fig.8 and Table 4.

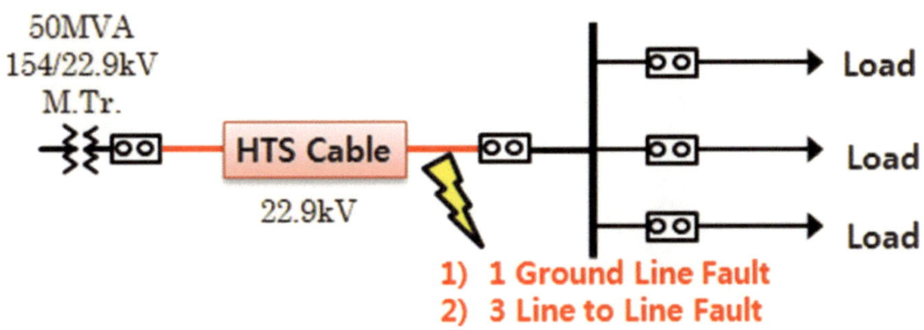

Fig. 8. Model distribution system

Items	Specification
Rated Voltage	22.9 kV
Rated Current	1,250 A
Capacity	50 MVA
Length	100 m
Cable Type	3 cores in one cryostat
Dielectric Type	Cold dielectric
Cable Size	Applicable for 175 mm duct
Response to Fault Current	There shall be no damage for the cable and cable system when the fault of 25kA is applied to the cable for 5 cycles.

Table 4. Ratings of modeled HTS cable

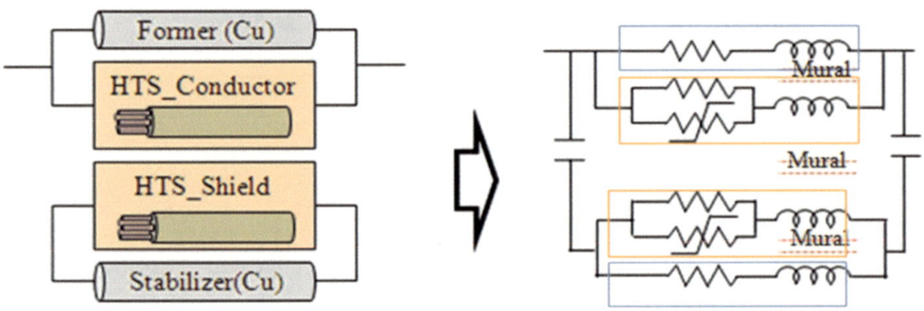

Fig. 9. CD type HTS cable modeling

To verify electrical characteristic more detail, each conductors and formers are modeled with EMTDC and compared with test results.

2.3.3.2 Normal operation characteristics –3 phase balanced case

When the operating current of HTS cable increased up to 2/3 of rated current, the conductor and shield current are measured[Fig 10]

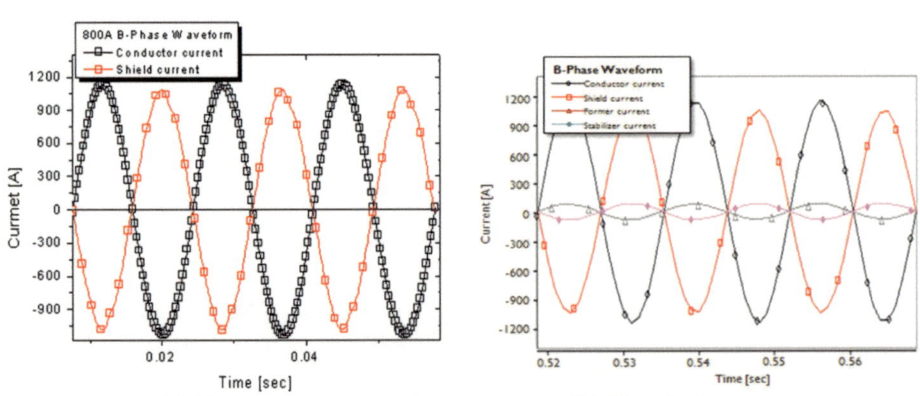

(a) Test (b) Simulation

Fig. 10. Test and simulation results (Balanced case 800Arms: conductor and shield current)

In a) and b), currents in conductor and shield are almost same and opposite phase. Errors of measured and simulated value are 1.7%(HTS conductor) and 0.7%(Shield), respectly. This errors are regarded as heat characteristics and AC loss effects of HTS cable.

Abnormal operation characteristics – 3 phase unbalanced case

Fig. represents the test and simulation results of 30% unbalanced case. Errors between test and simulation reaches 6.5% maximum.

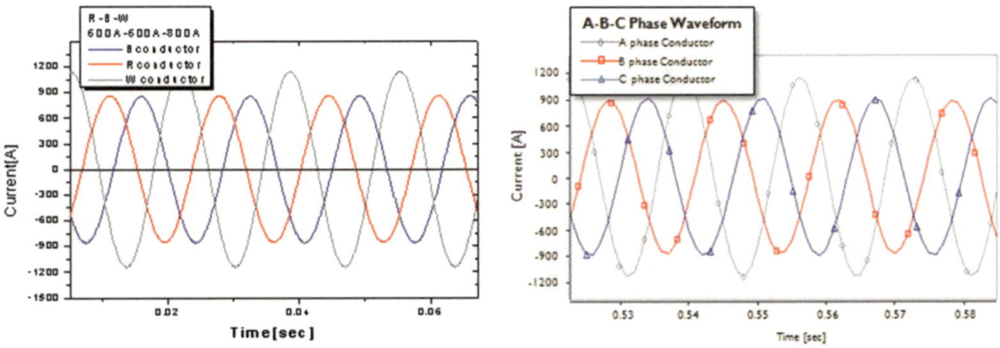

(a) Test (b) Simulation

Fig. 11. Test and simulation results (Unbalanced case 600/600/800Arms: conductor and shield current)

2.3.3.3 Abnormal operation characteristics – harmonics

Harmonics can increase AC loss of HTS cable due to hysteresis loss. Hysteresis loss model is as below equation.

$$P_{HY}=2\pi\omega\eta VB^{n}_{max} = kfB^{n}_{max} \ [W/m^3] \quad (1)$$

f : frequency [Hz]
B : flux density[Wb/m2]
n : exponential index on material [2.1]
V : volumn of material
k : total constant

In case of high THD, especially higher order harmonics are included dominantly, the hysteresis loss will be increased because it is proportional to frequency. Regarding harmonics, HTS cable system has to increase cooling capacity and/or decrease operating capacity of HTS cable.

2.3.3.4 Abnormal operation characteristics-fault currents and thermal characteristics

In abnormal operation status such as short curcuit current passing condition, superconducting cable has to pass large current securely. Usually, fault current rises 10 times more than normal current, this excessive current may over critical current (Ic) of superconductor. In this case, current quench may happen and very rapid temperature rise may take place and the HTS cable may be damaged . Therefore, various methods such as fast circuit breaker and/or parallel conductor(copper former) are applied to protect quench of HTS conductor.

In CD type HTS cable, most of fault currents are transferred from HTS conductor to former conductor because of superconductor resistance rise. When temperature is supposed as constant, HTS conductor resistance is calculated by next equation.

$$R_{HTS} = \frac{V_{HTS}}{I_{HTS}} = r_c \times \left(\frac{I_{HTS}}{I_{critical}} \right)^{n-1} \tag{1}$$

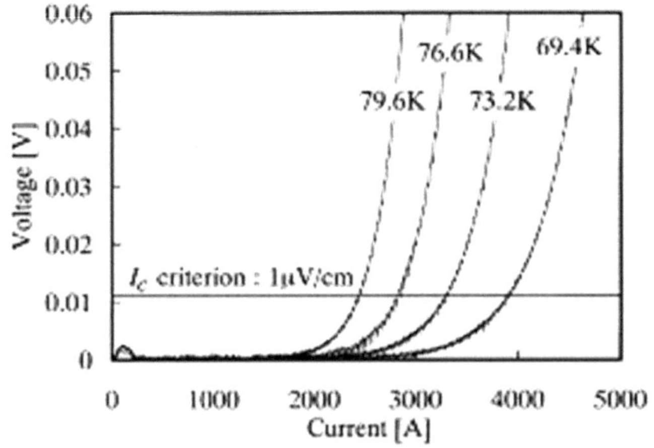

Fig. 12. V-I curve of 66kV HTS cable

During fault current, the internal heat dynamics can be approximately fomulated by heat insulated equation because electric dynamics ends within very short time(0.1 seconds) compare to heat dynamics.
Therefore, quench dynamics are represented next heat balnace differential equation.

$$C(T)\frac{\partial T}{\partial t} = \frac{\partial}{\partial x}\left(k(T)\frac{\partial T}{\partial x} \right) + Q(T) - W(T) \tag{2}$$

C(T) : heat capacity
The left side represent temperature rising rate of HTS cable, the first term of right side represent heat transfer to superconductor, and *k(T)* is heat transfer rate, *Q(T)* is internal heat generation due to current, *W(T)* is cooling heat.
Therefore,

$$Q(T) = \frac{\rho(T)}{A} I(t)^2 \tag{3}$$

I(t) is current, *ρ* is resistivity of tape, *A* is cross secion area.
If we suppose fault current flows within very short time, heat transfer and cooling effect can be disregarded. Therefore, equation (2) simplified as (4).

$$\int_{t_0}^{t} \frac{\rho_{Ag}(T)}{A_{Ag}} I^2 dt = \int_{T_0}^{T} m_{HTS} c_{p,HTS}(T) dT \tag{4}$$

In quench state, voltage of quench area will be increase and cable impedance(R+jX) is increased too.

Every nonconductors in cable acts heat resistances of heat tranfer. The heat resistance of each insulation can be calculated as follows.

$$T = \frac{\rho_{th}}{2\pi} ln\left(\frac{r_2}{r_1}\right) \tag{5}$$

T : Heat resistance of each insulation layer in unit length [K·m/W].

ρ_{th} : heat resistance of material

r_1, r_2 : inner and outer radius of insulator

Most of problem related cable rating is determined by passed time and modeled by heat balance equation. However, solving it is very difficult with numerical analysis. Therefore, in most calculation case, we define heat capacity of cable as equation (6) and use simple approach.

$$Q = V \cdot c \tag{6}$$

V = cable volumn[m^3]

c = heat capacity of material [J/m^3℃]

Next Figure represents and example of heat equivalent circuit between conductor and sheath of cable. Qc represents heat capacity of conductor and sheath. Heat capacity of dielectrics are calculated.

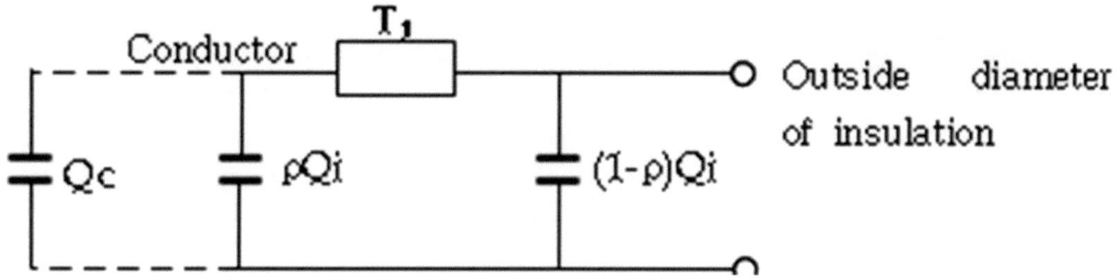

Fig. 13. Equivalent heat transfer circuit of HTS cable

T_1 : Total heat resistance of dielectric material

Qi : Total heat capacity of dielectric material

Qc : heat capacity of conductor

heat capacity coefficient ρ can be calculated equation (7)

$$\rho = \frac{1}{2\ln\left(\frac{D_i}{d_c}\right)} - \frac{1}{\left(\frac{D_i}{d_c}\right)^2 - 1} \tag{7}$$

D_i: Cable inner diameter

d_c : conductor diameter

2.3.3.5 Fault example - single line fault case

Fig. 14 shows the simulation results of single line to ground fault case on above distribution system

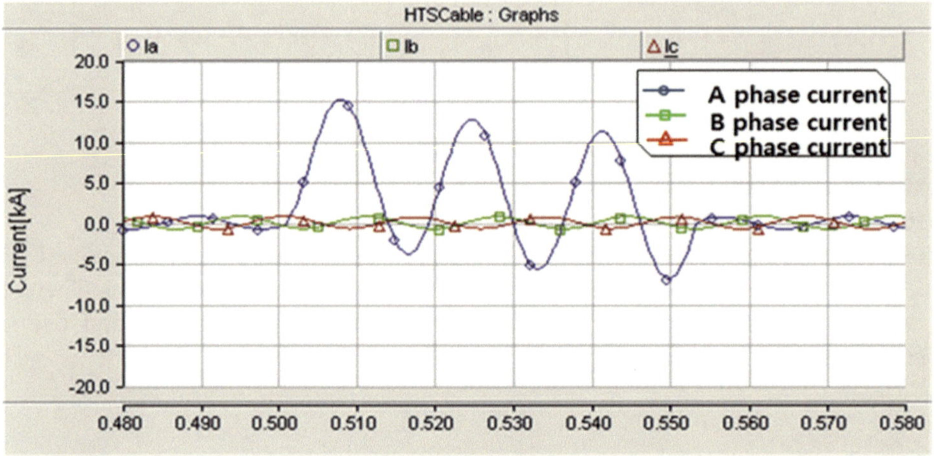

(a)

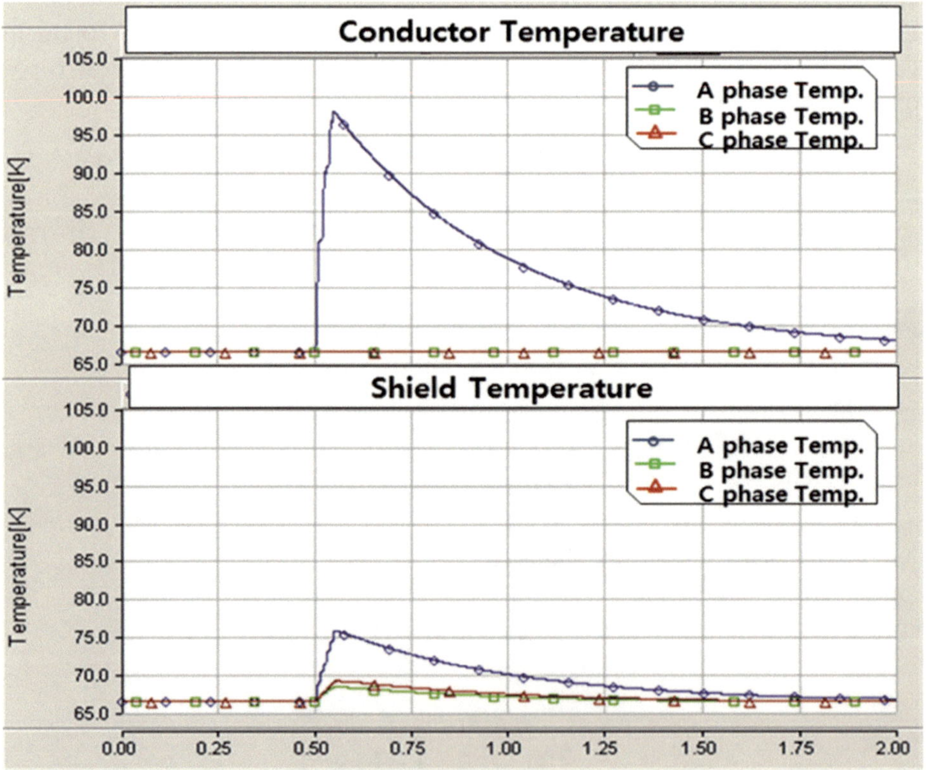

(b)

Fig. 14. Current and temperature of HTS cable in fault condition(SLG)
(a) fault current at Single line fault (b) temperature of conductor and shield

With the fault current of A phase, HTS conductor of phase A temperature rises from 67[K] to 97[K] during fault time. If quench temperature is 105[K] normally, there is little margin to this HTS cable system.

3. Superconducting Fault Current Limiter(SFCL)

3.1 Fault Current Limiter and SFCL

In electrical network, there are various faults, such as lightning, short circuits, grounding etc., which occurs large fault current. If these large currents are not properly controlled for power system security, there happens unexpected condition like fire, equipment and facility damage, and even blackout. Therefore, Circuit Breakers are installed and have the duty to cut off fault current, however, it takes minimum breaking time to cut, and sometimes fail to break.
Fault Current Limiter(FCL) is applied to limit very high current in high speed when faults occur. Different with normal reactor, normal impedance is very low and have designed impedance under faulted situation. Fault limiting speed is high enough that it can limit fault current within 1/4 cycle. Also, this function has to be recovered fast and automatically, too.
Various FCLs are developed and some of them are applied in power system. Most typical FCL is to change over circuit from low impedance circuit to high impedance circuit. Circuit breakers and/or power electronics devices are used to control FCL circuits. Fuse or snubber circuits are used to protect high recovery voltage. These FCLs are attractive as it implements normal conductor, however, there are weak points such as slow current limiting speed and big size in distribution and transmission level as well.
Superconducting fault current limiter (SFCL) has been known to provide the most promising solution of limiting the fault current in the power grid. It makes use of the characteristic of superconductor whose resistance is zero within critical temperature (Tc) and critical current (Ic). If fault current exceeds Ic, superconductor lose superconductivity and the resistance increase dramatically (called quench) and limit circuit current.

3.2 Classification of SFCL

Various types of SFCLs have been built and showed desired current limitation up to medium voltages. Some of them were actually field-tested in the electrical power grid. However, the SFCLs seem to be not near to commercial operation in the grid. This means that the SFCL is not ready to satisfy the utilities in various conditions. The conditions are dependent upon the application conditions, general purpose applications and special purpose ones.
We can classify these SFCLs as three types, which are resistance type(R-type), Inductance type (L-type) and saturable core type. R-type makes use of quench resistance of superconductor directly. L-type makes use of superconductor as trigger element for circuit inductance which limits fault current. Saturable core type makes use of superconductor magnet to saturate reactor iron core. In normal operation, this reactor has a little reactance in saturation state. However in fault state, fault current releases saturation state and increases impedance, therefore limits fault current.

3.2.1 R-type and L-type

The conceptual circuit of R-type and L-type SFCL is shown Fig. 15. In SFCL(Limiter), R_p is fault limiting resistance when R-type. In case of L-type, R_p will change as L_p (fault limiting inductance). If i_{ac} reaches critical current, R_{sc} should be quenched and its superconducting characteristics will be lost (resistance will be increased dramatically) , so fault current will be limited by Rp.

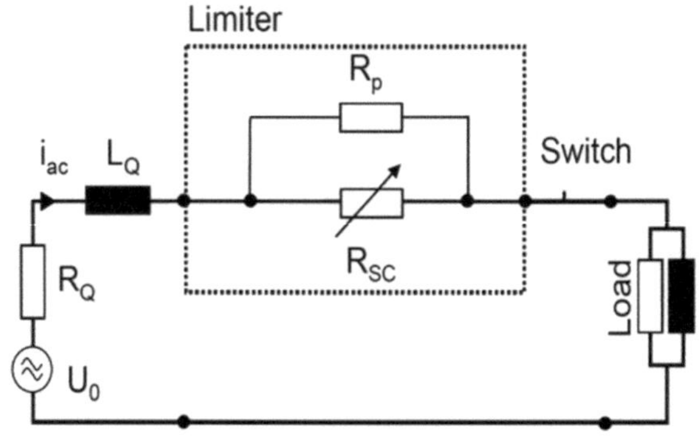

Fig. 15. R-type and L-type SFCL conceptual circuit

The mathematical model of SFCL is expressed as equation (8).

$$Z(t) = Z_s\left[1 - \exp\left(-\frac{t - t_0}{T_s}\right)\right] \tag{8}$$

T_s is time constant of impedance, t_0 is delay time of SFCL, Zs is impedance of SFCL.

$$Z_s = R_s + jX_s \tag{9}$$

$$R_s : \ R - Type$$

$$X_s : L - Type$$

By the equation (8), impedance dynamics of SFCL is as Fig. 16.

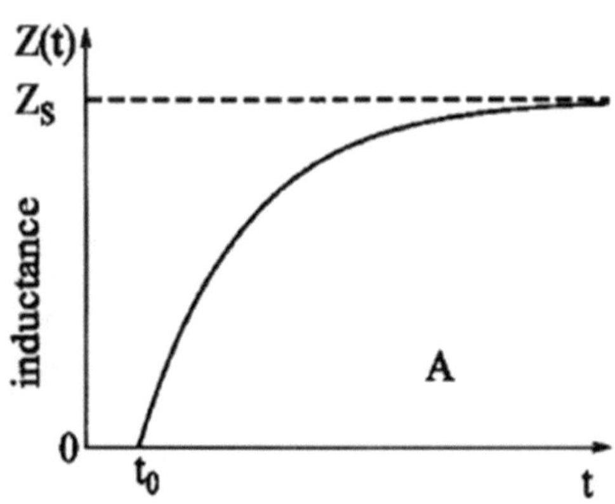

Fig. 16. Characteristics of SFCL impedance

R-type SFCL can limit peak current if proportional to Rs. L-type has slow damping characteristic because of transient DC component. The superconductor resistance value of SFCL (Rsc) is dependent to its type, it rise about 25 [pu] exponentally within 1[ms].

3.2.2 Saturable core type

The conceptual circuit diagram of saturable core type SFCL is shown Fig. 17. In normal state, two core fluxs are saturable with currents Io. When fault current iac flows, saturable fluxs are decreased and inductance of L1 and L2 increase along with B-H curve.

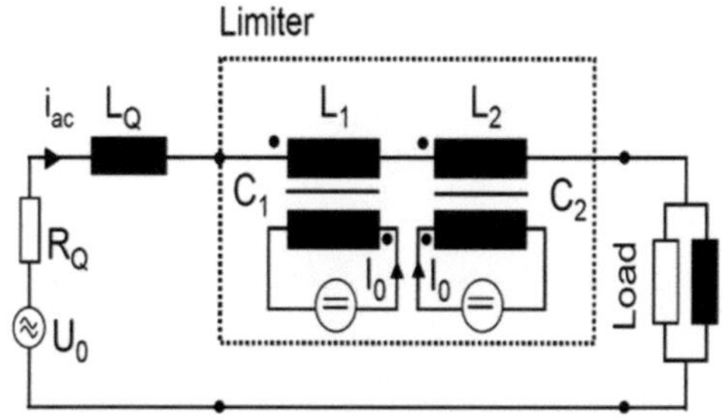

Fig. 17. Saturable core type SFCL conceptual circuit

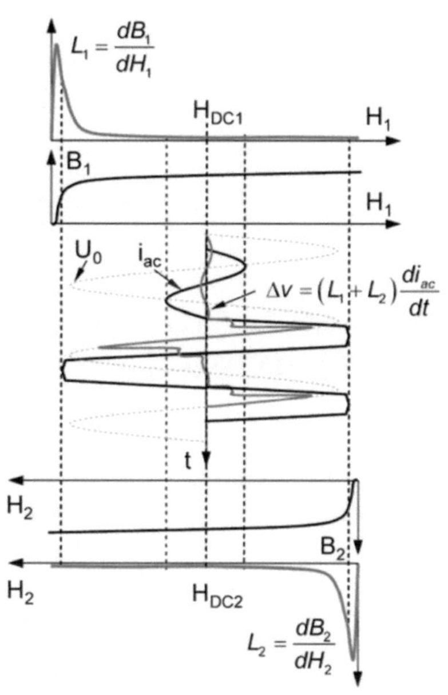

Fig. 18. Saturable core type characteristics

3.2.3 Hybrid type

Currently two types of SFCLs are widely developed at medium and high voltage scale, the resistive type and the saturable iron-core type SFCLs. Since a resistive SFCL component is limited in current and voltage ratings, inevitable is a large number of components to be assembled, so a large cryostat to cool them. Likewise, the saturable iron-core type carries large size iron cores.

To match these requirements, hybrid SFCL is developed for medium voltages class. The hybrid structure is composed of superconducting parts and conventional switches. This resulted in drastic reduction of superconductor volume, followed by smaller cryostat. The design also provides standing alone current limitation, reclosing capability, and other functions.

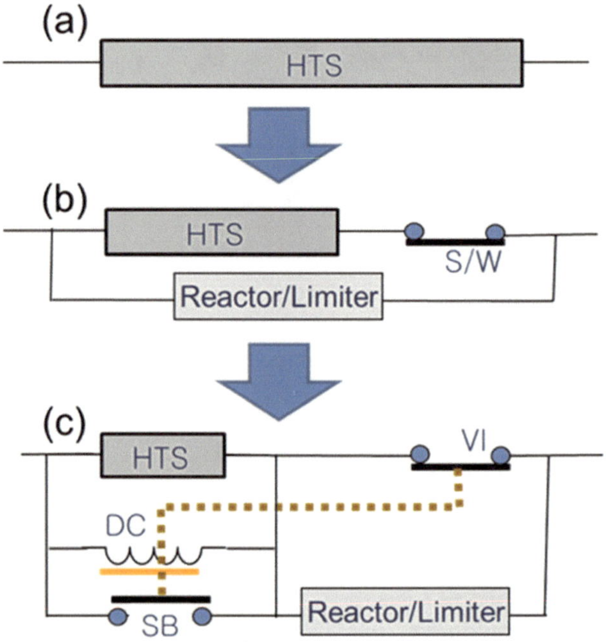

Fig. 19. Design innovation of resistive SFCLs. (a) conventional resistive type, (b) hybrid type with a conventional breaker, (c) hybrid type with a fast switch

3.3 Developed/Applied SFCLs

The first installed one is developed by ABB. After that, various SFCLs are developed for distribution and transmission application to protect bus and/or feeder from high fault currents . Fig. 20 shows recently developed and installed SFCLs for distribution level.

(a) (b) (c)

Fig. 20. Distribution class SFCLs, (a) Boxberg, Germany (b) Shandin, USA, (c) Kochang, Korea

place	developer	Voltage (kV)	Type	status
ABB P/P, Swiss	ABB	10.5	R-type	Operated 1997(6month)
Puji S/S, China	Innopower	10.5	Saturable Core	In operation (2008~)
SCE Shandin S/S USA	Zenergy Power	15	Saturable Core	In operation (2009~)
Tokyo Gas, Japan	Toshiba	6.6		In operation (2007~)
Lancashire, U.K	Nexans SC	12	R-type	In operation (2010~)
Boxberg P/P, Germany	Nexans SC	12	R-Type	In operation (2009~)
San Dionigi S/S Italy	CESI RICERCA	9	R-Type	In operation (2011~)
Kochang, Korea	KEPRI/LS	22.9	Hybrid	In operation (2009~)
SCE, USA	AMSC/Siemens	115	R-Type	In operation (2011~)
AEP, USA	ZenergyPower	138	Saturable Core	In operation (2011~)

Table 5. SFCL Developments for Transmission level

3.4 Applications of SFCL

The utilities used to require that the SFCL must be robust, reliable, of low cost, and (almost) maintenance-free for long time use. These would be universal conditions that any SFCL is expected to satisfy. In addition, there may be local conditions associated with the special purpose application of an SFCL by local demands. The local conditions may be specific size, cost, current limitation performance, reclosing capability, and so on.

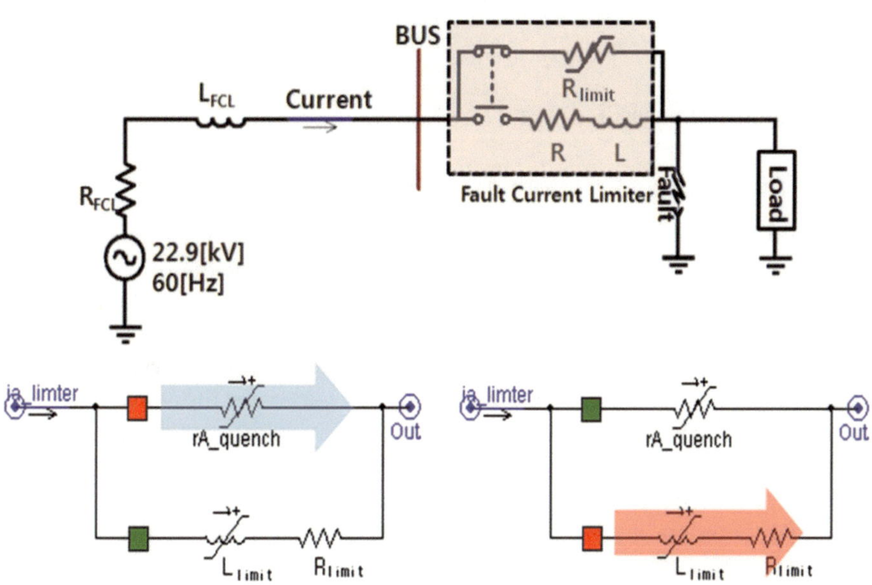

Fig. 21. SFCL sample system

SFCL has many good points, such as small size, faster fault current limiting, little parts, no power increase in fault circuit. Therefore, various applications are expected as belows, for example.

- Increase power transfer flexibility applied to bus-tie between distribution transformers
- Reduce voltage sag applied to sensitive load.
- Reduce ground fault current applied to neutral impedance for transformer

Below is case study result how SFCL is work in 22.9kV distribution system.

variables	L_{limit}	R_{limit}	rA_{quench}(normal)	**rA_{quench}(fault)**
Value	0.005[H]	1.0[Ω]	0[Ω]	$r_0 \times \left(\frac{I}{I_{crit}}\right)^{n-1}$

Table 6. Constants of sample SFCL

In this simulation, maximum quench resistance is 5 [Ω]. Fig 22 shows how SFCL limits fault current compare to non-SFCL circuit. The fault current could be reduced dramatically within 1/4 cycle by SFCL.

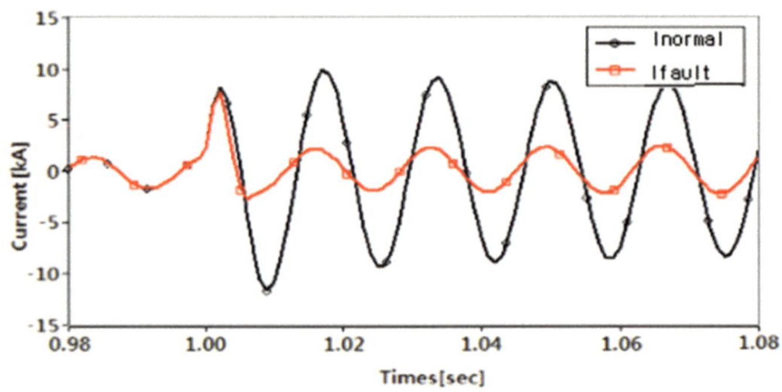

Fig. 22. Simulation result of SFCL dynamics

4. Dynamic Synchronous Condenser (DSC : SuperVar)

Synchronous Condensers (SC) are a good facility to support dynamic reactive power both capacitive and inductive area to improve system voltage characteristics. They are rotating machine rotating in synchronous speed. When we operate a synchronous generator connected power grid without driving motor, it is operated as synchronous condenser.

Its reactive power can be controlled with field current excitation. When overexcited, it generates capacitive reactive power. If under-excited, it generates inductive reactive power.

Today, this machine is not preferred because of high power loss and maintenance problem. Static reactive compensators such as STATCOM (Static Compensator) and/or SVC (Static Var Compensator) are preferred alternatives with rapid response and easy maintenance. However, synchronous condenser has excellent characteristic to support dynamic rating compare to above static compensators.

Dynamic Synchronous Condenser (DSC) has upgraded existing SC technology by using a conventional armature mated with a field winding made from High Temperature

Superconducting (HTS) wires. With the upgrading of field magnetic flux density as HTS conductor, it can provide up to 8 [pu] current for short periods to support transient VAR requirements. Key benefits of DSC are as follows:

- Fast response to transient voltage variation at both reactive power
- Low losses
- Simple installation (small footprint)
- Low maintenance
- No harmonic generation

4.2 Configuration

The major components of a DSC are shown in Figure 23. The field winding employs HTS conductor which is cooled with a cryocooler to about 35-40K. The cryocooler modules are located in a stationary frame and a fluid such as gaseous helium or liquid neon is employed to cool components on the rotor. The stator winding employs conventional copper windings.

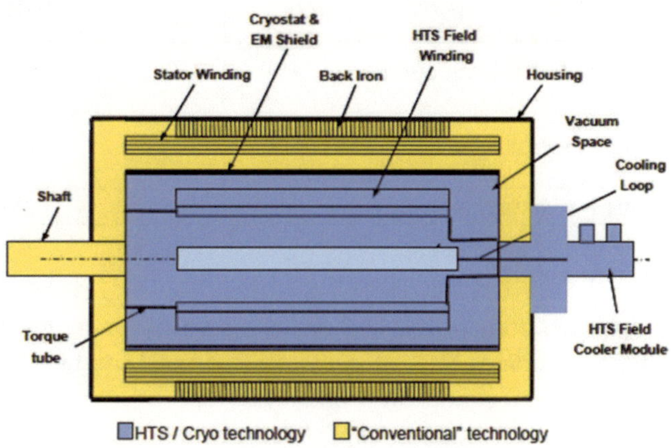

(a)

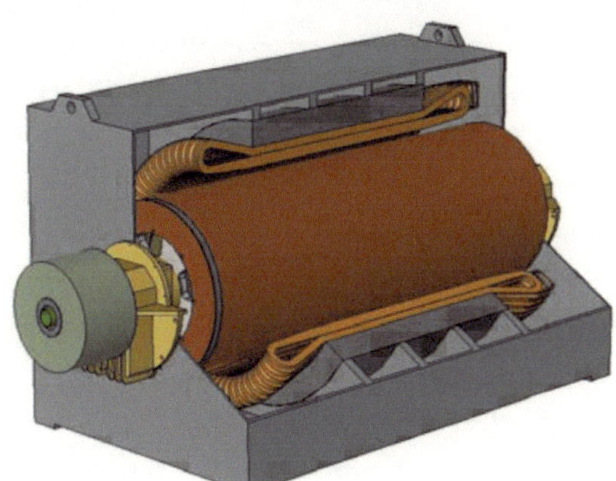

(b)

Fig. 23. Conceptual diagram of a DSC
(a) superconducting field winding in cryocooler , (b) DSC model picture

4.3 Electric characteristics and performance

The DSC has low synchronous reactance which increases power system stability and reactive power/voltage compensation compare to a conventional SC. The characteristics DSC are summarized below:

- With low synchronous reactance, DSC provides less voltage drop ratio between no-load and full-load operations
- The sub transient reactance (x_d'') of the machine is also low (0.11 pu) which lets the machine provide up to 8 pu first peak current for a terminal short circuit.

The major parameters of the machine are shown in table 7.

Parameters	value
Synchronous reactance (x_d)	0.5 pu
Transient reactance (x_d')	0.22 pu
Sub-transient reactance (x_d'')	0.11 pu
Armature short-current time constant (T_{sc})	0.045 sec
D-axis transient short circuit time constant (T_d')	7.31 hr
D-axis transient short circuit time constant (T_d'')	0.01 sec
Armature resistance (r_a)	0.007 pu

Table 7. DSC electric parameters

Figure 24 compares the efficiency of the DSC with a conventional synchronous condenser. The HTS field winding eliminates 50% of conventional machine field losses. Especially, It has good efficiency in light load condition.

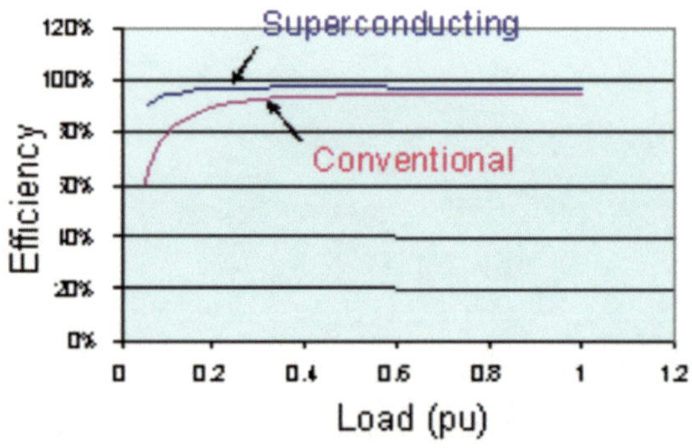

Fig. 24. DSC versus conventional machine efficiency

The DSC has no dynamic stability limit within its MVA rating. The machine can run stably without requiring any feedback control for dynamic voltage stabilization. This machine also has a superior dynamic stability during small oscillations and requires no field forcing for damping such oscillations. Figure 25 shows its damping of oscillations following a sudden change of load.

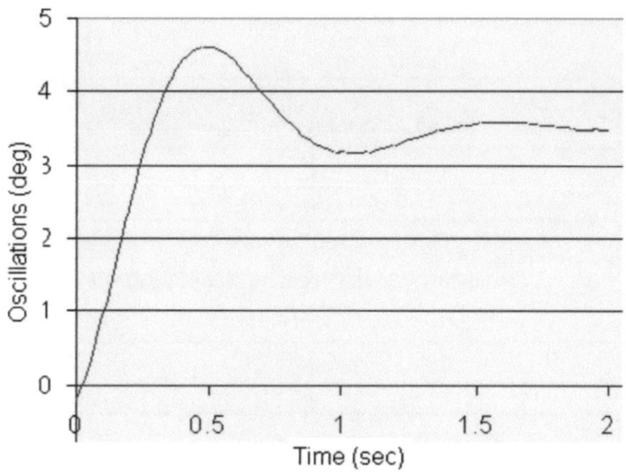

Fig. 25. DSC damping of low frequency oscillation following sudden load change

5. Application to power system

5.1 HTS cable

Before HTS cable application to power system, system planners have to understand the characteristics of power system and HTS cable. HTS cable system shall be applied special place in network which requires higher density power transmission.

There are several feasibility studies for HTS cable application. J. Jipping et al examined application validity of HTS cable for future load growth in a viewpoint for heat capacity and fault current. D. Politano et al examined technical economical efficiency for substitution high voltage transmission line for HTS cable. K. C. Seong et al examined transmission capability problem of power systems in a viewpoint for power flow and examined validity for HTS cable application. G.J.Lee et.el[ ] presented HTS cable application method to increase voltage stability limit. Recently, Ultera finished feasibility study of Amsterdam HTS project which will connect 6km, 50kV 250MVA HTS cable in 2013~2014 to increase inter-substation power transfer. Also, AMSC is planning to use DC HTS cable to interconnect North America network (Tres-Amigas Project).

For every application, total power system planning techniques are needed for the future's HTS cable implementation.

In this section, an example of HTS application study method shall be introduced to increase voltage stability limit. Fig 26 represents study procedure .

5.1.1 Case study

Sample system and verify initial transfer capacity

IEEE 39 bus system is considered for the sample system (Fig. 27). N-1 contingency is applied to estimate initial steady state transfer capacity. From the initial load condition (6098MW), maximum incremental transfer capacity applied N-1 contingency case is 3900MW. Therefore, system transfer capacity regarding security limit is 9,998MW.

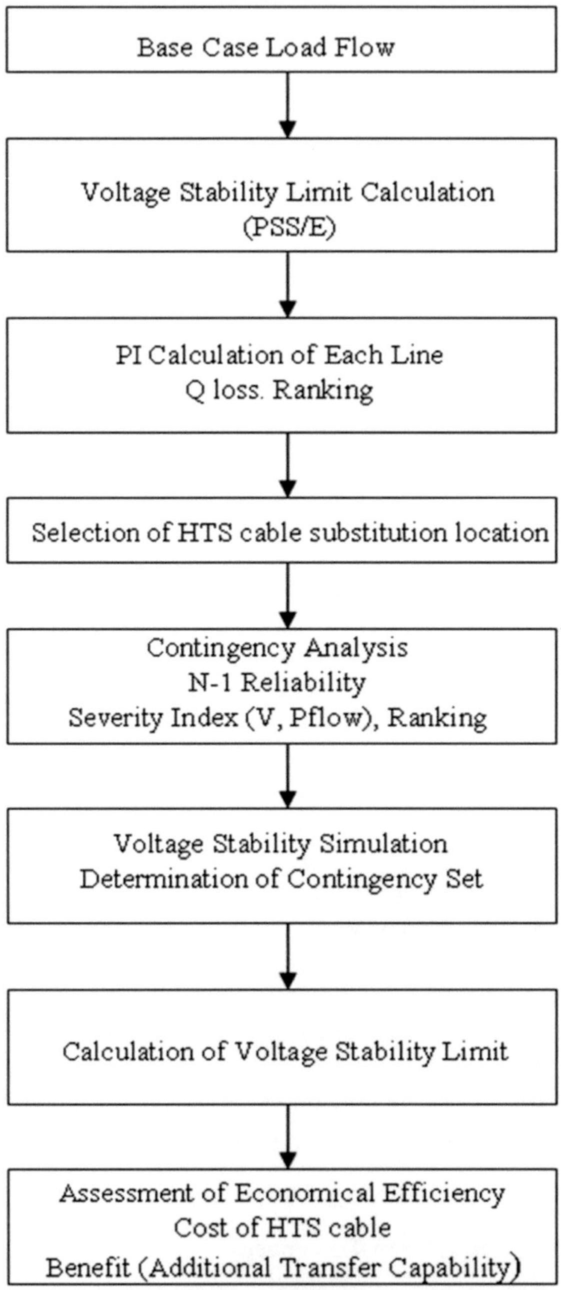

Fig. 26. Analysis procedure of HTS cable application

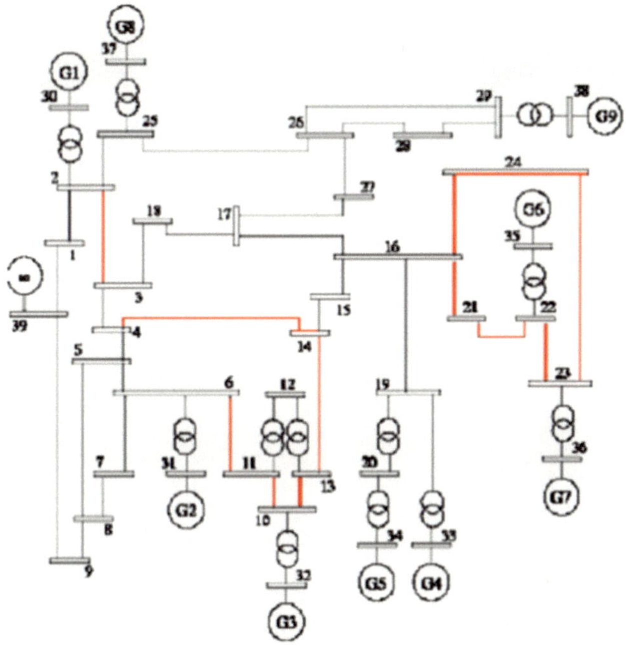

Fig. 27. IEEE 39 bus systems (HTS cable application: red line)

SI calculation of sample system

To consider power system reliability, N-1 contingency criteria was applied. Equation (3.1) and (3.2) shows the severity index (SI, over load index and voltage index) used in ranking.
① Over-load index
Equation 3.1 represents over-load index.

$$PI = \sum_{i=1}^{L}\left(\frac{P_i}{P_{\max,i}}\right)^2 \tag{10}$$

② Voltage index
Consumption of reactive power can be known by voltage ranker which represents increment of reactive power loss by increased load factor of line. Equation 3.2 represents voltage index.

$$PI = \sum_{i=1}^{L} X_i P_i^2 \tag{11}$$

where P_i is active power, X_i reactance, and $P_{\max,i}$ power ratings of i-line.
The results of SI on sample system results are shown in Table 3.4 and Table 3.5. As a result of calculation, the first two contingency cases of each SI are determined as the object cases of voltage stability calculation.

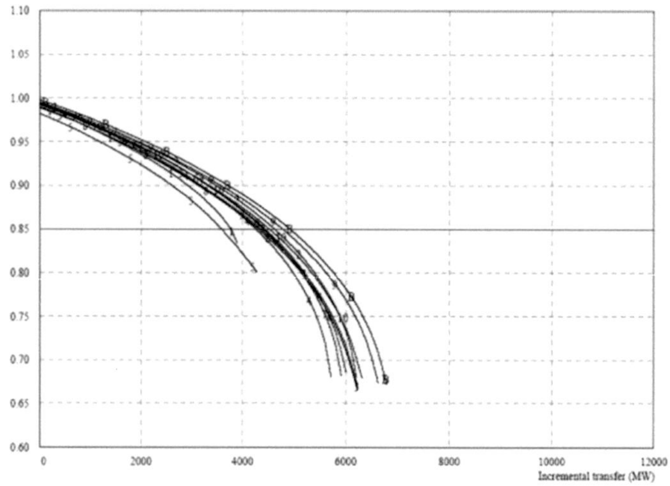

Contingencies

Mark	Label	Max incremental transfer
B	BASE CASE	6790.00
1	OVRLOD 1	3900.00
2	OVRLOD 2	6200.00
3	OVRLOD 3	6230.00
4	OVRLOD 4	5710.00
5	OVRLOD 5	4270.00
6	OVRLOD 6	6000.00
7	OVRLOD 7	5910.00
8	OVRLOD 8	6220.00
9	OVRLOD 9	6620.00
10	OVRLOD 10	6310.00

(a) before

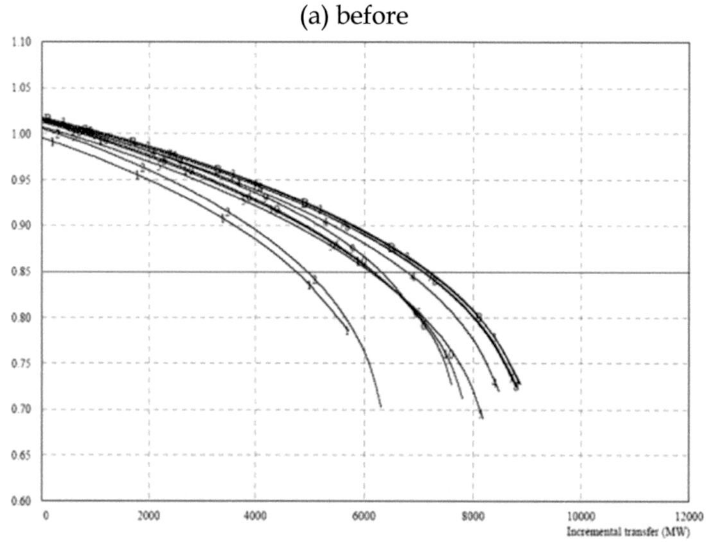

Contingencies

Mark	Label	Max incremental transfer
B	BASE CASE	8880.00
1	OVRLOD 1	5720.00
2	OVRLOD 2	6310.00
3	OVRLOD 3	8850.00
4	OVRLOD 4	8480.00
5	OVRLOD 5	8180.00
6	OVRLOD 6	7600.00
7	OVRLOD 7	8790.00
8	OVRLOD 8	8820.00
9	OVRLOD 9	6970.00
10	OVRLOD 10	7800.00

(b) after

Fig. 28. P-V curve (HTS cable application)

Ranking No.	Contingency Line		PI[p.u.]
	From Bus	To Bus	
1	21	22	10.8136-
2	23	24	8.6842
3	6	11	8.6463
4	13	14	8.6206
5	15	16	8.5228

Table 8. Performance index by line overload index

Ranking No.	Contingency Line		PI[p.u.]
	From Bus	To Bus	
1	28	29	10.8884
2	2	3	10.3888
3	16	21	10.2108
4	2	25	9.9931
5	6	7	9.8334

Table 9. Performance index by line voltage index of case I

Table 10 is the summary of the overloaded lines at severe contingency cases. HTS cable is applied as the order of severity of overloaded line. The replaced system is shown as Fig.29. Considered HTS cable constants are L = 0.10[uH/km], C=0.29[uF/km] respectly.
Incremented transfer capacity after HTS cable replacement is 8,880MW in base case and 5720MW in N-1 contingency case. Therefore, increased transfer capacity becomes 1820MW.

from	to	contingency	rating	flow	overload(%)
16	24	OVRLOD 1	600.0	630.4	105.0
22	23	OVRLOD 1	600.0	665.5	107.9
23	24	OVRLOD 1	600.0	945.9	157.5
16	21	OVRLOD 2	600.0	681.0	111.3
21	22	OVRLOD 2	900.0	955.9	104.2
4	14	OVRLOD 3	500.0	566.2	113.7
10	13	OVRLOD 3	600.0	620.8	102.3
13	14	OVRLOD 3	600.0	636.3	105.5
6	11	OVRLOD 4	480.0	636.8	132.3
10	11	OVRLOD 4	600.0	618.2	102.1

Table 10. Overloaded lines at N-1 contingency

5.2 SFCL

In power system, proper SFCL application places are considered as (a)~(c) points of Fig. 29. Point (a) is to limit fault current of distribution feeder. SFCL at (b) point reduces fault current impact of adjacent transformer in case of parallel operation and protects bus bar. Point (c) is general solution to reduce transformer secondary fault current and extend Circuit Breaker changing time when distribution system experiences high fault current.

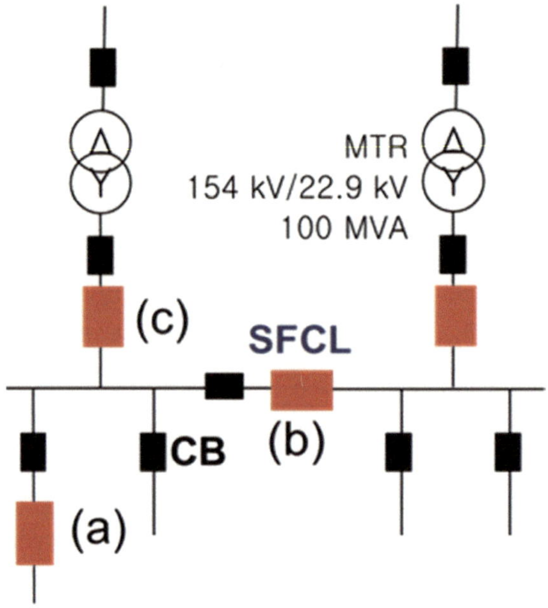

Fig. 29. SFCL application

6. Conclusion

The infrastructure of electric power system is based on conductor. With the change of power industry, such as Kyoto protocol and Energy crisis, superconducting technology is very promising one not only to increase efficiency of electricity but also to upgrade security of power system. Among various superconducting technology, most applicable ones -HTS cable, Fault current limiters, Dynamic SC are introduced and discussed how to apply.

Other superconducting facilities, like transformer, generator, SMES, Superconducting Flywheel, are in testing and will be implemented with the changes of power market needs.

However, the most critical obstacle of power system application is superconductor material and cooling system. Present HTS superconductors have to be improved much more than conventional ones, but still have difficulties in general use, such as extreme low temperature operation, hard manufacturing, AC loss and high cost. Cooling system is also hard task which have close relation of HTS failure due to quench mechanism. In operating point of view, monitoring and control to protect the local hot spot is another task to overcome.

More advanced superconductors and application methods are expected in power system usage in near future.

7. Acknowledgment

Thanks to support all referenced paper authors and researchers in the field of superconductor application in power system, especially Dr. OK-Bae Hyun and Si-Dol Hwang in KEPRI.

8. References

Jon Jipping, Andrea Mansoldo, "The impact of HTS cables on Power Flow distribution and Short-Circuit currents within a meshed network", *IEEE 2001 O-7803-7285-9/01*

M. Nassi, N. Kelley, P. Ladie, P. Coraro, G. Coletta and D. V. Dollen, "Qualification results of a 50m-115kV warm dielectric cable system", *IEEE Trans. on Applied Superconductivity*, Vol. 11, No. 1, 2001

G.J.LEE, J.P.LEE, S.D.Hwang, G.T.Heydt, "The Feasibility Study of High Temperature Superconducting Cable for Congestion Relaxation Regarding Quench effect", *0-7893-9156-X/05, IEEE General Meeting* 2005

Geunjoon LEE, Sanghan LEE, Songho-Son, Sidol Hwang, "Ground fault current variation of 22.9kV superconducting cable system", *KIEE Journal 56-6-1*, pp.993~999, 2007

Geun-Joon Lee, Sidol Hwang, Byungmo Yang, Hyunchul Lee, „An Electrical Characteristic Simulation and Test for the Steady and Transient state in the ww.9kV HTS cable Distribution", *KIEE Journal 58-12-3*, pp.2316~2321, 2009.

Geunjoon LEE, Jongbae LEE, Sidol Hwang, Song-ho Shon, "The Effects of Harmonic current in the operating characteristics of High Temperature Superconducting Cable", *KIEE journal, 56-12-2*, pp.2065~2071, 2007.

G.J. Lee, S.D. Hwang, H.C. Lee, "A Study on Cooperative Control Method in HTS Cable under Parallel Power System", *IEEE T&D Asia, Seoul* 2009

B. W. Lee, K. B. Park, J. Sim, I. S. Oh, H. G. Lee, H. R. Kim, and O. B. Hyun, "Design and Experiments of Novel Hybrid Type Superconducting Fault Current Limiters," *IEEE Trans. on Appl. Supercond.*, Vol 18, no. 2, (June 2008) pp. 624 - 627.

Ok-Bae Hyun, Jungwook Sim, Hye-Rim Kim, Kwon-Bae Park, Seong-Woo Yim, Il-Sung Oh, "Reliability Enhancement of the Fast Switch in a Hybrid Superconducting Fault Current Limiter by Using Power Electronic Switches," *IEEE Trans. on Appl. Superconductivity*, (presented at ASC2008, Chicago, USA), submitted for publication.

"The basic Study on Superconducting cable Application Technology on Electric Power System", *Report of Korea Industry and Resource Ministry (Chungbuk Provincial College, KEPRI)*, July, 2006

"A Study on Interconnection and Protection technology of superconducting cable for Distribution level power system application", *Report of Korea Knowledge and Economy Ministry (Chungbuk Provincial College, KEPRI, 02XKO1)*, September 2009.

Swarn Kalsi, David Madura, et.el. (2003)."Superconducting Dynamic Synchronous Condenser For Improved Grid Voltage Support", 2003 IEEE T&D Conference, Dallas, Texas, *IEEE Catalog No. 03CH37495C*, ISBN:0-7803-8111-4, 10 September 2003

Superconducting Fault Current Limiters: Technology Watch 2009. EPRI, Palo Alto, CA: 2009. 1017793.

S. Honjo, M. Shimodate, Y. Takahashi, T. Masuda, H.Yumura, C. Suzawa, S. Isojima and H. Suzuki, "Electric properties of a 66kV 3-core superconducting power cable", *IEEE Trans. on Applied Superconductivity*, Vol. 13, No. 2, pp. 1952-1955, 2003.

S. Mukoyama, H. Hirano, M. Yagi and A. Kikuchi, "Test result of a 30m high Temp. Superconducting power cable", *IEEE Trans. on Applied Superconductivity*, Vol. 13, No. 2, 2003

D. W. A. Willen et al, "Test results of full-scale HTS cable models and plants for a 36kV, 2kArms utility demonstration", *IEEE Trans. on Applied superconductivity*,Vol. 11, No. 1, pp. 2473-2576, 2001

J. Jipping, A. Mansoldo, C. Wakefield, "The impact of HTS cables on power flow Distribution and short-circuit currents within a meshed network", *IEEE/PES Transmission and Distribution Conference and Exposition*, pp. 736 – 741, 2001.

L. F. Martini, L. Bigoni, G. Cappai, R. Iorio, and S. Malgarotti, "Analysis on the impact of HTS cables and fault-current limiters on power systems", *IEEE Trans. On Applied Superconductivity*. Vol. 13, No. 2, pp. 1818-1821, 2003

D. Politano, M. Sjotrom, G. Schnyder, and J. Rhyner, "Technical and economical assessment of HTS cables", *IEEE Trans. on Applied Superconductivity*, Vol. 11, No. 1, 2367-2370, 2001.

K. C. Seong, S. B. Choi, J. W. Cho. H. J. Kim et al, "A study on the application effects of HTS power cable in Seoul", *IEEE Trans. on Applied Superconductivity*, Vol. 11, No. 1, pp. 2367-2370, 2001

K. W. Lue, G. C. Barber, J. A. Demko, M. J. Gouge, J. P. Stovall, R. L. Jughey and U. K. Sinha, "Fault current test of a 5-m HTS cable", *IEEE Trans. on Applied Superconductivity*, Vol. 11, No. 1, pp. 1785-1788, 2001

Anders, "Rating of Electric Power Cables in Unfavorable Thermal Environment", John Wiley & Sons

Guy Deutscher, "New Superconductors: From Granular to High Tc", *World Scientific*, 2006.

Donglu Shi, "High-Temperature Superconducting Materials Science and Engineering: New Concepts and Technology", Pergamon, 1995.

Design, Test and Demo of Saturable Core Reactor HTS FCL (DOE, Zenergy), 2009

H. Noji, K. Ikeda, K. Uto and T. Hamada, "Calculation of the total AC loss of high-Tc superconducting transmission cable", Physica C: Superconductivity Volumes 445-448, Pages 1066-1068, 1 October 2006

4

Current Distribution and Stability of a Hybrid Superconducting Conductors Made of LTS/HTS

Yinshun Wang
Key Laboratory of HV and EMC Beijing, State Key Laboratory for Alternate Electrical Power System with Renewable Energy Sources, North China Electric Power University, Beijing, China

1. Introduction

Although having made great progress in many applications, such as high magnetic field inserts in magnets at helium temperature and electrical engineering application in low magnetic fields at nitrogen temperature, the high temperature superconductor (HTS) is less commercially viable in mid- and large- scale magnets because of its high cost, low engineering critical current density, mechanical brittleness and low n value compared with conventional low temperature superconductors (LTS).

The superconductor with a high n value transfers quicker from superconducting state to the normal conducting state. From the standpoint of application, the transient characteristics strongly affect its stability. With a high current, in the low n value area, flux flow voltage becomes lower than in the high n value area. Generally, it is considered that quenching occurs at a weak point, which is defined as a low I_c and low n value area. However, when such transition is observed, it is predicted that the limit current of quenching will be reached sooner for the high n value than for the lower n value (Torii et al., 2001, Dutoit et al, 1999).

In general, the traditional superconductor has a higher n value than the Bi2223/Ag tape. In order to improve its stability, a LTS is always connected to a conventional conductor with low resistivity and high thermal conductivity, such as copper and aluminum, which then reduces its engineering critical current.

To enhance the performance of conventional composite NbTi superconductors with large current capacity (several tens of kA) utilized in large helical devices (LHD), a new LTS/HTS hybrid in which HTS is used as a part stabilizer in place of low-resistivity metals, was proposed (Wang et al, 2004; Gourab et al, 2006; Nagato et al, 2007). Thus its cryogenic stability against thermal disturbance, steady-state cold-end recovery currents and the minimum propagation currents (MPC) can be greatly improved because the HTS has low resistance and current diffusion which is faster than that in a pure conventional conductor matrix.

$$E = E_c\left(\frac{J}{J_c}\right)^n \tag{1}$$

Based on the power-law model (1) fitted in range of 0.1μV/cm ≤E≤1μV/cm, LTS has a higher n value (≥25) than HTS with a relative lower n value (<18) due to its intrinsic and

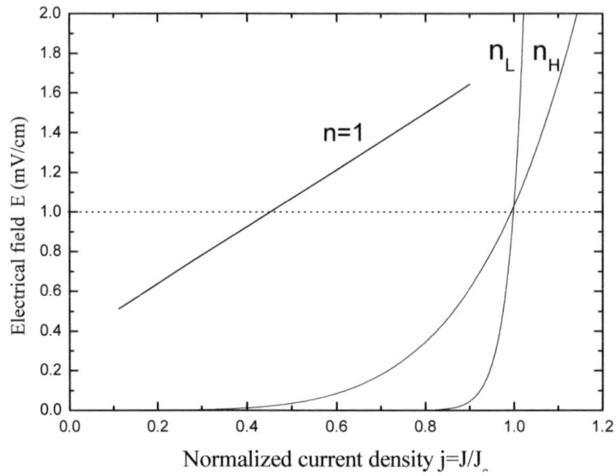

Fig. 1. Schematic E vs J plots of superconductors with n_H and $n_L(n_L>n_H)$, normal metal with n=1

granular properties (Yasahiko et al., 1995; Rimikis et al., 2000). According to different n values between LTS and HTS shown as Fig. 1, n=1 refers to the normal conductor according to the Ohm law. We firstly suggested a type of LTS/HTS hybrid composite conductor in 2004 in order to improve the stability of mid- and large scale superconducting magnets, in particular the cryo-cooled conduction superconducting magnet application.

Due to the different n values between LTS and HTS, the transport current flows initially through the LTS in the hybrid conductor. If there is a normal-transition in the LTS with some disturbance, the transport current will immediately transfer to the HTS, then the heat generation can be suppressed and full quench may be avoided. On the other hand, since the thermal capacity of HTS is two orders of magnitude higher than that of LTS, temperature rise can be smaller in the hybrid conductor than in the LTS. Therefore, the hybrid conductor can endure larger disturbances and maintain a higher transport temperature margin. In this chapter, we report on the current distribution and stability of a LTS/HTS hybrid conductor by simulation and experiment near in the range of 4.2K.

2. Numerical models of current distribution and stability

2.1 Current distribution

This kind of LTS/HTS hybrid conductor consists of soldering LTS wire and HTS tape together or by directly winding several LTS wires and HTS tapes together in parallel mode. The LTS/HTS superconductor is combination of LTS wire and HTS tapes shown in Fig.2.

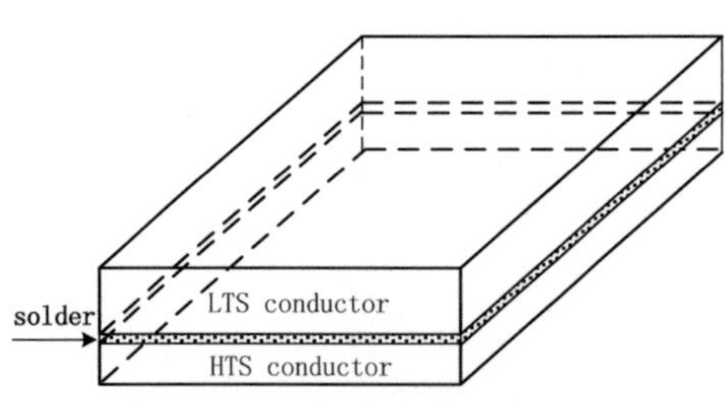

Fig. 2. Schematic view of LTS/HTS hybrid conductor with combination of LTS and HTS conductors

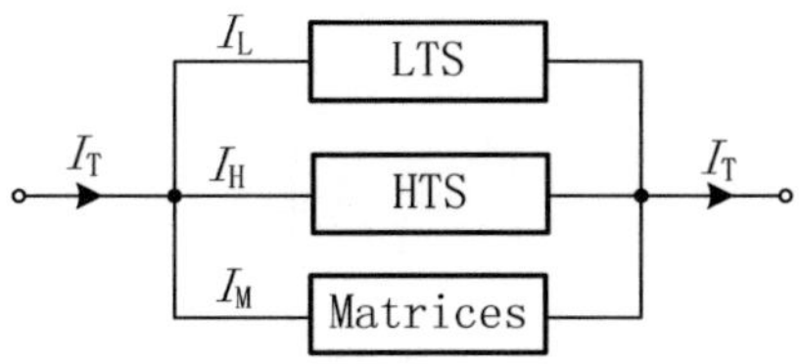

Fig. 3. Equivalent parallel circuit consisting of LTS/HTS hybrid conductor

According to its processing technology, the LTS/HTS hybrid conductor can be approximately considered to be equivalent parallel circuit consisting of LTS, HTS and metal matrix, shown in Fig. 3.

Let U_H, U_L, U_M be the voltages of the pure HTS, LTS conductors and the normal metal matrix including metal sheath, solder, etc, respectively; and J_H, J_L, J_M the corresponding branch current densities. Those parameters satisfy the following equations

$$U_H = U_c \left(\frac{J_H}{J_{cH}} \right)^{n_H} \tag{2}$$

$$U_L = U_c \left(\frac{J_L}{J_{cL}} \right)^{n_L} \tag{3}$$

$$U_M = I_M R_M \tag{4}$$

$$U_H = U_L = U_M \tag{5}$$

where $U_c=E_cL_0$, E_c is critical electric field ($E_c = E(I_c)$),and is usually equal to 1.0 μV/cm, L_0 is the length of the hybrid superconductor, n_H and n_L are the n indices of HTS and LTS, respectively; J_{cH} and J_{cL} are their critical current densities. R_M, the resistance of the matrices, is approximately given by

$$R_M = \rho_{avg} \frac{L_0}{S_M} \tag{6}$$

where ρ_{avg} and S_M are the effective resistivity and cross-sections of the matrices, estimation of ρ_{avg} is given , shown as Fig. 5

$$\frac{1}{\rho_{avg}} = \sum_{i=1}^{n} \frac{f_i}{\rho_i} \tag{7}$$

where f_i and ρ_i are volumetric ratio and resistivity of i-th components in matrices except for the LTS and HTS. Since the resistivity of superconductors is more at least one order than the metal conductor, it is reasonable to neglect the resistances of superconductors in this chapter.

Based on Eq. (2) through Eq. (6), following relations are found for unit length of the hybrid conductor

$$\begin{cases} I_T = I_H + I_L + I_M \\ \left(\frac{I_H}{I_{cH}}\right)^{n_H} = \left(\frac{I_L}{I_{cL}}\right)^{n_L} \\ \left(\frac{I_H}{I_{cH}}\right)^{n_H} \times 10^{-4} = R_{avg} I_M \end{cases} \tag{8}$$

where I_T is total transport current of hybrid conductor, and I_H, I_L and I_M the transport currents through HTS, LTS and matrices, respectively. The temperature dependence of critical currents of LTS and HTS in the hybrid superconductor can be approximately expressed as polynomial expressions with constant coefficients. Then the current distribution can be simulated according to Eq. (8).

2.2 Thermal stability

In order to conveniently analyze the thermal stability of the hybrid superconductor under the adiabatic condition, the heat source, made of heater, is located at the center of conductor with 200 mm length, and the length of heaters along the conductor is 10 mm, as schematically shown in Fig. 4.

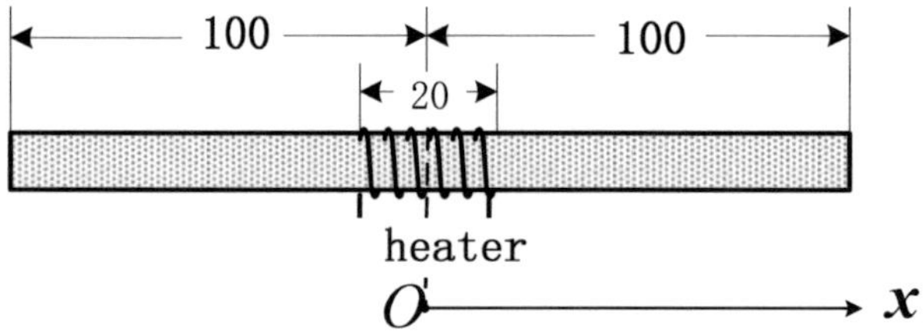

Fig. 4. Schematic view of heating on the hybrid conductor

The length of any segments is much larger than their cross-section and then the physical properties are assumed to be homogeneous over the cross-section. The numerical simulation may be simplified by choosing following one-dimensional, nonlinear, transient, heat balance equation (Wilson, 1983; Iwasa, 1994)

$$(\gamma C)_{avg} \frac{\partial T}{\partial t} = \frac{\partial}{\partial x}(k_{avg} \frac{\partial T}{\partial x}) + \frac{Q}{V} + \frac{G}{V_0} \tag{9}$$

where $(\gamma C)_{avg}$ is average heat capacity ($J \cdot m^{-3} \cdot K^{-1}$), k_{avg} the average thermal conductivity ($W \cdot m^{-1} \cdot K^{-1}$), Q the joule heat (W) generated in hybrid conductor, G the initial heat disturbance (W) applied by heater, V the total volume of the hybrid conductor and V_0 the volume of hybrid conductor surrounded by heater.

Both of average heat capacity and thermal conductivity are estimated according to Fig.5. Assuming that a composite conductor consists of n kinds of material, the heat capacity of each material is $(\gamma_i C_i)$ in which γ_i and C_i are mass density and heat specific, respectively, k_i and ρ_i its thermal conductivity and resistivity, the volumetric ratio of each component to total volume is f_i (i=1,2,⋯n). Fig. 5 is the schematic view of a composite conductor through which the heat Q and current I flow longitudinally. Fig.5 can be equivalent to "serial circuit" models for heat capacity $(\gamma C)_i$ and thermal conductivity k_i, but parallel circuit model for resistivity ρ_i. Then the maximum averages of heat capacity and thermal conductivity are respectively expressed by

$$(\gamma C)_{avg} = \sum_{i=1}^{n} f_i(\gamma_i C_i) \quad (10)$$

$$k_{avg} = \sum_{i=1}^{n} f_i k_i \quad (11)$$

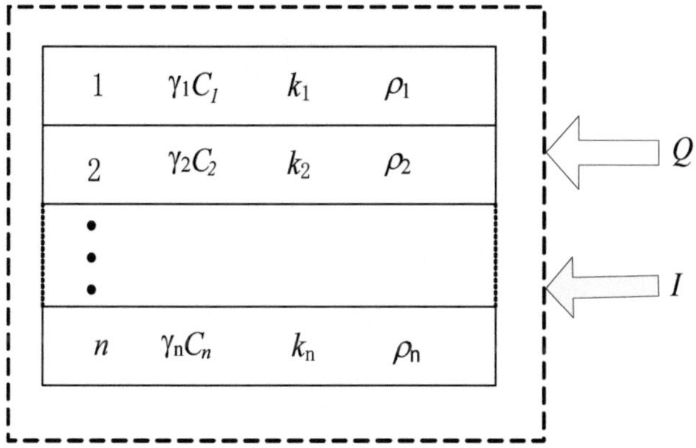

Fig. 5. Overview of a composite conductor with longitudinally flowing heat Q and current I

where f_i is volumetric ratio of each component in hybrid conductors, n the number of components, γ_i, C_i and k_i are mass density, heat specific and thermal conductivity of i-th components, respectively. In this chapter, 1 refers to copper, 2 represents NbTi, 3 indicates solder, 4 corresponds to YBCO CC for hybrid conductor made of NbTi/Cu conductor and YBCCO CC. On the other hand, 1 refers to copper, 2 represents NbTi, 3 indicates solder, 4, 5, 6 respectively refer to stainless steel, Bi2223 and silver for the one consisted of NbTi/Cu conductor and Bi2223/Ag tape.

In general, if a rectangular pulse disturbance is applied, the heat term G is given by

$$G = \begin{cases} I_g^2 R_g & (0 \le t \le t_g, 0 < x \le x_g) \\ 0 & (other\ conditions) \end{cases} \quad (12)$$

where t_g and x_g are the effective time and half length of the heater along the hybrid conductor located its center, here x_g=10 mm. I_g and R_g the current going through the heater and its resistance, respectively.

Let T_{cs} be the current sharing temperature, while $T>T_{cs}$, the Joul heat term Q in (W) is generated by

$$Q = I_M^2 R_M \tag{13}$$

When the hybrid conductor operates in superconducting state, I_M=0; If T_{cs} <T_{cL}, the current distributions are described by

$$I_M = \begin{cases} 0 & (T < T_{cs}) \\ I_T - (I_L + I_H) & (T_{cs} \le T < T_{cL}) \\ I_T - I_H & (T_{cL} \le T < T_{cH}) \\ I_T & (T \ge T_{cH}) \end{cases} \tag{14}$$

Substituting $\frac{\partial T}{\partial t} = \frac{\partial T}{\partial x}\frac{\partial x}{\partial t} = v\frac{\partial T}{\partial x}$ into Eq. (9), we have

$$k_{avg}\frac{\partial^2 T}{\partial x^2} + (\frac{\partial k_{avg}}{\partial x} - v(\gamma C)_{avg})\frac{\partial T}{\partial x} + \frac{Q}{V} + \frac{G}{V_0} = 0 \tag{15}$$

where v is the longitudinal quench propagation velocity (QPV) along the hybrid conductor length.
According to the one-dimensional model shown in Fig. 4, if the environment temperature is in liquid helium, the boundary and initial conditions are given by

$$\begin{cases} \left.\frac{\partial T}{\partial x}\right|_{x=0} = 0 \\ T|_{x=100} = 4.2 \\ T(x)|_{t=0} = 4.2 \end{cases} \tag{16}$$

Based on Eq.(13) through (16), the stability characteristics of the hybrid conductor, such as longitudinal quench propagation velocity (QPV) and minimum quench energy (MQE), can be simulated.

3. Simulation and results

In temperature T (<T_{cL}), the critical current of this kind of hybrid conductor is defined as

$$I_c(T) = I_{cH}(T) + I_{cL}(T) \tag{17}$$

For the sake of convenience, the normalized transport current $\alpha = I_T/I_c$ is defined and used thereafter. Table 1 lists geometrical and superconducting characteristics of composite NbTi/Cu, Bi2223/Ag by enforced stainless steel, YBCO coated conductor (YBCO CC) and the metal matrices in the hybrid conductor used in the numerical simulation in this paper. The critical currents are given in 4.2 K and magnetic field of 6 T with parallel to their wide surface for reducing their critical currents. In order to make the critical current of HTS comparable with the LTS as soon as possible, two commercial HTS tapes were selected in simulation and experiment (Section 4).

Material	Description	value
NbTi/Cu	Width/mm	4.3/4.5
	Thickness/mm	0.42/0.58
	Length/mm	1200
	Cross-section ratio of copper to superconductor	~1.38
	I_c@4.2 K and 6 T	960A
	n value@4.2K and 6 T	25
YBCO CC laminated on both sides with harden copper	Conductor Width/mm	4.41
	Conductor Thickness/mm	0.20
	Length/mm	1200
	YBCO width/μm	~1
	Copper stabilizer thickness/mm	~0.1
	Ic@ 4.2K and 6 T in parallel field (2 tapes)	2×350A=700A
	n value@4.2 K and 6 T	12
Bi2223/Ag tape enforced by stainless steel	Width/mm	4.3
	Thickness/mm	0.29
	Length/mm	1200
	Stainless-steel thickness/mm	0.05/each side
	Ratio of silver and stainless-steel to superconductor	~3
	I_c@4.2 K and 6 T in parallel field (2 tapes)	2×293=586A
	n value@4.2 K and 6 T	15
Solder (50Sn and 50Pb)	Width/mm	4.3
	Thickness/mm	<0.1

Table 1. Main parameters of superconductors and solders

3.1 Current distribution

3.1.1 Current distribution of hybrid conductor made of NbTi/Cu and YBCO CC

It is assumed that the hybrid conductor be soldered by solder Sn50-Pb50 and so that it is combination of NbTi/Cu with two YBCO CC. Under magnetic field of 6 T, dependence of critical currents of NbTi on temperature can be described by modified Morgan formulae

$$I_c(T) = 960 - 362(T-4.2) - 17.5(T-4.2)^2 + 1.85(T-4.2)^3 \quad (18)$$

In 4.2 K and 6 T, the critical current I_c is 960 A. With parallel magnetic field 6T, the relation of critical current with temperature in two YBCO CC is approximately described by

$$I_c(T) = I_c(0)\left(1 - \frac{T}{T_c}\right)^{1.2} \quad (19)$$

where $I_c(0)$ is critical current with T=0 K and B=6 T, $I_c(0)$=743 A, T_c =93 K is the critical temperature of YBCO CC, I_c=700 A is critical currents of two YBCO CC in 4.2 K and 6 T.

Neglecting magnetic field effect, the dependence of resistivity ($\Omega \cdot m$) in Cu and solder (50Sn-50Pb) on temperature are approximately described by

$$\rho_{Cu}(T)=\begin{cases}5.142\times10^{-15}T^3-1.1998\times10^{-14}T^2-1.714\times10^{-14}T+2.208\times10^{-10} & (4K<T\le 60K)\\ 6.856\times10^{-15}T^2+6.6846\times10^{-11}T-2.738\times10^{-9} & (60<T\le 300K)\end{cases} \quad (20)$$

And

$$\rho_{solder}(T)=\begin{cases}-2.15\times10^{-13}T^3+2.0\times10^{-11}T^2-1.2\times10^{-10}T+5.4\times10^{-9} & (4.2K<T\le 45K)\\ 4.85\times10^{-10}T-2.8\times10^{-10} & (45K<T\le 300K)\end{cases} \quad (21)$$

According to Eqs.(2)-(8) and (18)-(21), the current distribution among NbTi, YBCO CC and matrices can be numerically calculated with different transport current I and parallel magnetic field of 6T.

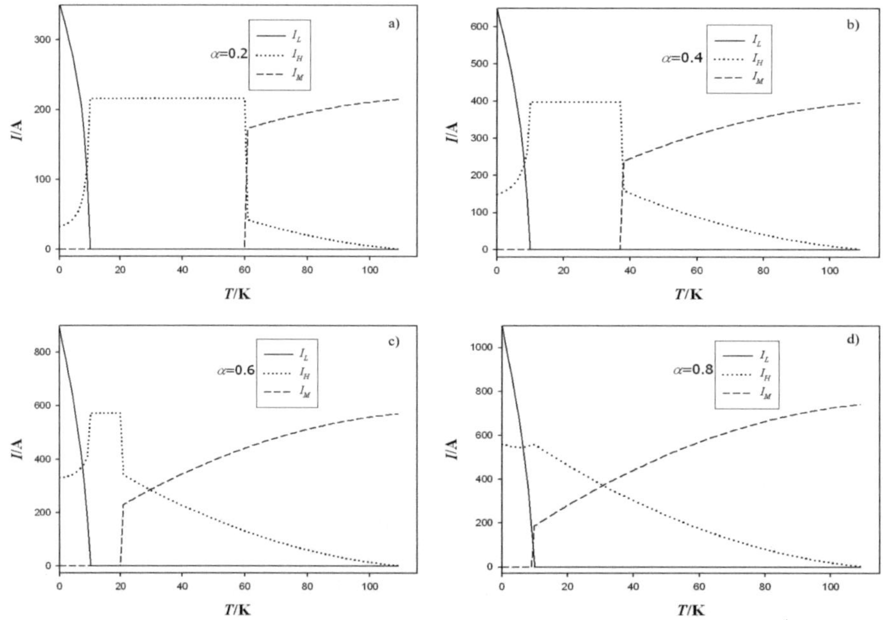

Fig. 6. The current distributions among NbTi, YBCO CC and the matrices in the hybrid superconductor vs temperature when the normalized transport current α=0.2, 0.4, 0.6, 0.8, respectively

In this simulation, n_H =12 and n_L =30 are adopted and assume that both of them are independence of temperature. Under conditions of normalized transport current α=0.2, 0.4, 0.6, 0.8, the temperature dependence of current distribution among three components are showed in Fig.6. With α=0.2, shown as in Fig. 6(a), the simulation results indicate that the current mainly flows in NbTi below 10 K, then transfers from NbTi to YBCO CC near above 10 K, and is totally transported by YBCO in range of 10 K through 60 K after NbTi quenching, then starts to transfer to the metal matrix gradually with temperature increasing. Finally, total current flows into the matrices after YBCO quenching completely. Figs.6(b), (c) and (d) show the larger normalized transport current, the lower temperature of current beginning to transfer from YBCO CC to the matrices. If $I_T<I_{cH}$, the current variation in the matrices is very flat, but there is a sudden increase which can be observed when $I_T>I_{cH}$, especially at $\alpha>0.5$.

When the hybrid conductor operates below 10 K, the current ratio of NbTi to YBCO CC decreases with increasing of the normalized current and temperature, as indicated in Fig.7.

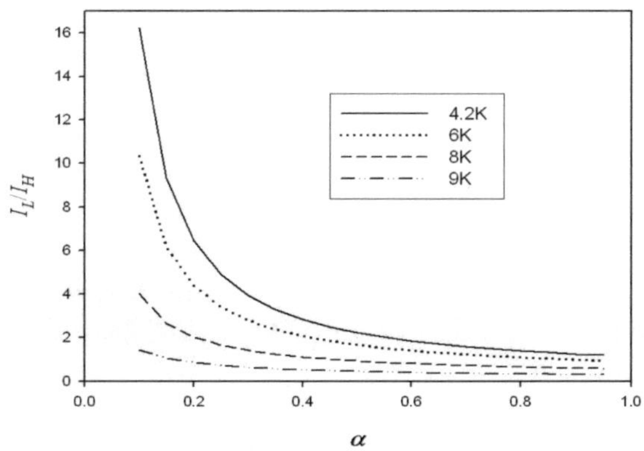

Fig. 7. The current ratio of NbTi to YBCO CC at different normalized transport currents

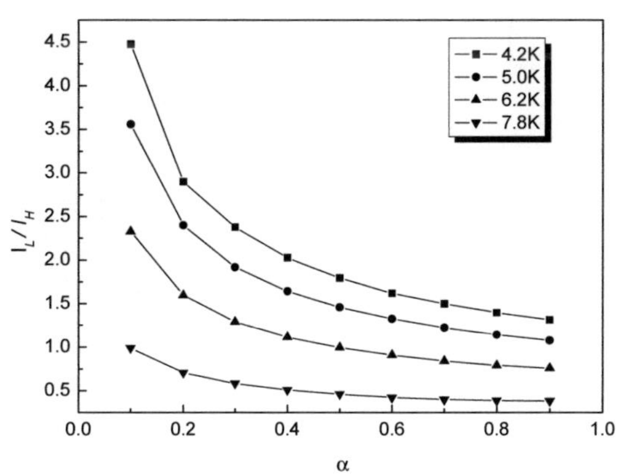

Fig. 8. Current distribution between NbTi/Cu conductor and Bi2223/Ag tapes

3.1.2 Current distribution of hybrid conductor made of NbTi/Cu and Bi2223/Ag

Same processing as the former hybrid conductor, this type of hybrid conductor is obtained by soldering NbTi/Cu with Bi2223/Ag. But there are much more components in matrices than the former one. Dependence of critical currents of two Bi2223/Ag tapes on temperature are approximately given by

$$I_c(T) = I_c(0)\left(1 - \frac{T}{T_c}\right)^{1.4} \tag{22}$$

where $I_c(0)$=620 A, T_c=110 K. The critical current is I_c=586 A with 6 T and 4.2 K. Unlike the former hybrid conductor, the matrices in this conductor include silver, stainless steel except for Copper and solder. The relations of resistivity (Ω·m) of silver and stainless steel with temperature can be expressed as polynomial terms

$$\rho_{Ag}(T)=\begin{cases}8.5\times10^{-12}e^{0.1037T} & (4K<T\le 40K)\\ -3.4\times10^{-16}T^3+8.5\times10^{-14}T^2+7.14\times10^{-11}T-2.428\times10^{-9} & (60K<T\le 300K)\end{cases} \tag{23}$$

and

$$\rho_{ss}(T)=1.05\times10^{-12}T^2+4.72\times10^{-10}T+4.8705\times10^{-7} \tag{24}$$

The current distributions are numerically calculated in a parallel magnetic field of 6 T by using Eqs.(2)-(7), (17), (19), (20)-(24)" with Q=0 (no disturbance) when the transport current I_T is smaller than its critical current in various temperatures below 8K. The results are shown in Fig.8, in which current distributions among NbTi, Bi2223 and matrices with different temperatures are indicated.

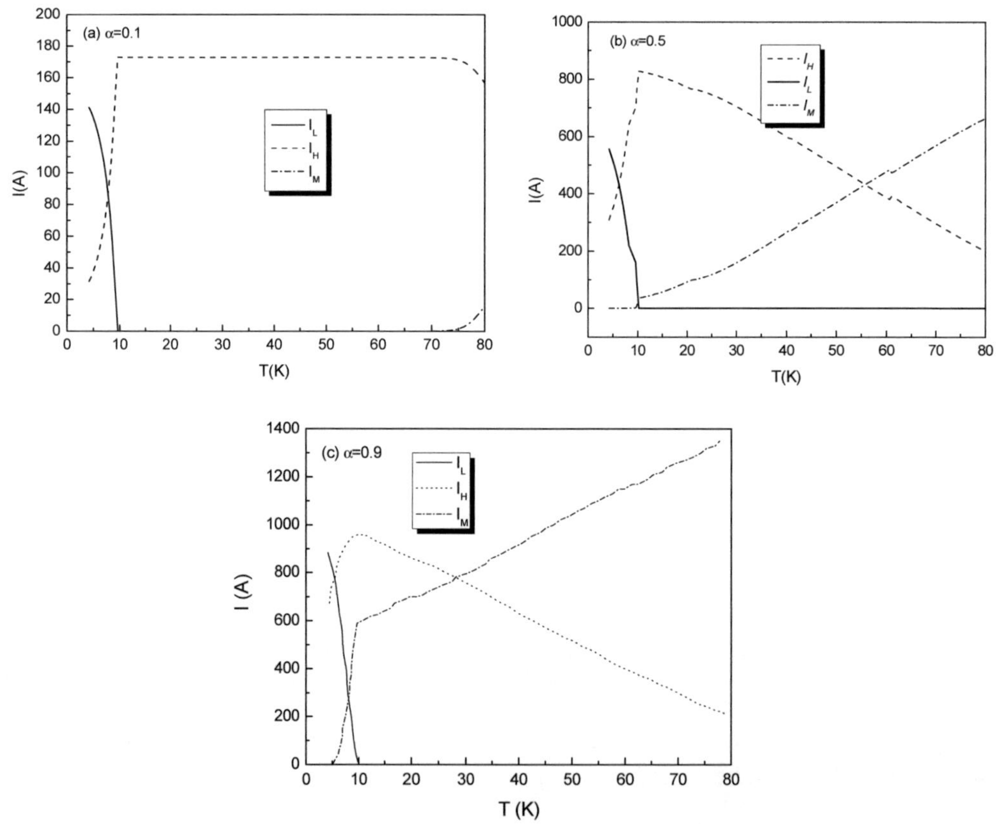

Fig. 9. Current distributions among NbTi, Bi2223 and matrices with different temperatures at α=0.1, 0.5 and 0.9

It is shown that I_L decreases while I_H increases with rise of temperature, which is due to the different n values in NbTi and Bi2223/Ag. Although the difference of both critical currents is not large enough at 4.2 K and 6 T, I_L is larger than I_H when α is smaller than 0.7. The value of I_H becomes close to I_L when α is close to 0.9 at about 5K, respectively. However, I_L approaches to I_H at 6.2K with α=0.6, and $I_L \leq I_H$ holds at 7.8 K for all values of α in the range of 0.1 through 0.9.

In order to consider the effect of the matrices on the current distribution with increasing temperature and transport current I_T, the current distributions among the NbTi conductor, Bi2223 tape and matrix were also simulated. Current distributions with different α 's are presented in Figs.9(a)-(c) in range of 4.2 K through 80 K.

Fig. 9(a) indicates that I_L decreases but I_H increases with rise of temperature when α= 0.1. I_H reaches a maximum then remains constant in the range of 9.5 K through 70 K while I_M is still zero for T≤70 K. If T>70 K, it gradually increases while I_H decreases. I_L and I_H intersect each other at 7.8 K, and I_L decreases to zero with temperature increasing to its critical temperature. However, I_M appears in temperature T=9.5K when transport current increases to α=0.5, and then it gradually increases with temperature, as shown in Fig. 9(b). At α=0.9, the current distribution is shown in Fig. 9(c). Similar to the case of α=0.5, I_H reaches its maximum at about 9.5 K then decreases gradually. But there is a difference in I_M's of Figs.9(b) and (c). In Fig.9(b), I_M gradually increases with temperature. However, it has a "knee point" at about 9.5K and its slope dramatically increases between 5.5 K and 9.5K, and then increases at a slower rate due to the large transport current and large n value of NbTi. Comparing with I_L, the rate of increase or decrease of I_H is lower since Bi2223 has a lower n value.

3.2 Stability of the hybrid conductor

3.2.1 Stability of hybrid made of NbTi/Cu and YBCO CC

In order to solve Eq. (15), the heat capacity γC (J m^{-3} K^{-1}) and thermal conductivity k(W m^{-1} K^{-1}) in all of the components including NbTi, YBCO and every kind of matrices must be known. All of them are listed as followings (Fujiwara et al, 1994)

i. NbTi

$$C_{NbTi}(T) = 0.152 + 2.10\times10^{-3}T^3 \tag{25}$$

With unit of J kg^{-1} K^{-1}, multiplied by mass density γ_{NbTi}=6550 (kg m^{-3}), (25) can be converted to volumetric heat capacity with unit (J m^{-3} K^{-1}).

$$k_{NbTi}(T) = \begin{cases} 0.38887\times T^{0.1532} - 0.371 & (4K < T \leq 6K) \\ 0.13957\times T^{0.782} - 0.4262 & (6K < T \leq 100K) \end{cases} \tag{26}$$

ii. YBCO CC

$$C_{YBCO}(T) = \begin{cases} 1.1 - 0.4T + 5\times10^{-2}T^2 - 4\times10^{-5}T^3 & (2K \leq T < 50K) \\ -125.2 + 5.6T - 1.9\times10^{-2}T^2 & (50 \leq T < 100K) \\ 80 + 2T - 3\times10^{-3}T^2 & (100K \leq T \leq 300) \end{cases} \tag{27}$$

With mass density γ_{YBCO}=6380(kg m^{-3})

$$k_{YBCO}(T)=\begin{cases}-3.5332+9.6273T-0.1282T^2+5\times10^{-4}T^3 & (2K<T\le 50K)\\ 208.45+0.2165T-5\times10^{-6}T^2 & (50<T<300K)\end{cases} \quad (28)$$

iii. Copper

$$C_{cu}(T)=7.582\times10^{-4}T^3 \qquad (4K<T\le 100K) \quad (29)$$

With mass density γ_{Cu}=8940 (kg m-3).

The expression for copper thermal conductivity is $k_{Cu}(T)=10^{f(T)}$, taking Y=Log(T), the f(T) is expressed by

$$f(T)=\begin{cases}1.9153+1.7202Y-0.4146Y^2 & (4K<T\le 15K)\\ 0.8611+3.7464Y-1.3762Y^2 & (15K<T\le 23.7K)\\ -13.9115+33.952Y-21.482Y^2+4.3295Y^3 & (23.7K<T\le 100K)\end{cases} \quad (30)$$

iv. Solder (Iwasa, 1994; Jack, W. Ekin, 2007)

$$C_{solder}(T)=\begin{cases}-3.27\times10^{-2}T^2+5.3731T-20.666 & (4.2K\le T<77K)\\ -1.0\times10^{-4}T^2+1.2582T+162.01 & (77K\le T<100K)\end{cases} \quad (31)$$

With mass density γ_{solder}=7310(kg m-3).

$$k(T)=\begin{cases}2\times10^{-3}T^3-0.2056T^2+6.4032T-4.008 & (4K\le T<45K)\\ 50.6 & (45K\le T<300K)\end{cases} \quad (31)$$

Substituting Eqs. (24)-(31) into Eq.(14) and considering the boundary condition (15) in adiabatic condition, the longitudinal quench propagation velocity (QPV) and the minimum quench energy (MQE) are numerically calculated by finite element method (FEM).

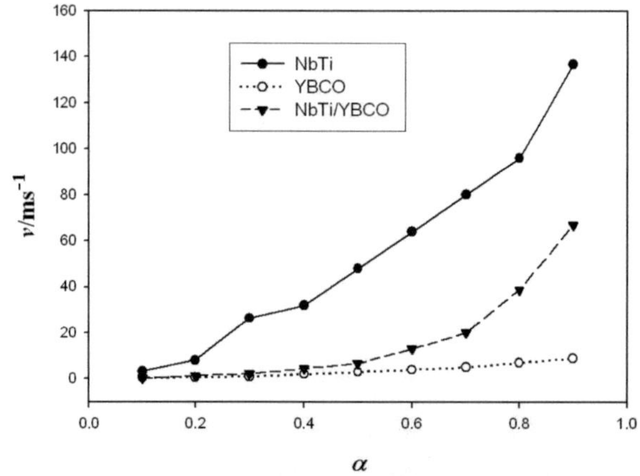

Fig. 10. The comparison of quench propagation velocities in NbTi, YBCO and NbTi/YBCO hybrid conductor with different transport currents.

The quench velocities of NbTi, YBCO and hybrid NbTi/YBCO conductor with different transport current ratios α are shown in Fig. 10, where the longitudinal QPV of NbTi is the maximum, the one in YBCO is minimum, and the QPV in hybrid NbTi/YBCO conductor is in the range of NbTi through YBCO. Therefore, the QPV of the hybrid conductor is neither faster than that of NbTi nor slower than that of YBCO CC, which is very useful for quench detection and protection of superconducting magnets.

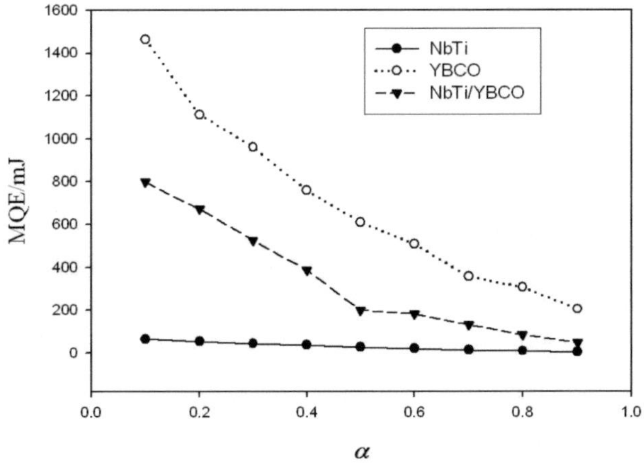

Fig. 11. The comparison of the MQE of NbTi, YBCO CC and NbTi/YBCO CC hybrid conductor with different transport currents.

Fig. 11. shows that the MQE of NbTi/YBCO CC hybrid conductor is in range of NbTi through YBCO CC with order of magnitude of mJ (several kJ·m^{-3}). When α<0.5, the MQE of the hybrid conductor is parallel to that of YBCO CC. An inflection point is observed in the curve at α=0.5, which is perhaps related to the transport current which exceeds the critical current of NbTi. When α>0.5, the MQE is close to that of NbTi. Generally, the MQE of NbTi/YBCO CC hybrid conductor is much larger than that of pure NbTi, which indicates that the stability of superconductor can be improved compared with pure NbTi/Cu wire.

3.2.2 Stability of hybrid made of NbTi/Cu and Bi2223/Ag tape

In section 3.1.1, the heat capacity and thermal conductivity of NbTi, YBCO, copper and solder are given, we list the heat capacity C (J·kg^{-1}·K^{-1}·) and thermal conductivity k(W·m^{-1}·K^{-1}) of other materials including Bi2223, stainless steel, silver.

i. Bi2223

$$C_{Bi2223}(T)=\begin{cases}4.5683\times10^{-3}T^3 & (T\le 10K)\\ -3.088+0.64996T+8.23239\times10^{-3}T^2+3.2406\times10^{-4}T^3 & (10K\le T\le 40K)\\ -58.32+3.18672T-7.8786\times10^{-3}T^2+6.5556\times10^{-6}T^3 & (40K\le T\le 300K)\end{cases} \quad (33)$$

With mass density γ_{Bi2223}=6500(kg·m^{-3}).

$$k_{Bi2223}(T)=\begin{cases}0.02T & (T\le 55K)\\ \begin{pmatrix}0.474+8.43\times10^{-3}T+3.25\times10^{-4}T^2-6.595\times10^{-6}T^3\\ +2.81\times10^{-8}T^4\end{pmatrix} & (55K<T\le 77K)\\ \begin{pmatrix}0.195+9.424\times10^{-3}T+3.4\times10^{-4}T^2-6.237\times10^{-6}T^3\\ +2.673\times10^{-8}T^4\end{pmatrix} & (77K<T<113K)\\ \begin{pmatrix}4.749-0.102T+9.901\times10^{-4}T^2-4.167\times10^{-6}T^3\\ +6.531\times10^{-9}T^4\end{pmatrix} & (113K\le T<200K)\end{cases} \tag{34}$$

ii. Silver

$$C_{Ag}(T)=\begin{cases}8.41\times10^{-4}T^3+5.10\times10^{-2}T^2-0.5566T+1.6341 & (4K<T\le 18K)\\ 2.341\times10^{-5}T^3-1.674\times10^{-2}T^2+3.8384T-50.775 & (18K<T\le 300K)\end{cases} \tag{35}$$

With mass density γ_{Ag}=10490 (kg m^{-3}).

$$k_{Ag}(T)=\begin{cases}43.343\times T^3-1227.2T^2+10513T-10254 & (4K<T\le 10K)\\ -0.6174T^3+72.264T^2-2816.5T+37594 & (10K<T\le 38K)\\ -0.0179T^3+3.6865T^2-253.64T+6292.2 & (38K<T\le 50K)\\ -1.3\times10^{-2}T^2-0.825T+640.3 & (50K<T\le 100K)\\ 420 & (100K<T\le 300K)\end{cases} \tag{36}$$

iii. Stainless steel (304L)

$$C_{ss}(T)=\begin{cases}2\times10^{-6}T^3+1.59\times10^{-2}T^2+0.2644T+0.4089 & (4K<T\le 30K)\\ 3.114\times10^{-4}T^2+0.7171T-1.7843 & (30K<T\le 300K)\end{cases} \tag{37}$$

With mass density γ_{ss}=7900(kg m^{-3}).

$$k_{ss}(T)=2.7041\times10^{-9}T^3-3.3219\times10^{-4}T^2+0.126T-0.1877 \quad (4K\le T\le 300K) \tag{38}$$

In this section, the longitudinal QPV and MQE are numerically simulated by solving Eq.(8) in combination with Eqs. (10)-(12), (16), (25), (26) and (29)-(38) under adiabatic conditions. Figs.12(a)-(c) show the temperature profiles at locations of 20 mm, 40 mm, 60 mm and 80 mm from the center with different external disturbances and transport currents. The variations of n values with temperature are not considered in simulation. The hybrid conductor was tested in magnetic field of 6 T which was applied for reducing its critical current in measurement. The dependence of n values on temperature is very difficult to obtain by experiment with supplying so high a transport current and background magnet. Therefore, we used the constant n values in simulation. Usually, the n values will decrease with increasing temperature, but we took n values approximately constant with fixed magnetic field regardless of influence of temperature in order to simplify calculation. In future, we should take NbTi and Bi2223 with small I_c in experiment within magnet bore in which temperature can be changed. Or the n values are measured by contact-free methods, like those methods adopted in HTS at 77 K (Wang et al, 2004; Fukumoto et al, 2004).

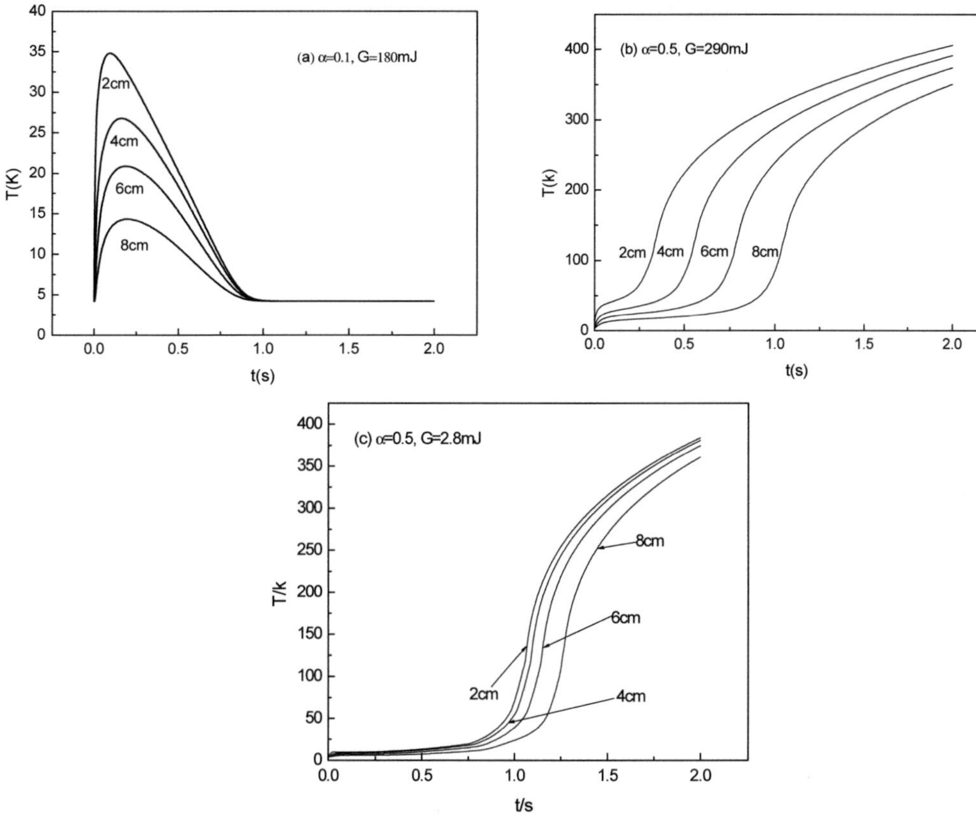

Fig. 12. Temperature profiles of the hybrid conductor with different transport current

When I_T is small (α=0.1) and there is a disturbance G=180mJ (~192kJ·m^{-3}), though the maximum temperature reaches to 35K, the quench doesn't propagate. Once the disturbance disappears, the hybrid conductor recovers to its original state at 4.2 K, as shown in Fig. 12(a). The reason is that the extra current in the NbTi transfers to the Bi2223 even though the temperature is far above the critical temperature of NbTi, but is still far below the critical current of Bi2223. There is no Joule heat generation in the hybrid conductor even though I_T is applied. For disturbances of G=290mJ (~310kJ/m^3) and 2.8mJ (~3kJ·m^{-3}) with transport currents α=0.3 and 0.5, the quench does propagate and the results are shown in Figs. 12(b) and (c) to indicate that the quench process can not recover.

Figs. 13 and 14 present the longitudinal QPV (V_Q) and MQE (Q_E)of three types of composite conductors (NbTi, hybrid NbTi/Bi2223 and Bi2223) with different normalized transport current factor α. The longitudinal QPV increases with increasing α. Among the three types of conductors, V_q in NbTi/Cu is the largest (~10^2 m/s), the one in Bi2223 is the lowest (~10^{-2}-10^{-1} m/s), but V_q in the hybrid NbTi/Bi2223 falls in the range of about 10^{-1} m/s through10 m/s for $\alpha \geq 0.4$. On the other hand, Q_E decreases with increasing α. Q_E of Bi2223 is the largest (~10^{3}mJ), the lowest in NbTi (~10^{-2}-1 mJ) and falls on the range of 1 mJ through 10^3 mJ in the hybrid conductor. In the case of $\alpha \leq 0.4$, Q_E in the NbTi/Bi2223 is more than 100mJ but significantly decreases while α is in the range of 0.4 through 0.5, and then it decreases gradually with increasing α. Nevertheless, Q_E of the hybrid conductor is at least more than one order of magnitude higher than the NbTi. Therefore, the stability of the hybrid conductor is improved greatly comparing with the NbTi conductor.

More exact simulation should be based on three-dimensional model in which the temperature distribution in cross-section can be numerically analyzed. This method will be used in future research.

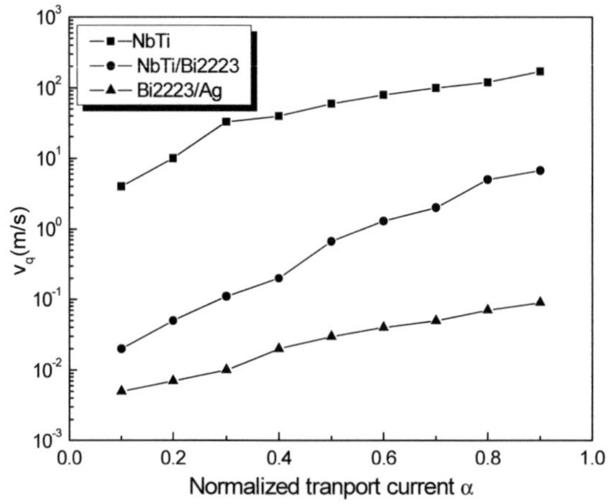

Fig. 13. Longitudinal QPV (Vq) of three types of conductors

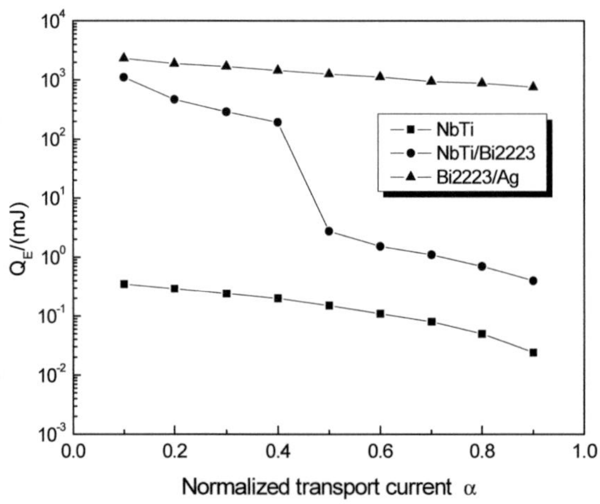

Fig. 14. MQE Q_E of three types of conductors

4. Experiment

The hybrid conductor was prepared by soldering two Bi2223/Ag tapes onto one NbTi/Cu conductor by use of Indium-Silver alloy solder under 200 °C in order to avoid degradation of Bi2223/Ag tape (Wang, 2009). One heater and two Rh-Fe thermometers were attached to the hybrid conductor. Next, the hybrid sample was wound by 10 layers of fiber glass tape and then immersed into epoxy resin in order to simulate the quasi-adiabatic environment.

The total length of the hybrid was 900 mm and was wound on a FRP bobbin with diameter of 70 mm. The main parameters of each conductor were also listed previously in Table 1 and the sample is shown as Fig.15.

A schematic diagram of the experimental set-up is illustrated in Fig.16. The hybrid sample was tested under a background field of 6 T provided by an NbTi NMR magnet with a core of diameter 88.6 mm and homogeneity of 1.7×10^{-7} in a 10 mm×10 mm spherical space, which ensured that the sample was located in the same field. The total length of the homogeneity region in axial orientation was 200 mm. The magnet was composed of 3 main coils and 2 compensated coils wound using NbTi/Cu composite wire. The heater, bifilar wound non-inductively by copper-manganese wire with a diameter of 0.1 mm, had a resistance of 69.7 Ω at 4.2 K.

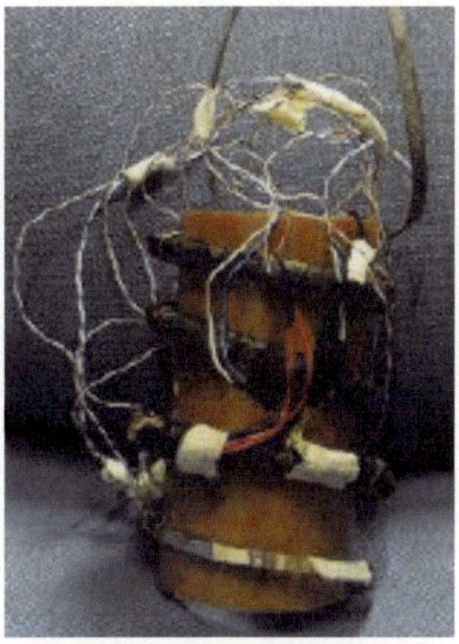

Fig. 15. Prepared sample

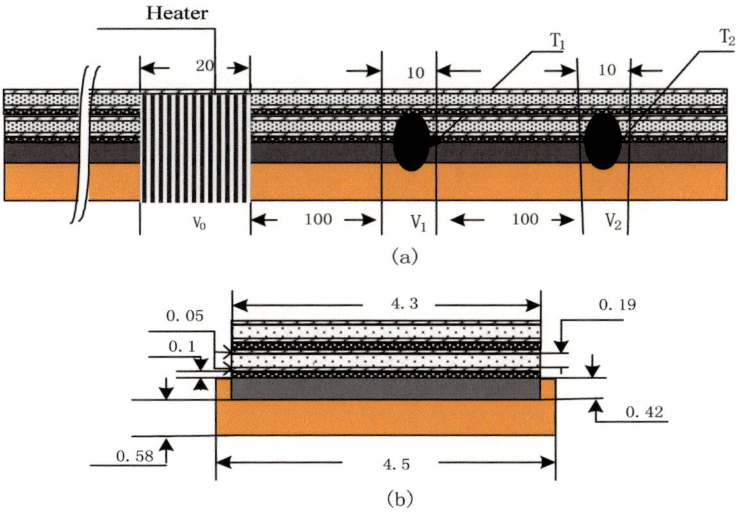

Fig. 16. Schematic of sample test arrangement. (a) and (b) are front-view and side-view of the hybrid conductor, respectively. T_1 and T_2 refer to temperature sensors. unit: mm.

The tests were performed in a 4.2 K helium bath and the magnet was excited with 6 T in all experiments. The quench voltage and temperature profiles were measured by triggering the heater with rectangular waveforms of different durations and amplitudes. Since 800 A was the limit of our power supply, the maximum transport current in sample was 800 A in this section.

5. Results and discussions

When a transport current of 400A was supplied in a background magnetic field of 6T, the quench voltage and temperature profiles are shown in Figs. 17 and 18, respectively,. The duration of power from the heater was 0.4s and the amplitude was 0.3A; therefore, the disturbance energy from the heater was 2.5J. When the heater was triggered, only the central part of the sample quenched, V_1 appeared slightly, but V_2 remained the same. The

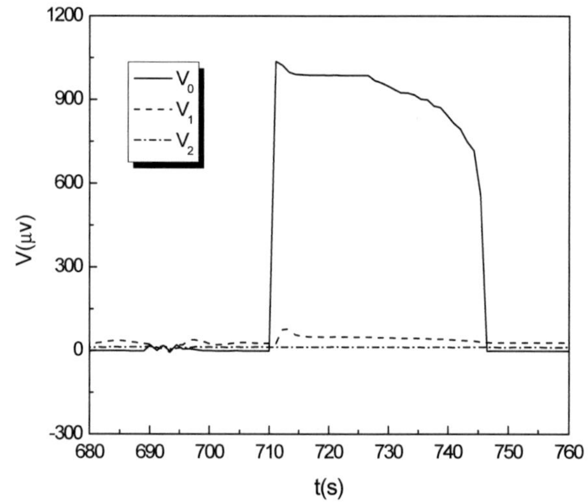

Fig. 17. Voltage profiles with impulse duration of 0.4 s and amplitude of 0.3 A

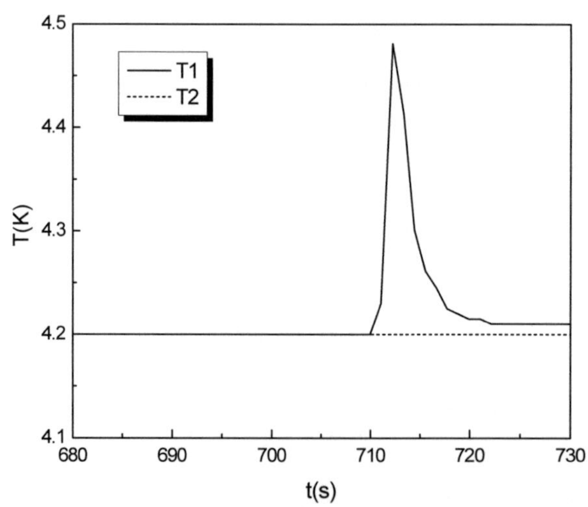

Fig. 18. Temperature profile with impulse duration of 0.4 s and amplitude of 0.3 A

temperature profiles with a peak of 4.4 K were different from voltages and the temperature of T_2 kept constant, which indicates that the quench recovered and there was no quench propagation during the triggering.

In order to measure the quench propagation, the transport current 800 A was applied, the triggering duration and amplitude were 59 ms and 0.1 A, respectively, i.e. the triggering energy is 41.12 mJ. The voltage and temperature profiles are presented in Figs. 19 and 20. When heater was triggered, three parts of the sample quenched, V_1 is slightly larger than V_2, the temperature profiles with maximum 11 K are similar to the voltages, which mean that the quench propagates. The order of the V_q is 10 m/s which is higher than the Bi2223/Ag tape (Dresner, 1993). Q_E has order of several tens of mJ, which are much larger than those of the NbTi/Cu conductor (Frederic et al, 2006). On contrary to 400 A, quench propagation does take place; Q_E and V_q are 41.12 mJ and 10 cm/s, respectively, which qualitatively agree with the simulated results in section 3.2.2 in case of normalized transport current α=0.5, though the experimental results are smaller than simulations. The differences between the experiment and simulation result from the assumptions of adiabatic conditions and constant n values in different temperatures. Practically, the quasi-adiabatic condition in the experiment is just an approximate and dependence of the n values on temperature and magnetic field should be included. In future, an experiment including normal zone propagation (NZP), V_q and Q_E should be performed by using cryo-cooler and LTS with lower critical current in order to obtain the quench parameters exactly. Furthermore, a three-dimensional model should be adopted. The stability of other types of hybrid conductor, such as LTS (NbTi, Nb_3Sn) /MgB_2, HTS(BSCCO, YBCO)/MgB_2 and Nb_3Sn/HTS, could be also needed to study by simulation and experiment.

Additionally, the variations of n values with temperature and magnetic field should be taken account into consideration and measured possibly by contact-free methods similar with those used in HTS tapes (Wang et al, 2004; Fukumoto et al, 2004). This work will need to conduct in near future.

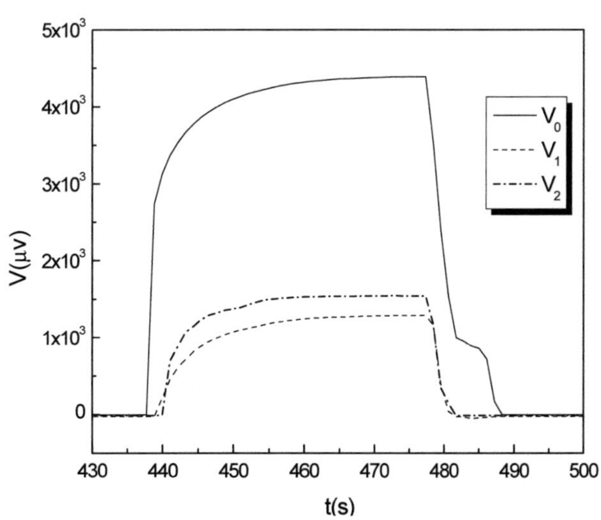

Fig. 19. Voltage profiles with impulse duration of 59 ms and amplitude of 0.1 A.

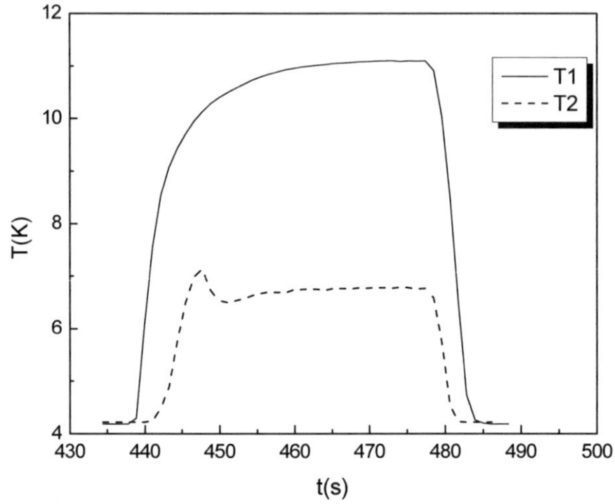

Fig. 20. Temperature profiles with impulse duration of 59 ms and amplitude of 0.1A

6. Conclusions

The current distribution and stability of LTS/HTS hybrid conductor, which is made of NbTi wire and YBCO coated-conductor, are numerically calculated. The results indicate that the current in LTS is larger than in HTS if both of them have the approximate critical currents and the current ratio of NbTi to YBCO CC decreases with increase of transport current and temperature when the hybrid conductor operates. On the other hand, the longitudinal quench propagation velocity is in the range of NbTi through HTS, which is very important for quench detection and protection of superconducting magnets. Finally, the MQE (Q_E) in the hybrid conductor is much higher than in NbTi wire and smaller than in YBCO CC conductor, which shows that the thermal stability of superconductor can be improved.

Based on the concept of a hybrid NbTi/Bi2223 conductor and power-law models, the current distribution was simulated numerically. Since NbTi has a higher n value than Bi2223, most of current initially flows through NbTi while the ratio of current in Bi2223 to that in NbTi increases with rise of temperature and transport current below their total critical current. The stability of the hybrid conductor was simulated using one-dimensional model. The results show that the V_q of the hybrid conductor is smaller, but the Q_E is bigger than NbTi conductor, which indicates that the stability of the hybrid superconducting conductor is improved. Simultaneously, a high engineering current density was also achieved. A short sample, made of Bi2223/Ag stainless-steel enforced multifilamentary tape and NbTi/Cu, was prepared and tested successfully at 4.2 K. The results are in qualitative agreement with the simulated ones.

With improving on their stability and engineering critical current compared with conventional LTS and HTS, the hybrid conductors have potential application in mid- and large scale magnet and particularly in the cryo-cooled conduction magnet application.

In future, the cryocooler-cooled conduction should be adopted in the experiments, and a three-dimensional model with n values depending on temperature and magnetic field and its orientation should be taken into account to improve the present numerical results. Stability in other types of hybrid conductor, such as (NbTi, $NbSn_3$)/MgB_2 and (NbTi, $NbSn_3$)/YBCO CC, should be also valuable for study in next step.

7. Acknowledgements

The author thanks Ms. Weiwei Zhou, Dr. Wei Pi, Prof. Xiaojin Guan and Dr. Hongwei Liu for their contributions to the research included in the chapter. This work was supported in part by the National Natural Science Foundation of China under grant No.51077051 and Specialized Research Fund for the Doctoral Program of Higher Education under grant No.D00033.

8. References

Dresner, L. (1993) Stability and protection of Ag/BSCCO magnets operated in the 20-40 K range. *Cryogenics*, Vol.33, pp 900-909

Dutoit, B; Sjoestroem, M. & Stavrev, S. (1999) Bi(2223) Ag sheathed tape Ic and exponent n characterization and modeling under DC applied magnetic field. *IEEE Trans. Appl. Supercond.*, Vol.9, No.2, pp. 809-812

Frederic, T, Frederic, A & Amaud, D. (2006) Investigation of the stability of Cu/NbTi multifilament composite wires. *IEEE Trans. Appl. Supercond.* Vol.16, No.2, pp. 1712-1716

Fukumoto, Y; Kiuchi, M. & Otabe, E. S. (2004) Evolution of E-J characteristics of YBCO coated–conductor by AC inductive method using third-harmonic voltage. *Physica C*, Vol. 412-414, pp 1036-1040

Fujiwara, T; Ohnishi, T; Noto, K; Sugita, K. & Yamamoto, J. (1994) Analysis on influence of temporal and spatial profiles of disturbance on stability of pooled-cooled superconductors. *IEEE Trans Appl. Supercond.*, Vol.4, No. 2, pp. 56-60.

Gourab, B.; Nagato, Y. & Tsutomu, H. (2006) Stability measurements of LTS/HTS hybrid superconductors. Fusion Eng. Des., Vol. 81, pp. 2485-2489

Iwasa, Y. (1994) *Case studies in superconducting magnet*. Plenum Press, New York and London.

Jack, W. Ekin. (2007) *Experimental Techniques for Low-Temperature Measurement*. Oxford University Press Inc., New York.

Rimikis, A.; Kimmich, R. & Schneider, Th. (2000) Investigation of n-values of composite superconductors. *IEEE Trans Appl. Supercond.*, Vol.10, No.1, pp.1239-1242

Torii, S.; Akita, S.; Iijima, Y.; Takeda, K. & Saitoh, T. (2001) Transport current properties of Y-Ba-cu-O tape above critical current region. *IEEE Trans Appl. Supercond.*, Vol.11, No.1, pp. 1844-1847

Wang, Y. S. ; Zhao, X and Han, J. J. (2004) A type of LTS/HTS composite superconducting wire or tape. Chinese patent (ZL200410048208.8 (In Chinese).

Wang, Y. S.; Zhang, F. Y. & Gao, Z. Y. (2009) Development of a high-temperature superconducting bus conductor with large current capacity. *Supercond. Sci. Technol.*, Vol.22, 055018 (5pp)

Wang, Y. S.; Lu, Y. & Xiao, L.Y. (2003) Index number (n) measurements on BSCCO tapes using a contact-free method. *Supercond. Sci. Technol.* Vol.16, pp. 628-63

Wilson, M. N. (1983). *Supercnducting Magnet*. Clarendon Press Oxford, London.

Yasahiko, I.& Hidefumi, K. (1995) Critical current density and n-value of NbTi wires at low field. *IEEE Trans Appl. Supercond.*, Vol.5, No.2, pp. 1201-1204

5

Magnetic Relaxation - Methods for Stabilization of Magnetization and Levitation Force

Boris Smolyak, Maksim Zakharov and German Ermakov
Institute of Thermal Physics Ural Branch of RAS
Russian Federation

1. Introduction

Bulk high-temperature superconductors (HTS) are used as current-carrying elements in various devices: electrical machines, magnetic suspension systems, strong magnetic field sources, etc. Supercurrents decay due to the relaxation of nonequilibrium magnetic structures. This phenomenon, which is known as magnetic flux creep or magnetic relaxation, degrades the characteristics of superconducting devices. A "giant flux creep" is observed in HTS. There is an extensive review on this phenomenon by Yeshurun et al. (1996), but the magnetic relaxation suppression was discussed only briefly in it. An overwhelming majority of studies dealing with applications of HTS also paid little attention to the problem of creep. In this chapter we describe the methods of influence on the relaxation rate both of local characteristics of the magnetic structure (vortex density and vortex density gradient) and averages over the volume of superconductor (magnetic flux, magnetic moment and levitation force). Particular emphasis is placed on the magnetization and the magnetic force whose stability is necessary for the normal operation of the majority of high-current superconducting devices.

Magnetic flux creep has its origin in motion vortex (flux lines) out of their pinning sites due to the thermal activation. The creep rate decreases when new or denser pinning sites are introduced into HTS sample. The overview of different techniques for producing pinning sites may be found in the review by Yeshurun et al. (1996). The dramatic decrease in the magnetic relaxation rate is observed if the temperature of the superconductor is reduced (Maley et al., 1990; Sun et al., 1990; Thompson et al., 1991). This effect known as "flux annealing" arises due to the transition of vortex system from the critical state having small activation energy to the subcritical state with relatively large activation energy. The "flux annealing" suppresses flux creep, but does not affect the magnetic structure. The induction gradient, which determines the supercurrent density and the superconductor magnetization, does not change after "annealing". However, this method is difficult to implement in technological applications. On the contrary, the exposure of ac magnetic fields strongly affects the nonequilibrium vortex configuration. The critical state in superconductor is completely destroyed at the certain amplitude of ac field (Fisher et al., 1997; Willemin et al., 1998). If the amplitude is less than it, the induction gradient is destroyed at the depth of ac field penetration (Fisher et at., 1997; Smolyak et al., 2007), and in the region bordering the penetration region gradient structure experiences strong relaxation which is not related to thermal activation (Brandt & Mikitik, 2003). After switching off ac field the remanent stationary magnetization is much smaller, but

it decays with time much slower than before the exposure of ac field. It was found that after the exposure of transverse ac field the remanent induction distribution does not change for a long time, i.e. the subcritical vortex configuration is formed (Fisher et al., 2005; Voloshin et al., 2007). However, the use of ac field to suppress creep in superconducting devices is not effective because the initial magnetization is highly reduced.
A classical paper on the flux creep (Beasly et al., 1969) probably was the first to note that the total magnetic flux in superconductor remains unchanged for a long time after the small reversal of external magnetic field. This effect was studied later in more detail, and it formed the basis of the reverse methods for the stabilization of magnetization (Kwasnitza & Widmer, 1991, 1993) and levitation force (Smolyak et al., 2000, 2002). The reversal leads to the internal magnetic relaxation (Smolyak et al., 2001) when the volume-averaged quantities do not change for a long time. The phenomenon of internal magnetic relaxation is considered in more detail below in the section 3.
Smolyak et al. (2006) studied the dependence of relaxation rate of magnetic force on the rigidity of constraints imposed on a "magnet-superconductor" system. The magnetic force in the suspension system decreases at maximum rate when HTS sample and magnet are rigidly fixed; that is, a rigid mechanical constraint is imposed on the suspension object (HTS sample or magnet). As the mechanical constraint is made weaker, the creep of magnetic force is retarded. The closer the suspension system to the "true" levitation (in which the mobility of the sample is determined predominantly by the magnetoelastic coupling), the slower the magnetic force decays with time. This effect is of great importance for levitation systems and discussed in the section 4.
A new effect has been described recently by Smolyak & Ermakov (2010a, 2010b). It was found that the magnetic relaxation is suppressed in HTS sample with a trapped magnetic flux when the sample approaches a ferromagnet. The local relaxation of induction is absent, too; that is, the flux distribution is rigid and does not vary with time. This effect is considered in the section 5.

2. Magnetization and magnetic force

Let a superconducting disk having a radius R and a thickness d be magnetized as it moves along the z-axis in a nonuniform magnetic field having an azimuthal symmetry (the side surface of the disk is parallel to the z-axis). The disk can perform reverse movements resulting to the azimuthal currents of density J_θ with alternating directions are induced in it. Assume that the critical state extends into the disk from its rim, i.e. the currents induced only by the radial vortex-density gradient:

$$J_\theta = -\frac{1}{\mu_0}\frac{dB_z}{dr}(r), \tag{1}$$

where B_z denotes the axial component of induction and μ_0 is the magnetic constant.
The disk magnetization along the z-axis may be written as:

$$M = \frac{1}{R_2}\int_0^R J_\theta(r) r^2 dr \,. \tag{2}$$

The force acting upon the disk along the z-axis:

$$F = \int_V J_\theta(r) B_r(r,z) dv , \tag{3}$$

where B_r is the radial component of the field induction; V is the disk volume. The density of ponderomotive forces $J_\theta B_r$ depends on the true value of field which is produced both by the external source and the currents influenced by the force. However, when a full force is calculated from Eq. (3), B_r may be assumed to mean the external field only (Landau et al., 1957).

Let us use the Bean's model of critical state according to which the critical current density J_c is constant throughout the volume. In Eq. (1) $J_\theta = J_c$ (if $dB_z/dr < 0$), $J_\theta = -J_c$ (if $dB_z/dr > 0$) and $J_\theta = 0$ (if $dB_z/dr = 0$). The disk will have a maximum magnetization if a unidirectional current flows in the whole volume of the disk ($0 \le r \le R$):

$$M_m = \frac{1}{3} JR \tag{4}$$

The subscript c at J is omitted because the current density decreases with time.

If the current flows in the region $r_1 \le r \le r_2$ ($r_2 < R$), the disk has a partial unipolar magnetization:

$$M = M_m \left(\frac{r_2^3 - r_1^3}{R^3} \right) . \tag{5}$$

If the currents producing opposite magnetic moments circulate in the sample, the disk has a bipolar magnetization:

$$M^* = M_m \left[2 \left(\frac{r^*}{R} \right)^3 - \left(\frac{r_1}{R} \right)^3 - 1 \right] , \tag{6}$$

where r^* is the boundary between regions passing counter currents ($r_1 < r^* < R$); the critical state occupies the region $r_1 \le r \le R$. Here and henceforth the quantities relating to the bipolar current structure are marked with an asterisk (e.g. $F(M^*) \equiv F^*$ etc.).

The magnetic field with azimuthal symmetry is usually created by disk or ring permanent magnets, and in some cases $B_r(r)$ can be approximated by a linear dependence. Then the expression for the force (3) may be written as (Smolyak et al., 2002):

$$F = \Phi_r M , \tag{7}$$

$$\Phi_r = 2\pi R \int_z^{z+d} B_r(R,z) dz , \tag{8}$$

where Φ_r is the radial magnetic flux piercing the disk rim. This flux can also be expressed as the axial induction gradient averaged over the disk volume V:

$$\Phi_r = \left\langle \frac{dB_z}{dz} \right\rangle V . \tag{9}$$

If the HTS sample does not move after the magnetization, the magnitude Φ_r in Eq. (7) does not change with time and, consequently, $F(t) \sim M(t)$. The force normalized to F_m determines the load factor:

$$w = \frac{F}{F_m} = \frac{M}{M_m}, \quad w^* = \frac{F^*}{F_m} = \frac{M^*}{M_m}, \tag{10}$$

where $F_m = \Phi_r M_m$, $F^* = \Phi_r M^*$.

3. Open and internal magnetic relaxation

3.1 Time evolution of current density

The flux creep develops when the critical state is established in the superconductor; i.e. the vortex-density gradient, or induction gradient, is formed. The critical gradient determining the critical current density is established in the result of the balance of opposing forces: pinning force holding vortices on the pinning sites and Lorentz force JB which drives vortices. The form of the induction distribution is determined by the dependence of pinning force on the vortex density. We use the Bean's model of critical state according to which the vortex density is distributed in superconductor with the same gradient, i.e. the critical current density J_c = const. The greater the pinning, the larger the value J_c. Magnetic relaxation was first studied in low-temperature superconductors (LTS). The experiment shows the trapped magnetic flux gradually leaves the sample. The explanation for this phenomenon was proposed by Anderson (1962) and Anderson & Kim (1964). They introduced the concept of thermal activation. The thermal fluctuations make the vortex surmount the pinning barrier and move in the direction of the Lorentz force to the region where their density is smaller. As a result, the induction gradient and the current density decrease with time. The creep effect in LTS is so small that it almost has no effect on the characteristics of LTS devices (the current density in superconductor is close to the value J_c for a long time).

The magnetization of high-temperature superconductors decreases immediately after magnetizing. Therefore, the critical state with current density J_c is only the initial state of the magnetic structure which relaxes rapidly so that the real current density is much less than J_c. The strong magnetic relaxation can also be described by the theory of Anderson-Kim in the first approximation. The decrease of the current density starting from the moment of time t_0 may be expressed in the term of relaxation coefficient:

$$\alpha(t) = \frac{J(t > t_0)}{J_0} = 1 - \frac{kT}{U_0} \ln \frac{t}{t_0}, \tag{11}$$

where t_0 is the relaxation observation start time; $J_0 \equiv J(t_0)$; $U_0 \equiv U(t_0)$ is the effective activation energy. The magnetization and the magnetic force also change with time. The linear dependence $M(J)$ (Eq. (4)) occurs when the current flows through the whole volume of the superconductor. If the critical state does not occupy the whole volume of the sample, the dependence $M(J)$ becomes nonlinear, because r_1, r_2 and r^* in Eqs. (5) and (6) also depend on J. In this case the variation of magnetization with time depends on the location of gradient zone in the sample. As it is shown below this location determines the type of magnetic relaxation and has the considerable effect on the variation rate of magnetization and force acting on the sample.

3.2 Open magnetic relaxation

Fig. 1 (*a*) and (*b*) present the radial magnetic flux distributions established in the disk when the induction of external field changes from B_{in} (field, in which the disk was cooled) to B_s.

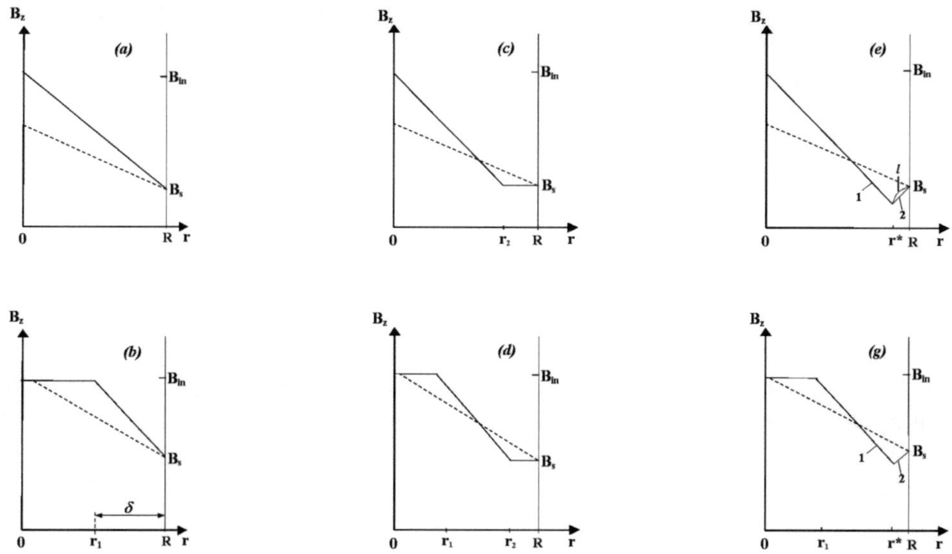

Fig. 1. One-gradient (*a*)–(*d*) and two-gradient (*e*), (*g*) magnetic flux distributions within the disk at the moment of time *t′* (—) and *t″* (---); for all distributions $t'' \gg t'$. The time $t'' \geq t_i$ for the distributions (*c*) and (*d*); $t'' \geq t_b$ for the distributions (*e*) and (*g*).

The (*a*) and (*b*) distributions exhibit the induction gradient in the whole volume of the disk, $0 \leq r \leq R$, and the ring layer, $r_1 \leq r \leq R$, respectively. In both cases the external boundary of the critical state is located on the disk surface *R* through which excess of vortices leaves the sample. The flux creep related to the vortex flow through the superconductor surface will be termed *an open magnetic relaxation.* Due to the creep the distribution slope and current density decrease with the relaxation coefficient $\alpha(t)$ (Eq. (11)).

The coefficient $\beta(t) = M(t>t_0)/M(t_0)$ will be taken to characterize the magnetization relaxation. For the partial penetration (Fig. 1 (*b*)) the magnetization is determined by Eq. (5), where $r_2 = R$ and $r_1 = R - \delta(t)$; $\delta(t)$ is the penetration depth of the critical state. Considering that in the Bean's model $\delta(t) = \delta(t_0) J_0 / J(t) = \delta(t_0) / \alpha(t)$ and using Eq. (10) and relations $\delta(t_0) \equiv \delta_0$, $\hat{\delta}_0 = \delta_0 / R$, $\hat{\delta}_0 = 1 - \sqrt[3]{1 - w_0}$, and also assuming the relaxation term in Eq. (11) is small as compared to unity, one may write (here and in the sections 3.3 and 3.4 we use the results received by Smolyak et al., 2002):

$$\beta(t) = \frac{M(t > t_0)}{M_0} \cong 1 - C_\delta \frac{kT}{U_0} \ln \frac{t}{t_0}, \tag{12}$$

$$C_\delta = \frac{3\hat{\delta}_0 - 2\hat{\delta}_0^2}{3 - 3\hat{\delta}_0 + \hat{\delta}_0^2}. \tag{13}$$

For the partial penetration of the critical state, the magnetization diminishes, similarly to the current density (Eq. (11)), by a logarithmic law, but at the smaller rate, because $C_\delta<1$. If $\hat{\delta}_0 \ll 1$, then $C_\delta \cong \hat{\delta}_0 \ll 1$, i.e. the relative variation of the magnetization 1 - β(*t*) is much less than the change of the current density $1-\alpha(t)$. When $\hat{\delta}_0 \to 1$ (the full penetration), $C_\delta \to 1$ and β(*t*) → $\alpha(t)$, i.e. the current density, the magnetization and the force have one and the same relaxation coefficient $\alpha(t) = J/J_0 = M/M_0 = F/F_0$.

3.3 Internal magnetic relaxation

3.3.1 Current zone removed from superconductor surface

Fig. 1 (*c*) and (*d*) present the flux distributions with the induction gradient in the region $0 \le r \le r_2$ (*c*) and $r_1 \le r \le R$ (*d*). These regions are separated from the superconductor surface *R* by the areas which are free of the vortex-gradient density. The induction distributions (*c*) and (*d*) may be obtained from the distributions (*a*) and (*b*) if an alternating magnetic field is applied to the latter for a short period of time. The induction gradients of the distributions (*c*) and (*d*) diminish thanks to the redistribution of vortices in the superconductor volume (We shall assume that the flux profile preserves its rectilinear behavior as in the case of distributions (*a*) and (*b*), Fig. 1). It may be shown that the current density in zone spaced from the superconductor surface has the same relaxation coefficient $\alpha(t)$ (Eq. (11)). However, oppositely to the open relaxation, the total magnetic flux remains unchanged in the sample. *An internal magnetic relaxation* takes place. The magnetization of the sample is constant, too, i.e. $\beta = M(t>t_0)/M(t_0) = 1$. The internal magnetic relaxation takes the time $t_0 \le t \le t_i$, where t_i denotes the time necessary for the emergence of boundary $r_2(t)$ on the superconductor surface *R*. This time may be found from Eq. (11):

$$t_i = t_0 \exp\left[(1-\alpha_i)\frac{U_0}{kT}\right], \tag{14}$$

where $\alpha_i \equiv \alpha(t_i)$ is the current relaxation coefficient at $r_2(t_i) = R$. The α_i value may be found from the condition of the full flux conservation. For the distribution in Fig. 1 (*c*) we have (considering Eq. (5), where $r_1 = 0$, and Eq. (10)):

$$\alpha_i = \left(\frac{r_{02}}{R}\right)^3 = w_0 . \tag{15}$$

For the distribution in Fig. 1 (*d*) we have:

$$\alpha_i = \frac{2\hat{\delta}_0}{3-\sqrt{4\frac{w_0}{\hat{\delta}_0}-3}}, \tag{16}$$

where $w_0 = \left(r_{02}^3 - r_{01}^3\right)/R^3$ and $\hat{\delta}_0 = \delta_0/R = (r_{02}-r_{01})/R$ (at the beginning of relaxation the size of the gradient zone $r_{02} - r_{01}$ (Fig. 1 (*d*)) is equal to the penetration depth $\delta(t_0)$ (Fig. 1 (*b*)), because the external field variation and pinning are the same in both cases).

3.3.2 Relaxation of opposite gradients

The magnetic structure with the opposite vortex-density gradients is established in the superconductor if the external field is reversed. Fig. 1 (*e*) and (*g*) present the induction distributions for the full and partial penetration of the critical state. The flux profile is a broken line comprising Sections *1* and *2*, in which the induction gradients are equal in the magnitude and opposite in the sign. The vortices diffuse during the creep to the region with the smaller vortex density, i.e. to the boundary r^* between the sections. We shall assume that the straight-line approximation is fulfilled for the both sections. Let the flux-flow density on the side of the Section *1* be larger than the density of the opposite flux flow. The excess vortices move from the Section *1* to the Section *2* and are distributed in the latter section with the same gradient as the one in the Section *1*. In another words, the Section *1* extends, while the Section *2* shrinks. (The arrangement of the gradients during the creep is similar to their rearrangement during the remagnetization of the superconductor: the new vortex distribution expands to the region with the different vortex-density gradient and "erases" the previous distribution. The only difference is that the speed of the gradient front depends on the external field variation rate in one case and on the creep velocity in another case.)

Let us consider the relaxation of distribution in Fig. 1 (*e*) when the sample has a bipolar magnetization (Eq. (6) when $r_1 = 0$). The dependences $r^*(t)$ and $M_m(t) \propto \alpha(t)$ determine the time evolution of the bipolar magnetization which proceeds with the relaxation coefficient $\beta^*(t) = M^*(t)/M_0^*$.

The bipolar magnetization is preserved in the sample for some time t_b only. Once this time has elapsed, the magnetization turns to the unipolar one and then the magnetic relaxation proceeds by the open type. It can be shown that during the time $t_0 \le t \le t_b$ the coefficient $\beta^*(t)$ changes from 1 to the value

$$\beta^*(t_b) = \frac{\alpha_b}{w_0^*} = \frac{1}{w_0^*}\left[4\sqrt{\frac{1+w_0^*}{2}} - 3\right]^{2/3}, \tag{17}$$

where $\alpha_b \equiv \alpha(t_b)$ is the current relaxation coefficient at the moment of time $t = t_b$ when the boundary r^* emerges to the superconductor surface ($r^*(t_b) = R$). Expanding Eq. (17) in the power series of $1 - w_0^*$ and keeping up to the third-order terms inclusive, we have:

$$\beta^*(t_b) \cong 1 + \frac{1}{36}\left(1 - w_0^*\right)^3, \tag{18}$$

where $w_0^* = M_0^*/M_{m0}$ is the load factor (Eq. (10)).

The relationship (17) shows that the magnetization rises slightly during the time t_b. The magnetic flux, which enters the sample during the time $t_b - t_0$, is very small. (This flux corresponds to the region limited by the initial distribution *2* and the surface *l* (see Fig. 1 (*e*)).) The lifetime of the bipolar magnetization may be calculated from Eq. (14) if t_b is substituted for t_i and α_b for α_i.

For the distribution in Fig. 1 (*g*), the relaxation pattern does not differ qualitatively from the relaxation of the distribution in Fig. 1 (*e*): the Section *1*, which contributes most to the magnetization, "swallows up" the section with the opposite magnetization. The full magnetic flux changes little in the sample. Using the flux conservation condition and Eqs. (6) and (10), it is possible to obtain an expression for α_b which is similar in its form to Eq. (16). Here w_0 should be replaced by $w_0^* = \left(2r_0^{*3} - r_{01}^3 - R^3\right)/R^3$ and $\hat{\delta}_0 = \delta_0/R = \left(2r_0^* - r_{01} - R\right)/R$ (as before δ_0 is equal to the penetration depth of the critical state in Fig. 1 (*b*)).

3.4 Results of experiment and calculation

To verify the theory, we use the experimental data on the relaxation of vertical magnetic force in the "magnet-superconductor" system. The experimental setup is described in the paper of Smolyak et al. (2002). The sample of melt-textured $YBa_2Cu_3O_7$ ceramics (disk 10 mm in diameter and 3.5 mm high) and the ring magnet are used in the experiments. The sample was cooled in the initial position in the field of magnet and then was moved to the suspension point in the forward or reverse stroke. During the forward stroke from the initial position to the suspension point the sample was magnetized by the current of one polarity (unipolar magnetization). During the reverse stroke (when the sample passed the suspension point, went a certain distance (reverse depth), and then returned to the suspension point), the opposite currents passed in the sample (bipolar magnetization). The magnetic force *F*, which was acting on the HTS sample in the suspension point, was measured as a function of time. The position of the sample at this point was fixed, i.e. the magnetized sample did not move relative to the magnet when the magnetic force was changing due to the flux creep. The loaded sample, or the suspension, with total weight *G*>*F* stood on the rest all the time except the measurement moments of the force *F* (the force *P* balancing the force difference *G* - *F* was applied to the suspension for a short time and the moment of separation of suspension from the rest was recorded). The initial force F_0 was measured after t_0 = 10 min (t_0 is the time elapsed from the moment the sample was placed at the suspension point).

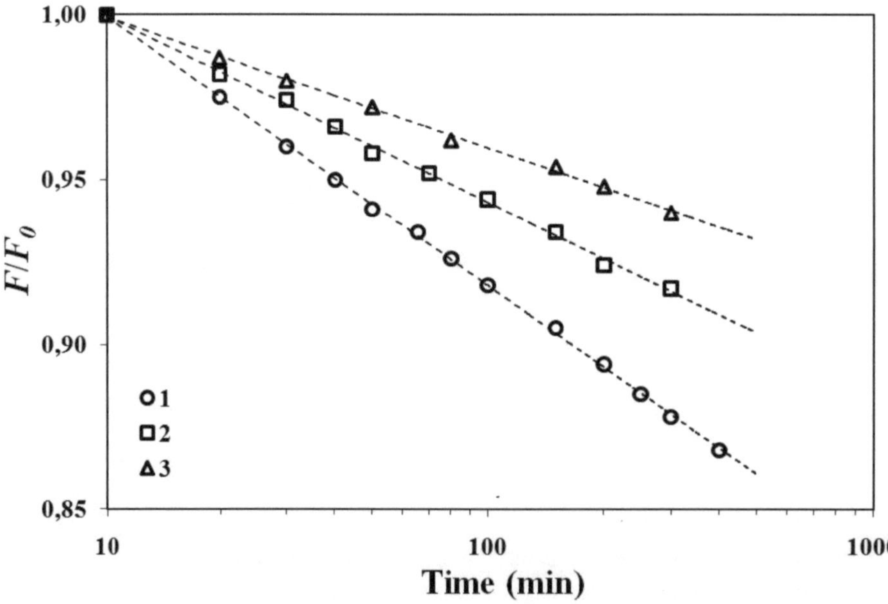

Fig. 2. Normalized magnetic force vs. time for the unipolar magnetization (open magnetic relaxation) at the different penetration depth of critical state: the initial force F_0 = 260 mN (dependence *1*), 205 mN (*2*) and 150 mN (*3*).

Fig. 2 presents the magnetic force vs. logarithmic time for the unipolar magnetization. The dependences *1-3* show the relative change of the force during the open magnetic relaxation (the flux distribution in Fig. 1 (*b*)) for different penetration depth of the critical state. These dependences also show the relative change of the sample magnetization because $F \propto M$. The slope of the dependences characterizes the logarithmic relaxation rate:

$$S_\beta = \frac{1}{F_0}\frac{dF}{d\ln t} = \frac{1}{M_0}\frac{dM}{d\ln t} = \frac{d\beta}{d\ln t}. \tag{19}$$

In the case of full penetration, the magnetization and the current-density relaxation coefficients are equal: $\beta(t) = a(t)$. From Eqs. (11) and (19) it follows that the quantity, which is inverse to the logarithmic relaxation rate, is the kT-normalized effective activation energy. In the case of partial penetration $\beta(t)$ is determined by Eq. (12). The activation energy is related to S_β as:

$$\frac{U_0}{kT} = \frac{C_\delta}{S_\beta}, \tag{20}$$

where C_δ is a correction factor (Eq. (13)). Considering open and internal magnetic relaxation we have made the assumption that the size and the location of current zone in the sample had no affect on the relaxation rate of current density. But the value of magnetization and the rate of its relaxation S_β depend on them. As follows from Eq. (20) the quantity C_δ/S_β should also be independent from the penetration depth of the critical state. The normalized penetration depth (calculated from expression $\hat{\delta} = 1 - \sqrt[3]{1 - w_0}$, where $w_0 = F_0/F_{m0}$) is equal to $\hat{\delta}_0 = 0.5$, 0.3, 0.2 at F_0 = 260, 205 and 150 mN respectively and $F_{m0} = 300$ mN. Calculating C_δ from Eq. (13) and determining the slopes S_β of the dependences *1-3* (Fig. 2), one may find that the effective activation energies are nearly equal, $U_0 \sim 15\ kT$, for all the three dependences.

However, the obtained values C_δ/S_β are very rough estimate of the effective activation energy. To calculate δ_0 and C_δ, we used the Bean's model which apparently could not describe correctly the expansion process of the critical state in the central region, i.e. for large δ_0. (The experiment shows that for $w_0 > 0.85$ the slopes of the relaxation dependences differ little from the slope of the dependence for the maximum magnetization ($w_0 = 1$). Therefore, the values δ_0/R and C_δ should be close to unity for the load coefficients, $0.85 < w_0 \le 1$.)

Fig. 3 presents the time dependence of the magnetic force for the bipolar magnetization which is imparted to the sample during the reverse stroke from the initial position to the suspension point. The force F^* is normalized to F_0 which acts at the same suspension point after the forward movement. The value of the force $F^*(t_0)$ depends on the reversal depth and, consequently, F_0^*/F_0 has different initial values (Fig. 3). The main specific feature of the dependences *1-3* consists in the presence of a plateau: the force relaxation is absent during a certain period of time. The stabilization time (the plateau) increases exponentially with the reversal depth (i.e. with decreasing F_0^*/F_0). The plateau is bounded by the dependence *4* which characterizes the relaxation of the force F acting at the same suspension point when the sample is magnetized without reversal. As soon as the force F^* reaches the said time boundary, it begins diminishing at the same rate as the force F: $dF^*/d\ln t = dF/d\ln t$.

The observed effect is described quite adequately in the terms of theory of the internal magnetic relaxation. A near-surface layer (a reverse-layer) with an opposite induction

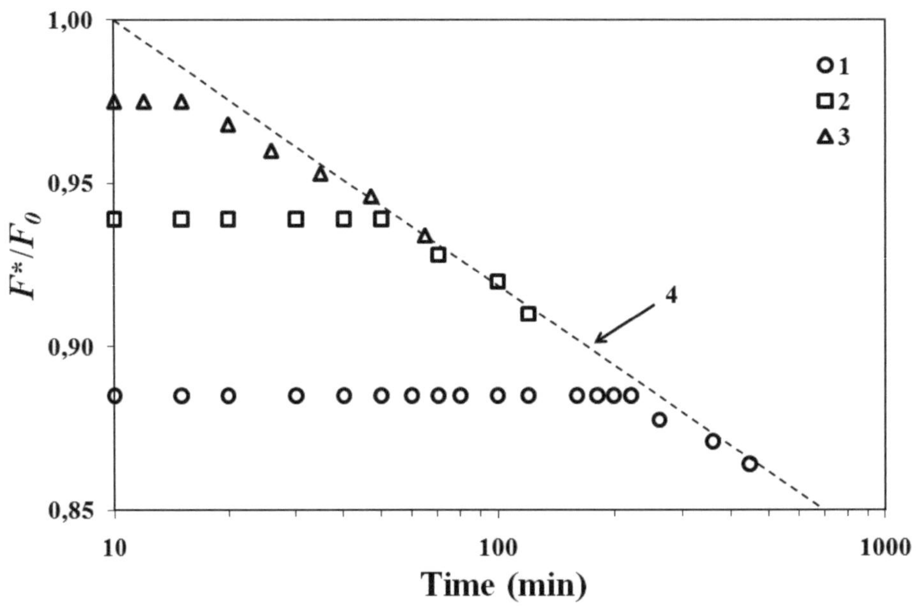

Fig. 3. Normalized magnetic force vs. time for the bipolar magnetization (internal magnetic relaxation) at the different depth of reveral: the initial force $F_0^* = 230$ mN (dependence *1*), 245 mN (*2*) and 255 mN (*3*); $F_0 = 260$ mN. The dashed line *4* is the approximation of data *1* in Fig. 2.

gradient appears in magnetized superconductor as a result of the small reversal of the external magnetic field. When a two-gradient distribution (Fig. 1 (*e*) and (*g*)) is relaxed, the vortices emerge from the volume to the reverse-layer rather than to the superconductor surface. The magnetic flux is redistributed inside the sample. It is known that a full force, which acts on a system of the closed-circuit currents in the magnetic field, can be expressed as the tensions operating at the boundary of the volume passing the currents. In another words, the force depends on the state of the field on the surface of the sample. Since the field does not change its state at the superconductor boundary during the time t_b , the force remains constant. The magnetic flux entering the reverse-layer on the side of the superconductor surface is very small. A relative change of the magnetization during the time t_b (the reverse-layer lifetime) is $1-\beta_b^* = -\left(1-w_0^*\right)^3 / 36$ (see Eq. (18)). In the experiment (Fig. 3) the load coefficient was $0.76 < w_0^* < 1$ which gives $\left|1-\beta_b^*\right| < 4\times10^{-4}$. If $F_0^* \cong 250$ mN, the force variation $F_0^* - F^*(t_b) = \left(1-\beta_b^*\right)F_0^*$ is less than 0.1 mN which is beyond the sensitivity limit of the experimental installation.

Let us estimate the time t_b taking into account that the critical state does not occupy the whole volume of the disk. For the partial penetration of the bipolar critical state (Fig. 1 (g)) the current relaxation coefficient at $t = t_b$ may be calculated using the expression $\alpha_b = 2\delta_0 \Big/ \left(3-\left(4\left(w_0^*/\hat{\delta}_0\right)-3\right)^{1/2}\right)$. In experiment (Fig. 3) the normalized penetration depth

$\hat{\delta}_0 = 0.5$. The load coefficient $w_0^* = 0.766$ (for the dependence *1*), 0.816 (*2*) and 0.95 (*3*). Given these w_0^* and $\hat{\delta}_0$ values, the aforementioned formula yields $\alpha_b = 0.813$ (*1*), 0.89 (*2*) and 0.95 (*3*). The time t_b may be estimated from Eq. (14) (if α_b is substituted for α_i and t_b for t_i). If $U_0/kT = 15$ and $t_0 = 10$ min, the calculated t_b is equal to 165 min (for the dependence *1*), 50 min (*2*) and 20 min (*3*). These values approach rather closely the values observed in the experiment (Fig. 3).

4. Magnetic relaxation in levitating and "fixed" superconductors

In the paper of Smolyak et al. (2006) it was noted the results of the experimental studies of magnetic force relaxation are contradictory. The direct measurements of the interaction force between magnet and superconductor (Moon et al., 1990; Riise et al., 1992; Smolyak et al., 2002) showed a considerable decrease of the force with time. However, in the experiments, where the drift of levitating HTS samples was observed, the levitation height did not change in the stationary magnetic field (Krasnyuk & Mitrofanov, 1990; Terentiev & Kuznetsov, 1992). We suggested that in the case of levitation the relaxation rate of magnetic force was much smaller than in the case of fixed position of superconductor and magnet (when the magnetic force acts on the superconductor, and the sample is fixed at the suspension point). The force stabilization in the levitation system must arise due to feedback. Let a superconductor be magnetized as it is moving to a magnet. The magnetization and the magnetic force F will increase until F balances the sample weight. Assume that the sample magnetization is maximal in the suspension point. Then the stability of the levitation is determined by the gradient function $\Phi_r(z)$ (Eq. (9)) which increases when the sample displaces from the suspension level. If the magnetization decreases due to the flux creep, the force F will also be reduced, and the sample moving slightly from the suspension level will be biased. Therefore, F will rise again and the sample will return to the level of suspension. As a result, the magnetization and the force are almost unchanged.

The feedback may be weakened (i.e. the magnetic bias reduces) by imposing the elastic mechanical constraint on the levitating sample. In this case, the relaxation rate of magnetization and force should increase. When the constraint is absolutely rigid, there is no magnetic bias, and the magnetization relaxation rate should be the largest.

In the experiments we used the same "magnet-HTS disk" system and the same method of magnetization as described in the section 3.4. The setup was upgraded to be able to measure the rate of relaxation when the sample is imposed absolutely rigid or elastic mechanical constraint with the stiffness coefficients 500 N/m or 15 N/m (the experimental details, see in the work of Smolyak et al. (2006).

Fig. *4* shows the dependences F(*lnt*) normalized to the initial force F_0 which were measured from the time $t_0 = 10$ min after the magnetization of the sample. The dependences are close to linear, and its slopes $S = (dF/d\ln t)/F_0$ characterize the logarithmic relaxation rate. The rate is maximum when the sample is fixed (dependence *1*). The relaxation slows down when the mechanical constraint is "softened" (dependences *2-4*). The closer the suspension system to the "true" levitation, in which the sample displacement is mainly determined by the magnetic coupling, the lower the rate of relaxation force.

To make a qualitative estimate of the experimental results, let us consider the magnetic force relaxation when the force F acts on the suspension with HTS, and at the same time the mechanical constraint is imposed on it. The magnetic force may be expressed as $F = \Phi_r M$

(Eq. (7)). For definiteness, consider the suspension of the superconductor above the magnet when the sample moving to the magnet from above is magnetized. Assume the critical state penetrates into the disk from the side surface, and the current density $J=(dB_z/dr)/\mu_0$ (Eq.(1)) is the same over the whole volume of the disk. In this case, the disk has the maximum magnetization $M = JR/3$ (Eq. (4)) (the subscript m at M is omitted). If the mechanical constraint is absolutely rigid, then J, M and F decrease with time with the same relaxation coefficient $\alpha(t)$ (Eq. (11)), i.e. $M(t)/M_0 = F(t)/F_0 = J(t)/J_0 = \alpha(t)$. If the constraint is elastic, then the current relaxation and decrease of F will cause the displacement of the suspension to the magnet. The field at the superconductor boundary grows up that leads to the formation of "fresh" critical state with the higher critical current density. The induction gradient, which is being destroyed by the flux creep, is restored.

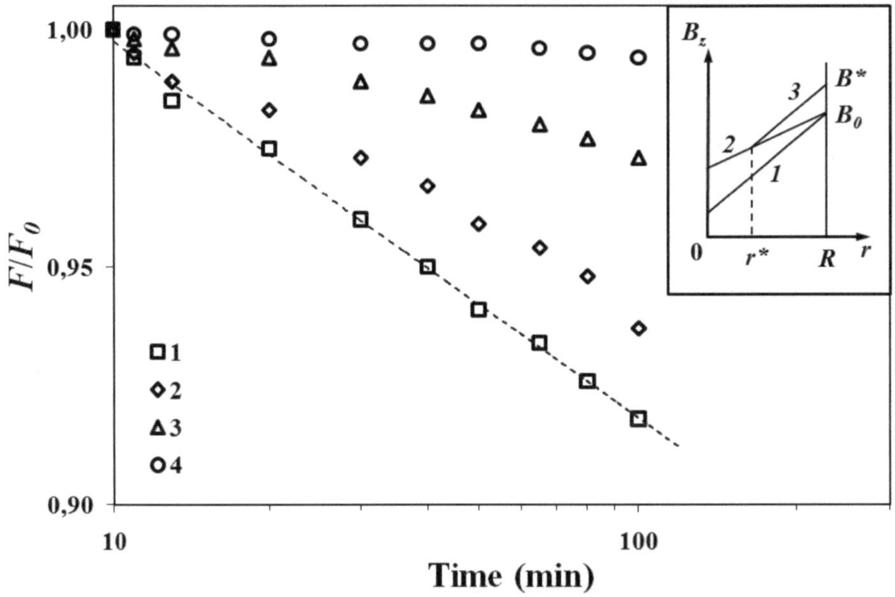

Fig. 4. The magnetic force relaxation depending on the rigidity of the constraint imposed on suspension: absolutely rigid (dependence *1*) and elastic constraint (*2-4*); rigidity of elastic constraint is much larger (*2*) and much smaller (*3, 4*) than the magnetic one (suspension under (*3*) and above (*4*) the magnet). The inset shows magnetic bias process (see the text for explanation).

The inset in Fig. 4 presents the induction distribution: the initial distribution at $t = t_0$ (*1*); the distribution at $t>t_0$ in the case of rigid constraint (*2*); the distribution at the same time $t>t_0$ in the case of elastic constraint (*3*). The critical state *3* penetrates to the depth $\delta = R - r^*$. Assume that the gradient dB_z/dr in this region is restored to its initial value $\mu_0 J_0$ and, consequently, the current density is reduced only in the region $r<r^*$ where $J(t)=\alpha(t)J_0$. The disk magnetization M^* consists of two components M' and M''. Using Eq. (5), we obtain $M'=\alpha(t)M_0\left(r^*/R\right)^3$ for region $0\le r\le r^*$ and $M''=M_0\left[1-\left(r^*/R\right)^3\right]$ for region $r^*\le r\le R$. Taking into account Eq. (11), the magnetization relaxation coefficient may be written as:

$$\beta^*(t) = \frac{M^*(t > t_0)}{M_0} = 1 - q\frac{kT}{U_0}\ln\frac{t}{t_0}, \tag{21}$$

where $M_0 = J_0R/3$, $q = [1 - (\delta/R)]^3$ is the factor of creep retardation. Assuming the function Φ_r in Eq. (7) varies slightly with the sample displacement, we have $F^*(t > t_0)/F_0 = \beta^*(t)$. The logarithmic relaxation rate $S^* = qkT/U_0$. Depending on the rigidity of mechanical constraint the factor q can range from unity ("fixed" superconductor) to zero (levitation). Using Eqs. (11) and (21) and relations $\delta = \Delta B_z / \mu_0 R(1-\alpha(t))$, $\Delta B_z = K_B \Delta z$, $\Delta z = \Delta P_m / k_m$, $\Delta P_m = \Delta F(1-\alpha(t))$ (where Δz is the suspension displacement, ΔB_z is the field variation on the boundary $r = R$, $K_B = dB_z/dz$ is the field gradient on the boundary, ΔF and ΔP_m are the variations of magnetic force and elastic mechanical force, k_m is the rigidity of mechanical constraint), the retardation factor q may be estimated from equation $Cq + q^{1/3} - 1 = 0$ where $C = F_0K_B/\mu_0 k_m R J_0$. Using $F_0 = 0.3$ N, K_B = 0.35 T/m, J_0 = 2×10^7 A/m^2, R = 5×10^{-3} m and rigidity $k_m = \infty$, 500 N/m and 15 N/m, the calculations yield the corresponding values of q = 1, 0.705 and 0.11, respectively.

From the experiment the values of q will be found by using the dependences *1-4* (Fig. 4). The slope of the dependence *1* determines the logarithmic relaxation rate in the absence of sample displacement, i.e. factor q = 1. Using this condition, we obtain the kT-normalized activation energy $U_0/kT \cong 29$. The dependences *2-4* show retarded relaxation with the rate $S^* = qkT/U_0$, which yields q = 0.724 (the suspension under the magnet with k_m = 500 N/m), $q = 0.31$ and 0.074 (the suspension under and above the magnet, respectively, with k_m = 15 N/m). The qualitative agreement between experimental and calculated results for the factor q is quite acceptable. The magnetic relaxation slows down when the suspension system is close to the "true" levitation, i.e. when the magnetic rigidity dF/dz is much greater than the rigidity of mechanical constraint (magnetic rigidity of the "magnet-superconductor" system was ~ 100 N/m). The different values S^*, when the suspension is under (dependence *3* (Fig.4)) and above (dependence *4*) the magnet, are probably due to the different values of magnetic rigidity which determines the sample displacement if k_m is small.

Fig. 5 illustrates the effect of retarded relaxation of the magnetic force F when the superconductor levitates. Image *1* in Fig. 5 presents two identical "magnet-loaded HTS sample" systems in the initial state when the samples are on the rest above the magnet, and the force F is absent (the supporting force is not shown). When the rest goes down, and the HTS sample approaches to the magnet, the magnetic force F appears and increases until it balances the body weight G at the suspension level. In the image *2* on the left the HTS sample levitates (the rest is removed), and on the right the HTS sample remains on the rest. This image corresponds to the initial moment $t = t_0$ that has passed since the establishing of $F = G$. The image *3* shows the same position as the image *2*, but for the moment $t \gg t_0$. During this time, the levitation height on the left remains the same since the force F has not changed. On the right the force F has decreased as a result of the magnetic relaxation. The image *4* shows the positions of the HTS samples after elimination of the right rest. The right HTS sample also levitates, but its levitation height is less than the left one. (The force F, which decreased as a result of flux creep, should increase again up to the magnitude G; the HTS sample should be biased, i.e. it should go down closer to the magnet.)

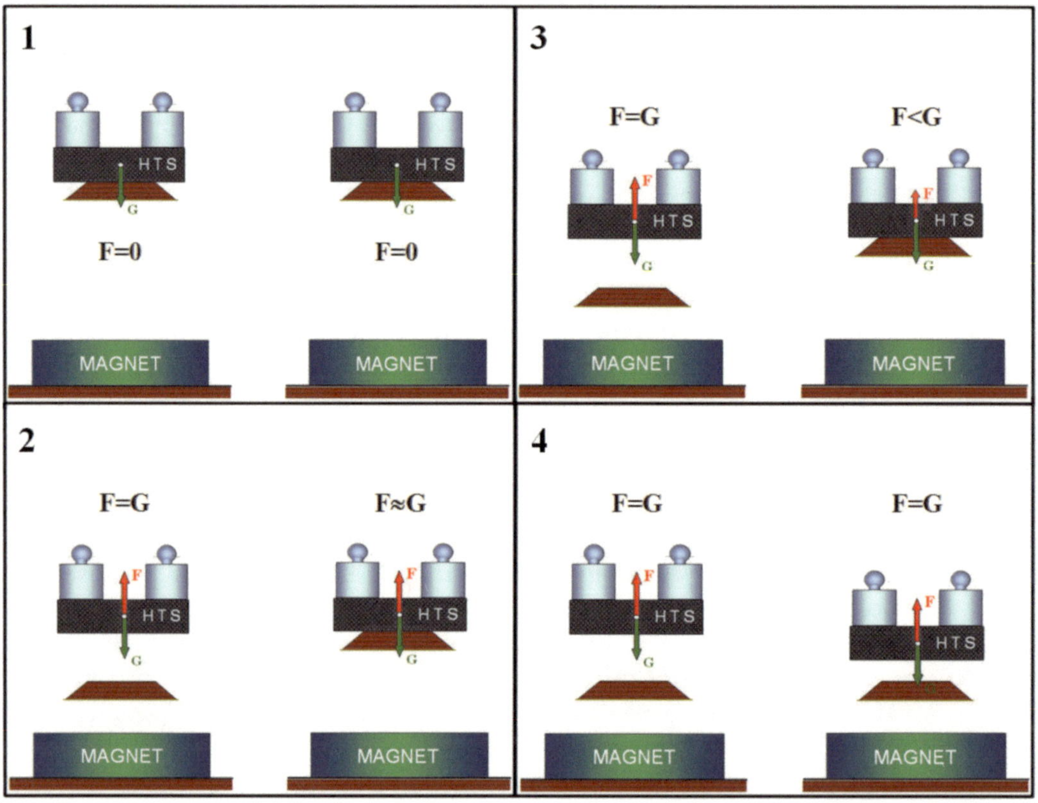

Fig. 5. The effect of retarded relaxation of the magnetic force in levitating superconductor.

5. Magnetic relaxation in superconductor placed near ferromagnet

A new effect was described by Smolyak & Ermakov (2010a, 2010b). It was found the magnetic relaxation is suppressed in HTS sample with a trapped magnetic flux when the sample approaches a ferromagnet. To have more precise idea of the conditions under which the suppression of relaxation is observed, we give a more detailed description of the experiment here.

5.1 Experimental details and results

The measurements were performed on a sample of melt-textured $YBa_2Cu_3O_7$ ceramics having the transition temperature $T_c \cong 91$ K and the transition width of less than 1 K. The sample was shaped as a disk 20 mm in diameter and 8.5 mm high. The *c*-axis was perpendicular to the disk plane. The Hall probes having the sensitive zone 1.5×0.5 mm^2 in size and the sensitivity of 130 μV/mT were attached to the base of the sample as sketched in the inset in Fig. 7. The probes detected the field component normal to the surface of the sample. The induction *B*, which determines the density of Abrikosov vortices, was measured simultaneously at five points on the surface of the sample as a function of time. (The vortices in Fig. 6 are shown conditionally as straight lines in the section of the sample. The arrow lines denote the magnetic field outside the superconductor.) The external magnetic field of the induction B_e was created by an electromagnet. Armco-iron plates (40 mm in diameter and 4 mm thick; the gap between the plates for placement of the sample was 10 mm) were also used in the experiments.

The experimental procedure was as follows. Three independent experiments on measurements of the local relaxation of the trapped magnetic flux were performed. Fig. 6 illustrates the magnetization conditions and the relative positions of the sample and the ferromagnet.

The experiment a. The HTS sample having the temperature $T>T_c$ was cooled in the external magnetic field B_e to 77 K, and then the field B_e was switched off. As a result, the sample trapped the magnetic flux (was magnetized).

The experiment b. The sample having the temperature $T>T_c$ was placed in the gap between the plates, and the external field B_e was applied to the "sample-ferromagnet" system. Then the sample was cooled, and the external field was turned off. In this experiment the sample trapped the magnetic flux when the sample and the ferromagnet were close together.

The experiment c. The sample was cooled in the field B_e, the external field was turned off, and then the sample was placed in the gap between the ferromagnetic plates. The final positions in the experiments *b* and *c* look identical, but in the experiment *b* the sample was magnetized in the presence of the ferromagnet, while in the experiment *c* the sample was first magnetized without the ferromagnet and then was brought close to it.

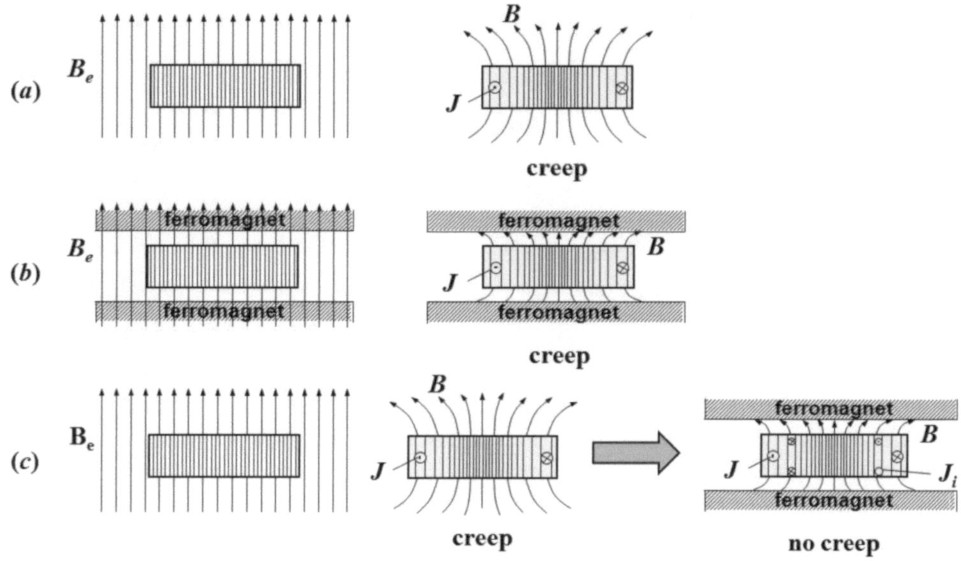

Fig. 6. Magnetizing conditions and relative position of the sample and the ferromagnet in the experiments (*a*), (*b*) and (*c*) (description of the experiments see the text). B_e , the external magnetic field; *B*, the induction of the trapped magnetic flux; *J* and J_i , the density of currents induced in the sample upon trapping of the flux and screening of the ferromagnet field, respectively.

5.2 Discussion

Fig. 7 depicts the profiles of the field which was trapped in the sample. The induction distributions on the surface of the sample was measured 2 min (the observation start point) and 100 min after the sample has been installed in the final position in the experiments *a*, *b* or *c*. The magnetic flux in the sample decreases in the experiments *a* and *b*. The flux value remains unchanged in the experiment *c*. The form of the distributions (the absence of the plateau) suggests that the critical state occupies the whole volume of the sample.

In the absence of the ferromagnet, experiment *a*, the induction near the edge reverses sign. (This feature was also observed in the experiments with slabs in perpendicular field by Abulafia et al. (1995) and Fisher et al. (2005)).

Fig. 8 presents the normalized induction at the center of the sample versus the logarithm of time. These dependences are linear, being a characteristic feature of the flux creep. The similar dependences with sharply different relaxation rates in the experiments *a*-*c* are observed for other regions of the sample.

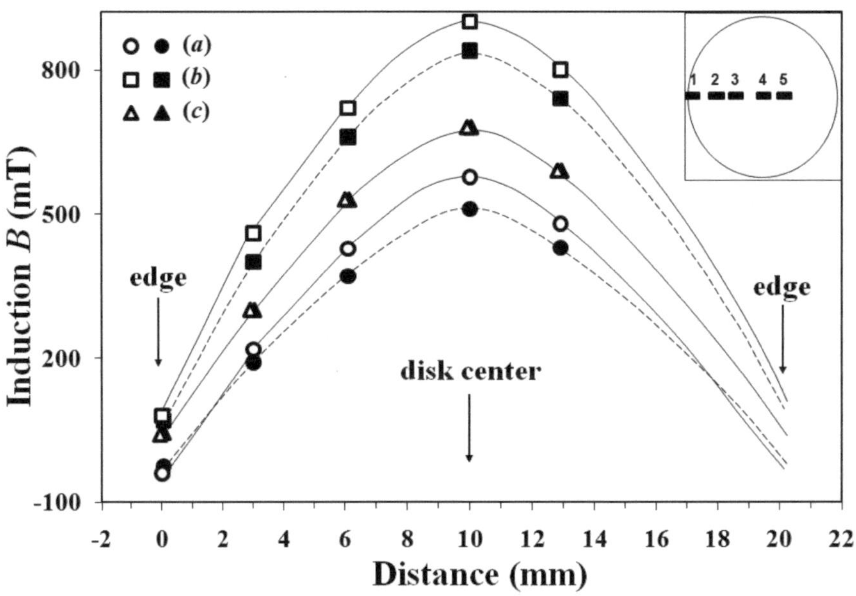

Fig. 7. Local induction *B* vs. Hall probe location measured on the surface of the sample in the experiments (*a*), (*b*) and (*c*): 2 min (open symbols) and 100 min (full symbols) after placing the sample in the final position. The solid and dashed lines serve as a guide for the eye. The inset shows the location of Hall probes.

The magnetic relaxation in the experiment *a* occurs in the absence of external effects on pinning and the nonequilibrium magnetic structure. Let us refer this relaxation to as "free". On the assumption that the current density is the same over the whole volume and diminishes at an equal rate everywhere, the local induction *B* is proportional to *J*. Therefore, the quantity $B(t)/B_0$ changes over time with the relaxation coefficient $\alpha(t)$ (Eq. (11)). The slope of the *a*-dependence, which determines the logarithmic relaxation rate, gives $1/S \sim 30$. This value is in agreement with known values of U_0/kT for melt-textured YbaCuO ceramics. The ferromagnet retards the flux creep in the superconductor. The magnitude of the retardation effect depends on the sequence of magnetization and approach of superconductor and ferromagnet. If they are brought close together before magnetization of the superconductor (experiment *b*, Fig. 6), the relaxation rate *S* is two times lower (*b*-dependence, Fig. 8) than the "free" relaxation rate (*a*-dependence). If they are brought close

together after magnetization (experiment *c*), the magnetic relaxation is almost fully suppressed (*c*-dependence).
This effect can be interpreted as follows. The driving force $f = JB$ depends on the magnetic field configuration, which determines the value and the direction of the current in the sample. When the field at the boundary increases, i.e. the magnetic flux enters the sample, the vortex density is larger near the boundary than in the bulk, and f acts on the vortices in the direction from the surface to the bulk of the sample. When the field at the boundary decreases (e.g. in the case of flux trapping), the vortex density gradient is directed from the bulk to the surface, and the driving force acts in the same direction.

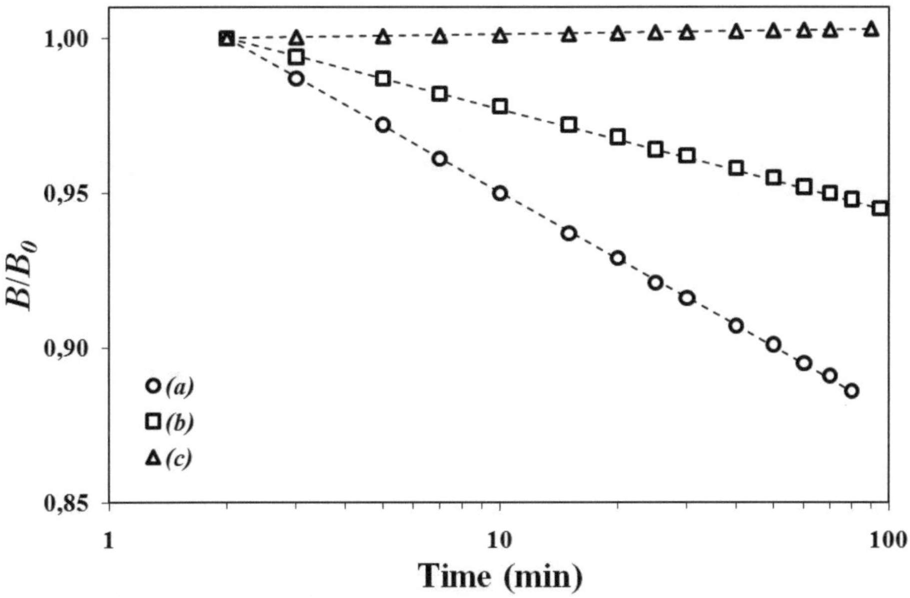

Fig. 8. Time dependence of the induction at the center of the sample normalized to the initial induction B_0 = 605 mT (experiment (*a*)), 867 mT (*b*) and 624 mT (*c*); t_0 = 2 min.

The azimuthal currents are induced in the disk sample located in the axial field. Depending on the way the external field changes upon magnetization of the sample, the critical state with forward, reverse, or counter circulation of currents is established. The magnetic field configuration, which is formed in the homogeneous external field in the absence of the ferromagnet, was calculated by Brandt (1996, 1998). The calculated configuration of the field and the direction of driving forces are shown in Fig. 9 on the left. "Free" magnetic relaxation corresponds to such direction of forces (experiment (*a*)). The current circulates in one direction in the whole volume of the sample. The driving force has two components. The radial force makes the vortices move from the center to the disk rim. The axial forces have the counter direction and do not contribute to the total force which moves vortices.
There are more complicated configurations of the flux lines in the experiments (*b*) and (*c*) because the magnetic field is produced by a screening current in the disk and by the ferromagnet. The sources of the ferromagnet field are domains oriented at the right angle to the plane of the disk. The distribution density of these domains in the disk plane

corresponds to the distribution of the local induction (Fig. 7). The ferromagnet field has the similar dome-shaped profile and has the same direction as the screening current field.

The critical state in the sample (experiment (*b*)) was established when the current in the electromagnet coil was cut off. In this case, the magnetization of the ferromagnet decreased (i.e. the number of oriented domains was reduced) from a maximum to a value corresponding to the distribution of the induction in final position in the experiment (*b*). The magnetic flux (produced by the coil and the domains, which were disorientated after the coil cutoff) left the sample through the base and the rim of the disk. As a result, the screening current circulating in one direction was excited in the sample. This state with unipolar current should undergo the magnetic relaxation. A slowdown of the creep rate in the experiment (*b*) with respect to the "free" relaxation can be due to an increase in the length of the vortices and their curvature. The effect of these factors on the total pinning force is discussed by Fisher et al. (2005) and Voloshin et al. (2007). Most likely, the mechanism of "external" pinning, which is connected with the interaction between vortices and the ferromagnetic domain structure (Garcia-Santiago et al., 2000; Helseth et al., 2002), is less probable. This effect is observed only when the superconductor and a ferromagnet are intimately in contact with each other.

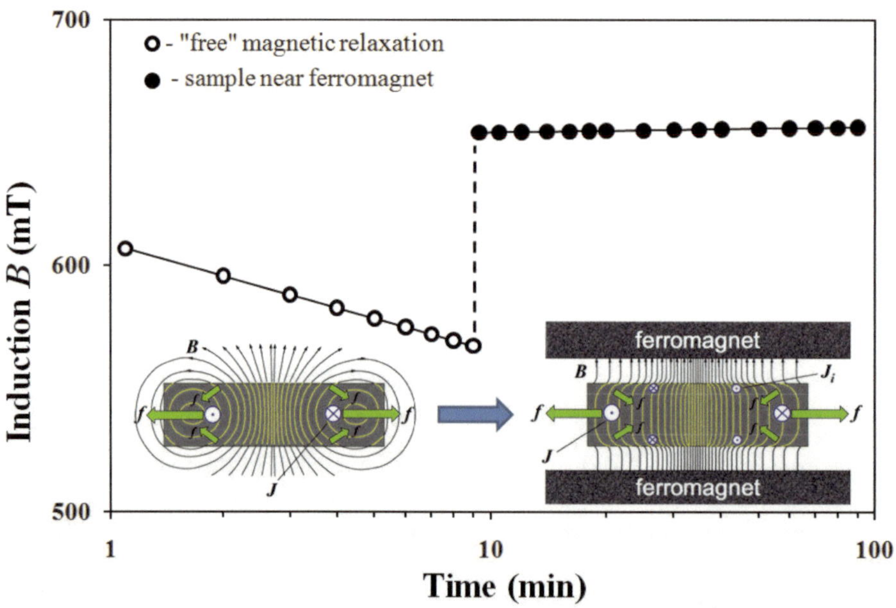

Fig. 9. The configuration of vortices and the direction of screening currents and driving forces acting on vortices, in the case of "free" magnetic relaxation (left image) and in the case the magnetized sample approaches to ferromagnet (right image). The relaxation dependences corresponding to the "free" magnetic relaxation and to the relaxation near ferromagnet are shown on the top.

The critical state in the experiment (*c*) was established when the sample with the trapped flux was placed between the ferromagnet surfaces; i.e. when the ferromagnet was brought into the magnetic field of the superconductor. Being magnetized, the ferromagnet produces its own magnetic field which penetrates into the sample and excites the currents circulating counter to the trapping current. It can be thought that the vortex density gradients, which are connected with the ferromagnetic field, generally appear on the plane surfaces of the disk; i.e. the reverse currents flow near the base of the disk. As a result, the critical state with a bipolar current structure is established in the sample (Fig. 9, on the right). This vortex configuration is more stable because the counter driving forces f act on the different sections of vortices. The ferromagnet field "supports" the nonequilibrium distribution of the trapped flux, leading to the formation of a "rigid" configuration of the magnetic field which remains unchanged with time (*c*-distribution, Fig. 7).

6. Conclusion

We have been considered the influence of the conditions of magnetization, the mobility of the samples in magnetic suspension system and the ferromagnetic medium on the relaxation rate of magnetization and magnetic force in bulk high-temperature superconductors.

(i) The features of open and internal magnetic relaxation have been discussed. It has been shown that both strong decrease in magnetization and force (open relaxation) and absence of any changes of these parameters (internal relaxation) could be observed in experiment. The magnetization of the sample and the magnetic force are stabilized thanks to the reversal of external magnetic field. A model is proposed for the internal magnetic relaxation which arises when the nonequilibrium region of vortex lattice is far from superconductor surface or is separated from it by the layer with an opposite vortex-density gradient.

(ii) It has been shown that in a "magnet-superconductor" system the creep rate depends on the rigidity of the constraints imposed on the system. The magnetization of the superconductor and the magnetic force decrease at a maximum rate when the HTS sample and the magnet are rigidly fixed. In the case of "true" levitation (when the mobility of the sample is determined predominantly by magnetic coupling) the magnetic force very slightly decreases with time. It is suggested that the force stabilization is related to magnetic bias feedback in the sample which restores the nonequilibrium structure broken by the magnetic flux creep.

(iii) It has been described the phenomenon of the retardation of magnetic relaxation in the HTS sample with a trapped magnetic flux when the sample approached a ferromagnet. The flux creep is fully suppressed when the superconducting sample first is magnetized and then the ferromagnet is brought into the magnetic field of the superconductor. It is supposed that the phenomenon results from the formation of stable vortex configuration in which counter Lorentz forces act upon the different regions of vortices.

7. References

Abulafia, Y.; Shaulov, A.; Wolfus, Y.; Prozorov, R.; Burlachkov, L.; Yeshurun, Y.; Majer, D.; Zeldov, E. & Vinokur, V.M. (1995). Local magnetic relaxation in high-temperature superconductors. *Phys. Rev. Lett.*, Vol.75, No.12, (September 1995), pp. 2404-2407, ISSN 0031-9007

Anderson, P.W. (1962). Theory of flux creep in hard superconductors. *Phys. Rev. Lett.*, Vol.9, No.7, (October 1962), pp. 309-311, ISSN 0031-9007

Anderson, P.W. & Kim, Y.B. (1964). Hard superconductivity: theory of the motion of Abrikosov flux lines. *Rev. Mod. Phys.*, Vol.36, No.1, (January 1964), pp. 39-43, ISSN 0034-6861

Beasly, M.R.; Labusch, R. & Webb, W.W. (1969). Flux creep in type-II superconductors. *Phys. Rev.*, Vol.181, No.2, (May 1969), pp. 682-700, ISSN 0143-0394

Brandt, E.H. (1996). Superconductors of finite thickness in a perpendicular magnetic field: strips and slabs. *Phys. Rev. B*, Vol.54, No.6, (August 1996), pp. 4246-4264, ISSN 1098-0121

Brandt, E.H. (1998). Superconductor disk and cylinders in an axial magnetic field. I. Flux penetration and magnetization curves. *Phys. Rev. B*, Vol.58, No.10, (September 1998), pp. 6506-6522, ISSN 1098-0121

Brandt, E.H. & Mikitik, G.P. (2003). Reversible magnetic behavior of superconductors forced by a small transverse ac magnetic field. *Journal of Low Temperature Physics*, Vol.131, No.5-6, (June 2003), pp. 1033-1042, ISSN 0022-2291

Fisher, L.M.; Kalinov, A.V.; Voloshin, I.F. & Yampol'skii, V.A. (2005). Suppression of magnetic relaxation processes in melt-textured $YBa_2Cu_3O_x$ superconductors by a transverse ac magnetic field. *Phys. Rev. B*, Vol.71, No.14, (April 2005), pp. 140503-(1-4), ISSN 1098-0121

Fisher, L.M.; Kalinov, A.V.; Savel'ev, S.E.; Voloshin, I.F.; Yampol'skii, V.A.; LeBlanc, M.A.R. & Hirscher, S. (1997). Collapse of the magnetic moment in a hard superconductor under the action of a transverse ac magnetic field. *Physica C*, Vol.278, No.3-4, (May 1997), pp. 169-179, ISSN 0921-4534

Garcia-Santiago, A.; Sanchez, F.; Varela, M. & Tejada, J. (2000). Enhanced pinning in a magnetic-superconducting bilayer. *Appl. Phys. Lett.*, Vol.77, No.18, (December 2000), pp. 2900-2902, ISSN 0003-6951

Helseth, L.E.; Goa, P.E., Hauglin, H.; Baziljevich, M. & Johansen, T.H. (2002). Interaction between a magnetic domain wall and a superconductor. *Phys. Rev. B*, Vol.65, No.13, (March 2002), pp. 132514-(1-4), ISSN 1098-0121

Krasnyuk, N.N. & Mitrofanov, M.P. (1990). Levitation of YbaCuO ceramics in magnetic field. *Superconductivity: Physics, chemistry, technique*, Vol.3, No.2, (February 1990), pp. 318-322, ISSN 0131-5366

Kwasnitza, K. & Widmer, Ch. (1991). Strong magnetic history dependence of magnetic relaxation in high-T_c superconductors. *Physica C*, Vol.184, No.4-6, (December 1991), pp. 341-352, ISSN 0921-4534

Kwasnitza, K. & Widmer, Ch. (1993). Methods for reduction of flux creep in high and low T_c type II superconductors. *Cryogenics*, Vol.33, No.3, (March 1993), pp. 378-381, ISSN 0011-2275

Landau, L.D.; Lifshitz, E.M. & Pitaevskii, L.P. (1984). *Course of theoretical physics, vol.8 – Electrodynamics of continuous media* (2nd edition), Pergamon Press, ISBN 0080302750, New York

Maley, M.P.; Willis, J.O.; Lessure, H. & McHenry, M.E. (1990). Dependence of flux-creep activation energy upon current density in grain-aligned $YBa_2Cu_3O_{7-x}$. *Phys. Rev. B*, Vol.42, No.4, (August 1990), pp. 2639-2642, ISSN 1098-0121

Moon, F.C.; Chang, P.-Z.; Hojaji, H.; Barkatt, A. & Thorpe, A.N. (1990). Levitation forces, relaxation and magnetic stiffness of melt-quenched $YBa_2Cu_3O_x$. *Japanese Journal of Applied Physics*. Vol.29, No.7, (July 1990), pp. 1257-1258, ISSN 0021-4922

Riise, A.B.; Johansen, T.H.; Bratsberg, H. & Yang, Z.J. (1992). Logarithmic relaxation in the levitation force in a magnet-high T_c superconductor system. *Appl. Phys. Lett.* Vol.60, No.18, (May 1992), pp. 2294-2296, ISSN 0003-6951

Smolyak, B.M.; Perelshtein, G.N.; Ermakov, G.V. & Postrekhin, E.V. (2000). Stopping of levitation force relaxation in superconductors: the flux-locking effect. *Physica C*, Vol.341-348, No.PART 3, (November 2000), pp. 1129-1130, ISSN 0921-4534

Smolyak, B.M.; Perelshtein, G.N. & Ermakov, G.V. (2001). Internal magnetic relaxation in levitating superconductors. *Technical Physics Letters*, Vol.27, No.8, (August 2001), pp. 674-676, ISSN 1063-7850

Smolyak, B.M.; Perelshtein, G.N. & Ermakov, G.V. (2002). Effects of relaxation in levitating superconductors. *Cryogenics*, Vol.42, No.10, (October 2002), pp. 635-644, ISSN 0011-2275

Smolyak, B.M.; Perelshtein, G.N. & Ermakov, G.V. (2006). Retarded magnetic relaxation in levitated superconductors. *Technical Physics Letters*, Vol.32, No.2, (February 2006), pp. 98-100, ISSN 1063-7850

Smolyak, B.M.; Ermakov, G.V. & Chubraeva, L.I. (2007). The effect of ac magnetic fields on the lifting power of levitating superconductors. *Superconductor Science and Technology*, Vol.20, No.4, (April 2007), pp. 406-411, ISSN 0953-2048

Smolyak, B.M. & Ermakov, G.V. (2010). Elimination of magnetic relaxation in superconductors on approaching a ferromagnet. *Physica C*, Vol.470, No.3, (February 2010), pp. 218-220, ISSN 0921-4534

Smolyak, B.M. & Ermakov, G.V. (2010). Suppression of magnetic relaxation in a high-temperature superconductor placed near a ferromagnet. *Technical Physics Letters*, Vol.36, No.5, (May 2010), pp. 461-463, ISSN 1063-7850

Sun, J.Z.; Lairson, B.; Eom, C.B.; Bravman, J. & Geballe, T.H. (1990). Elimination of current dissipation in high transition temperature superconductors. *Science*, Vol.247, No.4940, (January 1990), pp. 307-309, ISSN 0036-8075

Terentiev, A.N. & Kuznetsov, A.A. (1992). Drift of levitated YBCO superconductor induced by both a variable magnetic field and a vibration. *Physica C*, Vol.195, No.1-2, (May 1992), pp. 41-46, ISSN 0921-4534

Thompson, J.R.; Sun, Y.R.; Malozemoff, A.P.; Christen, D.K.; Kerchner, H.R.; Ossandon, J.G.; Marwick, A.D. & Holtzberg, F. (1991). Reduced flux motion via flux creep annealing in high-J_c single-crystal $Y_1Ba_2Cu_3O_7$. *Appl. Phys. Lett.*, Vol.59, No.20, (November 1991), pp. 2612-2614, ISSN 0003-6951

Voloshin, I.F.; Kalinov, A.V.; Fisher, L.M. & Yampol'skii, V.A. (2007). Suppression of magnetic relaxation by a transverse alternating magnetic field. *Journal of Experimental and Theoretical Physics*, Vol.105, No.1, (July 2007), pp. 278-282, ISSN 1063-7761

Willemin, M.; Rossel, C.; Hofer, J.; Keller, H.; Erb, A. & Walker, E. (1998). Strong shift of the irreversibility line in high-T_c superconductors upon vortex shaking with an oscillating magnetic field. *Phys. Rev. B*, Vol.58, No.10, (September 1998), pp. R5940-R5943, ISSN 1098-0121

Yeshurun, Y.; Malozemoff, A.P. & Shaulov, A. (1996). Magnetic relaxation in high-temperature superconductors. *Rev. Mod. Phys.*, Vol.68, No.3, (July 1996), pp. 911-949, ISSN 0034-6861

6

3-D Finite-Element Modelling of a Maglev System using Bulk High-Tc Superconductor and its Application

Guang-Tong Ma[1,2], Jia-Su Wang[1] and Su-Yu Wang[1]
[1]*Applied Superconductivity Laboratory of Southwest Jiaotong University,*
[2]*Traction Power State Key Laboratory of Southwest Jiaotong University,*
P. R. China

1. Introduction

Maglev using bulk High-Tc superconductor can realize stable levitation without any active control (Brandt, 1989), and this facsinating property can reduce remarkably the complexity of the levitation system and therefore exhibits promising application in several fields such as maglev bearing (Hull, 2000; Ma, et al., 2003) and maglev transit (Wang, et al., 2002 ; Wang, J. & Wang, S., 2005; Schultz, et al., 2005; Sotelo, et al., 2010). To understand the electromagnetic interaction between the bulk high-Tc superconductor and its applied fields generated by various permanent magnetic devices and to provide a numerical tool to conduct the design for practical application, many methods have been proposed to numerically estimate the characteristics of the magnetic force of the bulk high-Tc superconductor.

The earliest method (Davis, et al., 1988) after the discovery of the high-Tc superconductor was basically established on the traditional mirror-image-model which uses Bean's critical model (Bean, 1964). However, this model and the later frozen-image model (Kordyuk, 1998) can not reflect the important hysteresis property of the levitation force (Hull & Cansiz, 1999). Though this demerit can be overcome to some extent by introducing additional image dipoles (Yang & Zheng, 2007), this kind of model is essentially a phenomenological one and its applicable scope is also confined to miniature scale systems due to the essential dipole approximation of the levitated body in deducing the model.

Based on the principle of minimum energy, the current distribution in the high-Tc superconductor can be acquired by an iterative process, and then the magnetic force of the high-Tc superconductor can be calculated by Lorentz equation (Sanchez & Navau, 2001; Navau & Sanchez, 2001; Sanchez, et al., 2006). However, up to date, this method is used only to investigate the axisymmetric system with cylindrical high-Tc superconductor and permanent magnet (Navau & Sanchez, 2001) or 2-D translational symmetry system with rectangular high-Tc superconductor and permanent magnetic array (Sanchez, et al., 2006).

According to the state variables used in the governing equations, the methods describing the electromagnetic property of the high-Tc superconductor based on Maxwell's equations can be classified into three types, i.e., **A**-V-method (Prigozhin, 1997), T-method (Hashizume, et al., 1991), and H-method (Pecher, et al., 2003). Most previous work (Qin, et al., 2002; Uesaka, et al., 1993; Yoshida, et al., 1994; Luo, et al., 1999; Alonso, et al., 2004;

Gou, et al.,2007a; Gou, et al.,2007b) using these methods was to calculate the levitation force of a bulk high-Tc superconductor by aid of the assumption that the induced current only flows in the *ab*-plane of the high-Tc superconductor. In this case, only the component of the state variable along the *c*-axis of the high-Tc superconductor is considered in the governing equation and the number of degrees of freedom is therefore reduced and the problem to be solved is actually a 2-D one. This assumption is acceptable when the studied problem has an axisymmetric geometry and the movement of the bulk high-Tc superconductor is restricted to the vertical direction above the center of the magnetic device because in this situation, the induced current in the high-Tc superconductor due to the variation of the applied field will flow along the plane parallel to the *ab*-plane. As a result, the numerical results of the levitation force compare well with the measured data (Uesaka, et al., 1993; Alonso, et al., 2004; Gou, et al.,2007a), but for other situations, e.g., a bulk high-Tc superconductor is above a rectangular permanent magnet or a magnetic rail, the assumption is no longer satisfied and modelling high-Tc superconductor in full 3-D case is thereby required. Based on the **A**-*V*-method, the 3-D model has been proposed to numerically estimate the characteristics of the levitation force as well as the lateral force of a levitating transporter using bulk high-Tc superconductor (Ueda, et al., 2006), and also the dynamic behavior of the levitation system composed of a rectangular high-Tc superconductor and permanent magnet (Alloui, et al., 2009). However, though an anisotropic critical model is employed in the numerical iterative proces (Ueda, et al., 2006), none of them has considered the special electromagnetic anisotropic behavior of the high-Tc superconductor in deducing the governing equations. Those models are hence still an anisotropic 3-D model. For the maglev transit using bulk high-Tc superconductor above a magnetic rail, a 3-D model considering the anisotropic behavior is reported (Lu, et al., 2008). In their model, in order to describe the anisotropic behavior, the bulk high-Tc superconductor is split into two different parts: One is a homogeneous part that is identical to an anisotropic superconductor, and the other is considered as a conductor whose conductivity only along the *c*-axis is not equal to zero. However, the reasonable explanation of this superposition approach to describe the anisotropic behavior in physics is still a question needed to be answered at present.

In this chapter, we report a 3-D finite-element model using current potential **T** as state variable. In this model, the anisotropic behavior of the high-Tc superconductor is contained in the 3-D governing equations by considering a tensor resistivity, and the finite-element technique is empolyed to numerically solve the mathmatical formulations on a VC++ software platform. The numerical results of both levitation force and lateral force were validated by the measured data. Lastly, one example using this 3-D finite-element model to optimize the magnetic rail is introduced to present its viable use for practical design of maglev system using bulk high-Tc superconductor.

2. Mathematical formulations

2.1 Formulations to model the anisotropy in high-Tc superconductor

The special microstructure, which consists of the alternating stack of superconductive CuO_2 layers and almost insulating block layers, results in a remarkable anisotropic behavior in the present high-Tc superconductor (Dinger, et al., 1987). Due to this anisotropic behavior arising from the intrinsic pinning and other defects in the high-Tc superconductor, the flux-line curvature will always occur when the high-Tc superconductor is placed in a magnetic field.

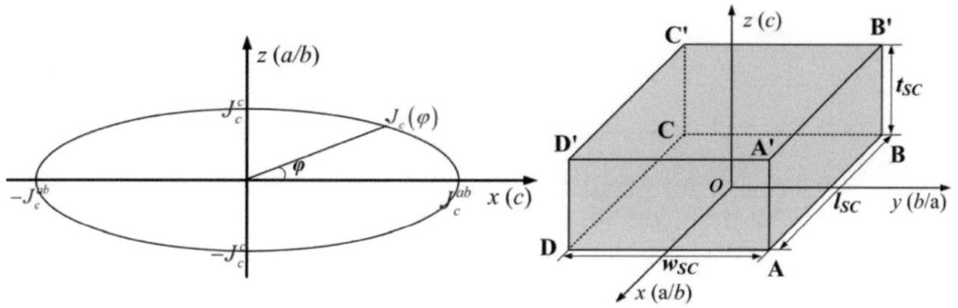

Fig. 1. Schematic drawing of the elliptical model (left) and the Cartesian coordinate system with a bulk high-Tc superconductor whose *c*-axis is parallel to the *z*-axis (right).

Consequently, the critical current density J_c of the high-Tc superconductor is anisotropic and strongly dependent on the orientation of the applied field (Mikitik & Brandt, 2000). Namely,the value of the critical current density flowing in the *ab*-plane J_c^{ab} is larger than that along the *c*-axis J_c^c (Matsushita, 2007).
Many methods have been proposed to formulate the dependence of the critical current density J_c on the angle φ between the orientation of the local applied field and *c*-axis. However, a few of them are difficult to be employed in the present calculation due to its complexity in determining the parameters involved in the formulation (Mikitik & Brandt, 2000) or its applicable field range is beyond our focus (Sawamura & Tsuchimoto, 2000). Another model reported in (Yang, et al., 1999) is simple to be employed and has been valiadated in an axisymmetric levitation system. This model is desribed in the following:
When the out-of-plane anisotropic ratio *a* of the critical current density J_c is defined as $a = J_c^{ab} / J_c^c$, the $J_c(\varphi)$ relation can be rewritten as follows:

$$J_c(\varphi) = J_c^{ab}\left(\cos^2\varphi + \sin^2\varphi/\alpha\right) \tag{1}$$

Here, we will introduce an elliptical model, which has been used to investigate the electromagnetic problem involving the anisotropic ferromagnetic material (Napoli & Paggi, 1983), to describe the angular-dependence property of the critical current density J_c in the high-Tc superconductor. According to the schematic drawing shown in Fig. 1, the elliptic model can be expressed by the following equation:

$$\left(\frac{J_{cx}}{J_c^{ab}}\right)^2 + \left(\frac{J_{cz}}{J_c^c}\right)^2 = 1 \tag{2}$$

where J_{cx} and J_{cz} are the induced current densities in the *ab*-plane and parallel to the *c*-axis, respectively. According to (2), the $J_c(\varphi)$ relation with respect to J_c^{ab} and *a* can be expressed as:

$$J_c(\varphi) = \sqrt{J_{cx}^2 + J_{cz}^2} = J_c^{ab}\sqrt{\cos^2\varphi + \left(\sin\varphi/\alpha\right)^2} \tag{3}$$

The resistivity of the high-Tc superconductor is also anisotropic (Wu, et al., 1991) and can be represented by a tensor while modelling the high-Tc superconductor. The tensor of the resistivity of the high-Tc superconductor can be reduced to a following diagonal matrix when only the out-of-plane anisotropy is considered,

$$\overline{\overline{\rho_s}} = \begin{bmatrix} \rho_{ab} & 0 & 0 \\ 0 & \rho_{ab} & 0 \\ 0 & 0 & \rho_c \end{bmatrix} = \begin{bmatrix} \rho_{ab} & 0 & 0 \\ 0 & \rho_{ab} & 0 \\ 0 & 0 & \alpha\rho_{ab} \end{bmatrix}$$

where ρ_{ab} and ρ_c are the resistivity in the *ab*-plane and along the *c*-axis, respectively. It is important to remark here that, for the identical electrical field criterion E_c to determine the critical current density J_c, the anisotropic ratio of the resistivity ρ and the critical current density J_c are reciprocal to each other. The above tensor expression of the resistivity illustrates a possible way to establish the governing equations of the high-Tc superconductor including its anisotropic behavior.

2.2 3-D governing equations

The *T*-method is adopted in our model because it has the merits that, the number of unknown and space needed to be meshed can be remarkably reduced because the variable *T* can be defined as zero outside the conductor (Miya & Hashizume, 1988).
As a conducting material, Maxwell's equations are also valid to describe the electromagnetic phenomena in the high-Tc superconductor. Thus, we have,

$$\nabla \times \mathbf{H} = \mathbf{J} \tag{4}$$

$$\nabla \times \mathbf{E} = -\frac{\partial \mathbf{B}}{\partial t} \tag{5}$$

According to (4), the current density **J** is a divergence free vector with a quasi-static approximation in the low-frequency problem, i.e., $\nabla \bullet \mathbf{J} = 0$, and thus, a current vector potential **T** can be introduced and defined as,

$$\mathbf{J} = \nabla \times \mathbf{T} \tag{6}$$

The Coulomb gauge is applied to vector **T** to guarantee the uniqueness of the solution, i.e., $\nabla \bullet \mathbf{T} = 0$. Applying Helmholtz's theorem to vector **T** yields the following equation:

$$\begin{aligned} C(P)\mathbf{T}(P) = & \frac{1}{4\pi}\int_V \left(\nabla' \bullet \mathbf{T}(P')\right)\nabla' \frac{1}{R(P,P')} dV' - \frac{1}{4\pi}\int_S \left(\mathbf{n}' \bullet \mathbf{T}(P')\right)\nabla' \frac{1}{R(P,P')} dS' \\ & + \frac{1}{4\pi}\int_V \left(\nabla' \times \mathbf{T}(P')\right) \times \nabla' \frac{1}{R(P,P')} dV' - \frac{1}{4\pi}\int_S \left(\mathbf{n}' \times \mathbf{T}(P')\right) \times \nabla' \frac{1}{R(P,P')} dS' \end{aligned} \tag{7}$$

where $R(P, P')$ is the distance between the source point P' and the field point P, the superscript ' refers to the quantity at the source point, $\mathbf{n}'$ is a unit vector out of the surface S', and the coefficient $C(P)$ takes the following values (Hashizume, et al., 1991) :

$$C(P) = \begin{cases} 1 & P \in V' \ (excluding\ S') \\ 1/2 & P \in S' \\ 0 & elsewhere \end{cases}$$

According to the physical fact that the normal component of **J** must be zero on the surface of the high-Tc superconductor, i.e., J_n = 0, **T** has the following boundary condition (Miya & Hashizume, 1988):

$$\mathbf{n}' \times \mathbf{T} = 0 \tag{8}$$

Therefore, only the normal component T_n exists on all the surfaces of the high-Tc superconductor, e.g., for the surface of AA'D'D as shown in Fig.1, vector **T** is reduced to a scalar T_x. Furthermore, T is zero on all edges of the rectangular high-Tc superconductor bulk, e.g., for edge AA' as shown in Fig.1, when it is regarded as a part of surface AA'D'D, $T_y = T_z = 0$, whereas when it is considered as a part of surface AA'B'B, $T_x = T_z = 0$. Thus, we have $T_x = T_y = T_z = 0$ on all the edges.
Equation (7) is reduced to the following form when the Coulomb gauge and boundary condition (8) are considered,

$$C(P)\mathbf{T}(P) = \frac{1}{4\pi}\int_V \left(\nabla' \times \mathbf{T}(P')\right) \times \nabla' \frac{1}{R(P,P')} dV' - \frac{1}{4\pi}\int_S \left(\mathbf{n}' \cdot \mathbf{T}(P')\right) \nabla' \frac{1}{R(P,P')} dS' \tag{9}$$

The *B-H* constitutive law of the high-Tc superconductor can be assumed to be linear as that in vacuum with a good approximation because its applicable conditions (Brandt, 1996) can be easily satisfied in a levitation system using bulk Y-Ba-Cu-O due to its small lower critical field B_{c1} (Krusin-Elbaum, et al., 1989) and large applied field as well as geometry. Thus,

$$\mathbf{B} = \mu_0 \mathbf{H} \tag{10}$$

The induced field $\mathbf{B}_s$ produced by the induced current in the high-Tc superconductor can be expressed in terms of the vector **T** in the following equation when we combine (9) with Biot-Savart's law (Miya & Hashizume, 1988),

$$\mathbf{B}_s = \mu_0 C(P)\mathbf{T}(P) + \frac{\mu_0}{4\pi}\int_S \left(\mathbf{n}' \cdot \mathbf{T}(P')\right) \nabla' \frac{1}{R(P,P')} dS' \tag{11}$$

When an equivalent conductivity σ_s, which is nonlinear and dependent on the local electrical field of the high-Tc superconductor, is introduced, the traditional Ohm's law in the high-Tc superconductor has the following form:

$$\mathbf{J} = \sigma_s\left(|\mathbf{E}|\right)\mathbf{E} \tag{12}$$

By substituting (6) and (12) into (5) and considering $\mathbf{B} = \mathbf{B}_e + \mathbf{B}_s$ where $\mathbf{B}_e$ is the applied field, yields the following equation:

$$\nabla \times \frac{1}{\sigma_s}(\nabla \times \mathbf{T}) + \frac{\partial(\mathbf{B}_e + \mathbf{B}_s)}{\partial t} = 0 \tag{13}$$

The governing equation of the high-Tc superconductor based on the variable T is finally derived from (11), and (13) as follows,

$$\nabla \times \frac{1}{\sigma_s}(\nabla \times \mathbf{T}) + \mu_0 C(P)\frac{\partial \mathbf{T}}{\partial t} + \frac{\mu_0}{4\pi}\int_S \frac{\partial(\mathbf{n}' \cdot \mathbf{T}')}{\partial t} \nabla' \frac{1}{R(P,P')} dS' + \frac{\partial \mathbf{B}_e}{\partial t} = 0 \tag{14}$$

When we replace σ_s by a tensor resistivity $\overline{\overline{\rho_s}}$, (14) and consider that,

$$\overline{\overline{\rho_s}}(\nabla\times\mathbf{T})=\begin{bmatrix}\rho_{ab} & 0 & 0\\ 0 & \rho_{ab} & 0\\ 0 & 0 & \alpha\rho_{ab}\end{bmatrix}\begin{bmatrix}\left(\frac{\partial T_z}{\partial y}-\frac{\partial T_y}{\partial z}\right)\hat{\boldsymbol{x}}\\ \left(\frac{\partial T_x}{\partial z}-\frac{\partial T_z}{\partial x}\right)\hat{\boldsymbol{y}}\\ \left(\frac{\partial T_y}{\partial x}-\frac{\partial T_x}{\partial y}\right)\hat{\boldsymbol{z}}\end{bmatrix}=\rho_{ab}\left[\left(\frac{\partial T_z}{\partial y}-\frac{\partial T_y}{\partial z}\right)\hat{\boldsymbol{x}}+\left(\frac{\partial T_x}{\partial z}-\frac{\partial T_z}{\partial x}\right)\hat{\boldsymbol{y}}+\alpha\left(\frac{\partial T_y}{\partial x}-\frac{\partial T_x}{\partial y}\right)\hat{\boldsymbol{z}}\right] \quad (15)$$

The following equality for the first term in the left side of (14) is satisfied,

$$\begin{aligned}\nabla\times\frac{1}{\sigma_s}(\nabla\times\mathbf{T})=\nabla\times\overline{\overline{\rho_s}}(\nabla\times\mathbf{T})=\rho_{ab}&\left[\alpha\frac{\partial^2 T_y}{\partial x\partial y}-\alpha\frac{\partial^2 T_x}{\partial y^2}-\frac{\partial^2 T_x}{\partial z^2}+\frac{\partial^2 T_z}{\partial x\partial z}\right]\hat{\boldsymbol{x}}\\ +\rho_{ab}\left[\frac{\partial^2 T_z}{\partial y\partial z}-\frac{\partial^2 T_y}{\partial z^2}-\alpha\frac{\partial^2 T_y}{\partial x^2}+\alpha\frac{\partial^2 T_x}{\partial y\partial x}\right]\hat{\boldsymbol{y}}&+\rho_{ab}\left[\frac{\partial^2 T_x}{\partial z\partial x}-\frac{\partial^2 T_z}{\partial x^2}-\frac{\partial^2 T_z}{\partial y^2}+\frac{\partial^2 T_y}{\partial z\partial y}\right]\hat{\boldsymbol{z}}\end{aligned} \quad (16)$$

According to the Coulomb gauge, we have $\nabla(\nabla\bullet\mathbf{T})=0$� i.e.,

$$\left(\frac{\partial^2 T_x}{\partial x^2}+\frac{\partial^2 T_y}{\partial y\partial x}+\frac{\partial^2 T_z}{\partial z\partial x}\right)\hat{\boldsymbol{x}}+\left(\frac{\partial^2 T_x}{\partial x\partial y}+\frac{\partial^2 T_y}{\partial y^2}+\frac{\partial^2 T_z}{\partial z\partial y}\right)\hat{\boldsymbol{y}}+\left(\frac{\partial^2 T_x}{\partial x\partial z}+\frac{\partial^2 T_y}{\partial y\partial z}+\frac{\partial^2 T_z}{\partial z^2}\right)\hat{\boldsymbol{z}}=0 \quad (17)$$

The following identities can be derived from (17),

$$\frac{\partial^2 T_z}{\partial z\partial x}=-\frac{\partial^2 T_x}{\partial x^2}-\frac{\partial^2 T_y}{\partial y\partial x},\quad \frac{\partial^2 T_z}{\partial z\partial y}=-\frac{\partial^2 T_x}{\partial x\partial y}-\frac{\partial^2 T_y}{\partial y^2},\quad \frac{\partial^2 T_x}{\partial x\partial z}+\frac{\partial^2 T_y}{\partial y\partial z}=-\frac{\partial^2 T_z}{\partial z^2} \quad (18)$$

Equality (16) can be further written in the following way when (18) is taken into account,

$$\begin{aligned}\nabla\times\overline{\overline{\rho_s}}(\nabla\times\mathbf{T})=\rho_{ab}&\left[-\frac{\partial^2 T_x}{\partial x^2}-\alpha\frac{\partial^2 T_x}{\partial y^2}-\frac{\partial^2 T_x}{\partial z^2}+(\alpha-1)\frac{\partial^2 T_y}{\partial x\partial y}\right]\hat{\boldsymbol{x}}\\ &+\rho_{ab}\left[-\alpha\frac{\partial^2 T_y}{\partial x^2}-\frac{\partial^2 T_y}{\partial y^2}-\frac{\partial^2 T_y}{\partial z^2}+(\alpha-1)\frac{\partial^2 T_x}{\partial x\partial y}\right]\hat{\boldsymbol{y}}+\rho_{ab}\left(-\frac{\partial^2 T_z}{\partial x^2}-\frac{\partial^2 T_z}{\partial y^2}-\frac{\partial^2 T_z}{\partial z^2}\right)\hat{\boldsymbol{z}}.\end{aligned} \quad (19)$$

Besides,

$$\nabla'\frac{1}{R(P,P')}=\frac{\partial}{\partial x'}\left(\frac{1}{R(P,P')}\right)\hat{\boldsymbol{x}}+\frac{\partial}{\partial y'}\left(\frac{1}{R(P,P')}\right)\hat{\boldsymbol{y}}+\frac{\partial}{\partial z'}\left(\frac{1}{R(P,P')}\right)\hat{\boldsymbol{z}}, \quad (20)$$

and

$$\mathbf{B}_e=B_{ex}\hat{\boldsymbol{x}}+B_{ey}\hat{\boldsymbol{y}}+B_{ez}\hat{\boldsymbol{z}}. \quad (21)$$

Finally, the following 3-D governing equations are obtained once (19) to (21) are substituted into (14) and ρ_{ab} is replaced by its reciprocal σ_{ab}.

$$\frac{1}{\sigma_{ab}}\left(-\frac{\partial^2 T_x}{\partial x^2}-\alpha\frac{\partial^2 T_x}{\partial y^2}-\frac{\partial^2 T_x}{\partial z^2}+(\alpha-1)\frac{\partial^2 T_y}{\partial x\partial y}\right)+\mu_0 C(P)\frac{\partial T_x}{\partial t}+\frac{\mu_0}{4\pi}\int_S\frac{\partial(\mathbf{n}'\cdot\mathbf{T}')}{\partial t}\frac{\partial}{\partial x'}\left(\frac{1}{R(P,P')}\right)dS'+\frac{\partial B_{ex}}{\partial t}=0 \quad (22)$$

$$\frac{1}{\sigma_{ab}}\left(-\alpha\frac{\partial^2 T_y}{\partial x^2}-\frac{\partial^2 T_y}{\partial y^2}-\frac{\partial^2 T_y}{\partial z^2}+(\alpha-1)\frac{\partial^2 T_x}{\partial x\partial y}\right)+\mu_0 C(P)\frac{\partial T_y}{\partial t}+\frac{\mu_0}{4\pi}\int_S\frac{\partial(\mathbf{n}'\cdot\mathbf{T}')}{\partial t}\frac{\partial}{\partial y'}\left(\frac{1}{R(P,P')}\right)dS'+\frac{\partial B_{ey}}{\partial t}=0 \quad (23)$$

$$\frac{1}{\sigma_{ab}}\left(-\frac{\partial^2 T_z}{\partial x^2}-\frac{\partial^2 T_z}{\partial y^2}-\frac{\partial^2 T_z}{\partial z^2}\right)+\mu_0 C(P)\frac{\partial T_z}{\partial t}+\frac{\mu_0}{4\pi}\int_S\frac{\partial(\mathbf{n}'\cdot\mathbf{T}')}{\partial t}\frac{\partial}{\partial z'}\left(\frac{1}{R(P,P')}\right)dS'+\frac{\partial B_{ez}}{\partial t}=0 \quad (24)$$

where σ_{ab} is the conductivity in the *ab*-plane. It is worth noting that, compared with the traditional *T*-Ω method, the complexity of the governing equations is reduced due to the omission of the variable Ω (Miya & Hashizume, 1988), and thus the number of varibles in the problem is three with unknown T_x, T_y, and T_z.

2.3 Nonlinear E-J relations

Typically, there are three different models to address the highly nonlinear relationship between **E** and **J** of the high-Tc superconductor, i.e., Bean's critical current model (Bean, 1964), power law model (Rhyner, 1993) and flux flow and creep model (Yamafuji & Mawatari, 1992). For Bean's critical current model, it fails to investigate problems such as force relaxation (Luo, et al., 1999) and drift under vibration (Gou, et al., 2007) due to the assumption that the current flowing in the superconductor is constant with time and the lack of material related parameters in its model. In addition, Bean's critical current model can be considered as an infinite case of the power law model (Rhyner, 1993). This model is thereby not employed in the following calculation.

When an index *n* is introduced and defined as $n=U_0/k\Theta$ where U_0 is the pinning potential of the superconductor at an absolute temperature Θ and *k* is the Boltzmann constant, the power law model is expressed as:

$$\mathbf{E}=E_c\left(|\mathbf{J}|/J_c\right)^n\mathbf{J}/|\mathbf{J}| \quad (25)$$

According to Anderson theory (Anderson, 1962), the responding behavior between **E** and **J** due to flux flow and creep model phenomena in the superconductor can be described as:

$$\mathbf{E}=\begin{cases}2\rho_c J_c\sinh(\frac{U_0}{\kappa\Theta}\frac{|\mathbf{J}|}{J_c})\exp(-\frac{U_0}{\kappa\Theta})\frac{\mathbf{J}}{|\mathbf{J}|} & 0\le|\mathbf{J}|\le J_c\\ E_c+\rho_f J_c(\frac{|\mathbf{J}|}{J_c}-1)\frac{\mathbf{J}}{|\mathbf{J}|} & |\mathbf{J}|>J_c\end{cases} \quad (26)$$

where ρ_c and ρ_f are the creep and flow resistivities respectively, and $E_c=\rho_c J_c\left[1-\exp(-2U_0/\kappa\Theta)\right]\approx\rho_c J_c$.

3. Numerical solution using finite-element technique

The previous work in handling linear eddy current problem has proven that Boundary Element Method is an effective method to numerically solve the governing equations including both differential and integral terms (Miya & Hashizume, 1988). Unfortunately, for the case studied here, the high-Tc superconductor is a highly nonlinear media, and this special property leads to difficulty when Boundary Element Method is employed to handle the 3-D governing equations. Consequently, the finite element method is a frequent choice to discretize the governing equation of the high-Tc superconductor in space (Pecher, et al., 2003; Uesaka, et al., 1993; Yoshida, et al., 1994; Luo, et al., 1999; Alonso, et al., 2004; Gou, et al.,2007a; Gou, et al.,2007b; Lu, et al., 2008). In particular, one format of the finite difference method, named Crank-Nicolson-θ method, is employed to numerically perform the time domain of the 3-D governing equations.

3.1 Finite-element matrices

The governing equations (22) to (24) incorporated with the boundary condition are numerically solved by finite-element technique via Galerkin's method. The tetrahedral element is chosen to mesh the domain of the high-Tc superconductor. The final algebraic equations corresponding to (22) to (24) can be compactly expressed by the following matrix equation,

$$\left[\mathbf{K}_i(\sigma_{ab})\right]\{T_i\}+\left[\mathbf{K}_{12}(\sigma_{ab})\right]\{T_{3-i}\}+[\mathbf{Q}_{0i}]\left\{\frac{\partial T_i}{\partial t}\right\}+[\mathbf{Q}_{1i}]\left\{\frac{\partial(\mathbf{n}'\cdot\mathbf{T}')}{\partial t}\right\}=\{\mathbf{L}_i\}\ (i=1,2,3) \tag{27}$$

where

$$\left[\mathbf{K}_i(\sigma_{ab})\right]=\sum_e\left[\mathbf{K}_i(\sigma^e_{ab})\right]_e=\sum_e\frac{1}{\sigma^e_{ab}}\int_{V_e}\left(\alpha_i\left[\frac{\partial N}{\partial x}\right]^{\mathrm{T}}_e\left[\frac{\partial N}{\partial x}\right]_e+\alpha_{3-i}\left[\frac{\partial N}{\partial y}\right]^{\mathrm{T}}_e\left[\frac{\partial N}{\partial y}\right]_e+\left[\frac{\partial N}{\partial z}\right]^{\mathrm{T}}_e\left[\frac{\partial N}{\partial z}\right]_e\right)dV$$

$$\left[\mathbf{K}_{12}(\sigma_{ab})\right]=\sum_e\left[\mathbf{K}_{12}(\sigma^e_{ab})\right]_e=\frac{1}{2}(1-\alpha)\sum_e\frac{1}{\sigma^e_{ab}}\int_{V_e}\left(\frac{\partial[N]^{\mathrm{T}}_e}{\partial y}\frac{\partial[N]_e}{\partial x}+\frac{\partial[N]^{\mathrm{T}}_e}{\partial x}\frac{\partial[N]_e}{\partial y}\right)dV$$

i	K_i	T_i	T_{3-i}	Q_{0i}	Q_{1i}	L_i	α_i	α_{3-i}	u'_i	B_{ei}
1	K_x	T_x	T_y	Q_{0x}	Q_{1x}	L_x	1	α	x'	B_{ex}
2	K_y	T_y	T_x	Q_{0y}	Q_{1y}	L_y	α	1	y'	B_{ey}
3	K_z	T_z	0	Q_{0z}	Q_{1z}	L_z	1	1	z'	B_{ez}

Table 1. Specification of the variable parameters in (27)

$$[\mathbf{Q}_{0i}]=\sum_e[\mathbf{Q}_{0i}]_e=\mu_0\sum_e\int_{V_e}[N]^{\mathrm{T}}_e[C(P)N]_e dV$$

$$[\mathbf{Q}_{1i}]=\sum_e[\mathbf{Q}_{1i}]_e=\frac{\mu_0}{4\pi}\sum_e\int_{V_e}[N]^{\mathrm{T}}_e\left(\int_S[N]_e\frac{\partial}{\partial u'_i}\frac{1}{R(P,P')}dS'\right)dV$$

$$\{\mathbf{L}_i\}=\sum_e\{\mathbf{L}_i\}_e=-\sum_e\int_{V_e}[N]^{\mathrm{T}}_e\frac{\partial B_{ei}}{\partial t}dV$$

N is the shape function of the linear tetrahedral nodal element. The subscript e represents its detailed formula in each element. The conductivity is different for each element and it is therefore represented by σ_{ab}^{e} here. All the other parameters with a subscript including i can be determined by Table 1.

The finite-element matrices (27) for three components are integrated into one matrix in the numerical program, i.e.,

$$\left[\mathbf{K}(\sigma_{ab})\right]\{T\}+[\mathbf{Q}_0]\left\{\frac{\partial T}{\partial t}\right\}+[\mathbf{Q}_1]\left\{\frac{\partial(\mathbf{n}'\bullet\mathbf{T}')}{\partial t}\right\}=\{\mathbf{L}\} \tag{28}$$

where

$$\left[\mathbf{K}(\sigma_{ab})\right]=\sum_{i=1}^{2}\left[\left[\mathbf{K}_i(\sigma_{ab})\right]+\left[\mathbf{K}_{12}(\sigma_{ab})\right]\right]+\left[\mathbf{K}_3(\sigma_{ab})\right]$$

$$[\mathbf{Q}_0]=\sum_{i=1}^{3}[\mathbf{Q}_{0i}] \qquad [\mathbf{Q}_1]=\sum_{i=1}^{3}[\mathbf{Q}_{1i}] \qquad \{\mathbf{L}\}=\sum_{i=1}^{3}\{\mathbf{L}_i\}$$

After applying Crank-Nicolson-θ method to (28) for time discretization, the matrix at nth time step can be written as:

$$\left(\frac{[\mathbf{Q}_0]+[\mathbf{Q}_1]}{\Delta t}+\theta\left[\mathbf{K}(\sigma_{ab})\right]\right)\left\{T^n\right\}=\theta\left\{\mathbf{L}^n\right\}+(1-\theta)\left\{\mathbf{L}^{n-1}\right\}+\left(\frac{[\mathbf{Q}_0]+[\mathbf{Q}_1]}{\Delta t}-(1-\theta)\left[\mathbf{K}(\sigma_{ab})\right]\right)\left\{T^{n-1}\right\} \tag{29}$$

The disadvantage of the T-method is that the coefficient of (29) is a dense matrix. In the previous work, an over-relaxation iterative solution approach has been proposed to handle this dense matrix on a linear eddy current problem (Takagi, et al., 1988). However, to the situation we are facing, the nonlinearity of the E–J characteristic brings an additional iterative procedure in determining the conductivity of the high-Tc superconductor. Therefore, if the over-relaxation approach is employed in our computation, there would be two iterative procedures in the numerical program, and this would give rise to numerical instability in the calculation as well as the complexity of the numerical program. The dense feature of the coefficient matrix of (29) arises from the dense matrix $[Q_1]$, which is related to the integral term of (27). Therefore, the coefficient matrix will be a sparse one if $[Q_1]$ is transferred to the right side of (29), and this operation gives the following form,

$$\left(\frac{[\mathbf{Q}_0]}{\Delta t}+\theta\left[\mathbf{K}(\sigma_{ab})\right]\right)\left\{T^n\right\}=\theta\left\{\mathbf{L}^n\right\}+(1-\theta)\left\{\mathbf{L}^{n-1}\right\}+\left(\frac{[\mathbf{Q}_0]}{\Delta t}-(1-\theta)\left[\mathbf{K}(\sigma_{ab})\right]\right)\left\{T^{n-1}\right\}-[\mathbf{Q}_1]\left\{T^{n-1}-T^{n-2}\right\} \tag{30}$$

In (30), the dense matrix $[Q_1]$ is related to the difference of unknown variables between the last and the previous two time steps. The value of the unknown at the adjacent time step will approach to each other as the continuous decrease of the time step size Δt. Therefore, in order to obtain a higher precision, we should assign a sufficient small time step size.

3.2 Nonlinear equation solution approach

The common Newton-Raphson method is employed to solve the nonlinear equations obtained from (30). Basically, the nonlinearity of the equations is eliminated by introducing a linear residual, and the corresponding linear equations can be integrated after calculating

the Jacobi matrix for each element. To improve the stability of the calculation, a relaxation coefficient is also introduced as suggested in (Grilli, et al., 2005). The relaxation coefficient is assigned with an initial value before calculation. During the calculation, it will be reduced to a smaller one once the convergence can not be achieved within a threshold of the total iterative steps, and in this case the current time step is recalculated with the new value.
Because of the significant increase of the order of the coefficient matrix when the problem is extended from 2-D to 3-D and also the element of the matrix with the anisotropy consideration involved (an additional second derivative in term of x and y appears in (22) and (23) when the anisotropic behavior is taken into account), the Incomplete Cholesky-Conjugate Gradient method (Kershaw, 1978), which is regarded as an effective approach to solve linear algebraic equations with large symmetric positive definite matrix, is employed to solve the linear equations at each iterative step in an accelerated form. In this accelerated form, an accelerated factor is introduced in the incomplete decomposition process (Fujiwara, et al., 1993) after the coefficient matrix and column vector are preconditioned by a method proposed in (Cui, 1989).

3.3 Numerical procedure

The above-discussed numerical method is implemented by a self-written numerical program based on a VC++ platform with the following typical steps:
Step 1: Initial σ_{ab} in all elements is assumed to be the same and sufficiently large value before the first time step.
Step 2: Calculating $[\mathbf{K}(\sigma_{ab})]$ with the present σ_{ab} in each element at the current time step, and vector $\mathbf{T}$ is obtained by solving (30).
Step 3: The σ_{ab} at the kth iterative step (σ_{ab}^k) is calculated in each element according to the following method using either power law model or flux flow and creep model.
If $J_c(\varphi)$ is expressed by (1),

$$\sigma_{ab}^k = |\mathbf{J}_k| / |\mathbf{E}_k| \Big/ \left(\cos^2\varphi + \sin^2\varphi/\alpha\right) \tag{31}$$

If $J_c(\varphi)$ is expressed by (3),

$$\sigma_{ab}^k = |\mathbf{J}_k| / |\mathbf{E}_k| \Big/ \sqrt{\cos^2\varphi + (\sin\varphi/\alpha)^2} \tag{32}$$

At the end, σ_{ab} in each element is replaced by its new value σ_{ab}^k.
Step 4: Repeating steps 2 and 3 until the residual becomes less than a prescribed tolerance ε, i.e.,

$$\left[\boldsymbol{b} - \boldsymbol{f}\left(\mathbf{T}^{(k)}\right)\right]^2 < \varepsilon \tag{33}$$

where $\boldsymbol{b}$ is a column vector corresponding to the right side of (30), and $f(\mathbf{T}^{(k)})$ is the result of the left side of (30) at the kth iterative step.
Step 5: The current density $\mathbf{J}$ is obtained by (6), and then the magnetic force is calculated by Lorentz equation,

$$\mathbf{F} = \int_V \mathbf{J} \times \mathbf{B}_e dV \tag{34}$$

Step 6: $t = t + \Delta t$ until the maximum number of time step is achieved, and the steps 2-5 are repeated.

3.4 Numerical precision

Based on a levitation system composed of a bulk high-Tc superconductor (single-domain with a cubical shape) and magnetic rail (assembled by the permanent magnets with opposite magnetization direction as shown in the inset of Fig. 2) , the dependence of the computational precision of the levitation force on both mesh density and time step is discussed in this section. In the calculation, the bulk high-Tc superconductor was downward in a speed of 1 mm/s from the filed cooling position to the nearest gap, and then upward to its original position. On the basis of the geometrical and material parameters listed in Table 2 and power law model, the levitation force of the bulk high-Tc superconductor was calculated in this vertical down-and-up movement with different mesh densities and time steps.

w_{SC} (mm)	l_{SC} (mm)	t_{SC} (mm)	J_c^{ab} (A/m^2)	E_c (V/m)	U_0 (EV)	a
10	10	10	2.5×10^8	1×10^{-4}	0.1	3

Table 2. Parameters of high-Tc superconductor used in the calculation for investigating the numerical precision.

Fig. 2 shows the dependence of the levitation force versus gap curve and the maximum levitation force F_{max} at the nearest gap on the total finite element node. This figure clearly illustrates that the levitation force and its hysteresis behavior will approach to a saturated state with the continuous increment of the total finite element node, or in other words, the continuous increment of the mesh density. It can be seen from the inset that the F_{max} continuously increases with the total finite element node but the increased rate is gradually

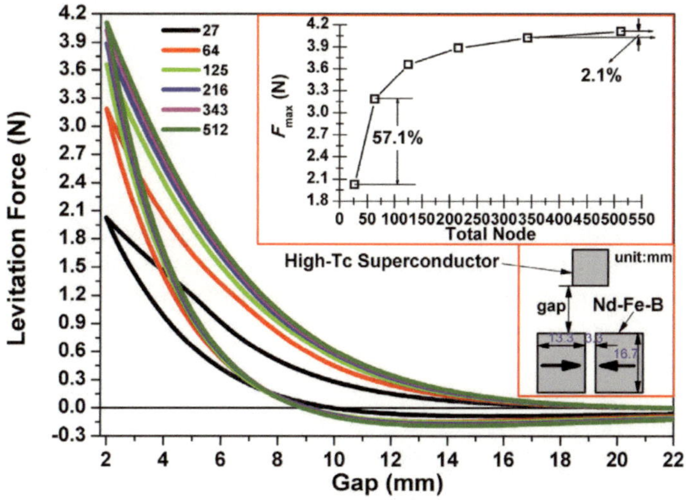

Fig. 2. Levitation force versus gap with different total nodes and the maximum levitation force F_{max} as a function of the total finite element node (inset).

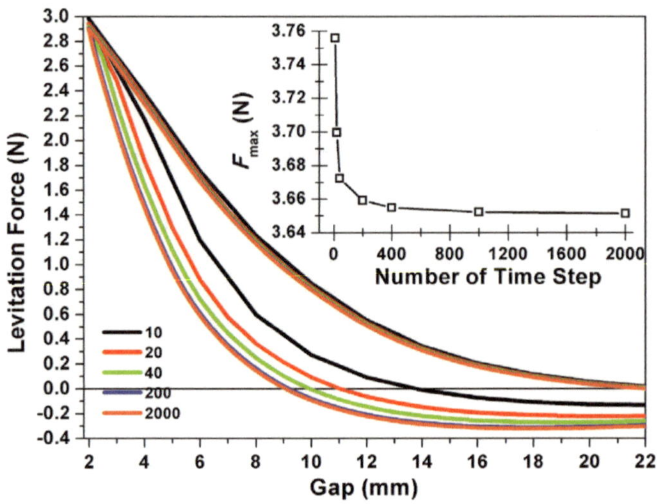

Fig. 3. Levitation force versus gap with different number of time step and the maximum levitation force F_{max} as a function of the number of time step (inset).

reduced, e.g., the rate of the increment is about 57.1% when the total finite element node is raised from 27 to 64, but it is only 2.1% when the node is raised from 343 to 512. This indicates that the computed result is consistent with the finite-element thoery, i.e., the computed value will approach to the real value with the continuous increment of mesh density. Fig. 3 plots the levitation force versus gap curve with different number of time steps and the dependence of the F_{max} on the number of time step. It is obvious that the levitation force and its hysteresis behavior will also approach to a saturated state with the increment of the number of time step, e.g., the reduced rate of the F_{max} is less than 0.024% when the number of time step is raised from 1000 to 2000 and the value of the two cases is almost identical as shown in the inset of Fig. 3. This indicates that, the approach used to handle the coupling problem of the governing equations is applicable when the number of time step adopted is sufficiently large.

4. Experimental validation

The experimental validation is of great importance to support the development of any theoretical model and to confirm the practical application of the theoretical mode to the real world. Compared with the previous validations of the model, we will validate the 3-D finite-element model in a more comprehensive way, i.e., in which two different types of motion are considered, i.e., vertical movement (perpendicular to the surface of the magnetic rail) and transverse movement (parallel to the surface of the magnetic rail), are considered, and the associated magnetic force computed with our model are compared with the experimental data.

w_{SC} (mm)	l_{SC} (mm)	t_{SC} (mm)	J_c^{ab} (A/m^2)	E_c (V/m)	U_0 (eV)	ρ_f (Ωm)	a
30	30	15	1.6×10^8	1×10^{-4}	0.1	5×10^{-10}	3

Table 3. Parameters of the high-Tc superconductor used in the calculation for verifying the 3-D finite-element model.

4.1 Brief introduction of the experiment

The self-developed maglev measurement system was used to measure the magnetic force of the bulk high-Tc superconductor (Wang, 2010). The Y-Ba-Cu-O sample used here is of single-domain with a rectangular shape, and its geometrical parameters and photo as well as its schematic drawing are presented in Table 3 and Fig. 4 respectively. The geometric parameters and photos of the two magnetic rail demonstrators (Rail_1 and Rail_2) used to generate the applied field for high-Tc superconductor are also shown in Fig. 4.

In the experiment, the sample was firstly bath-cooled in a liquid nitrogen vessel for several minutes to to insure the superconductive state was established at a certain position above the center of the magnetic rail, and then, for the vertical movement, the sample was put downward to an expected nearest gap (distance from the bottom of the high-Tc superconductor to the upper surface of the magnetic rail) and then upward to its original position in the levitation case, whereas in the suspension case, the movement was opposite; For the transverse movement, the sample was put downward (levitation case) or upward (suspension case) to its levitation height, and then it was forced to move along the transverse direction parallel to the surface of the magnetic rail (y-axis shown in Fig. 1) after a 30 seconds' relaxation at the levitation height to reduce the influence of the force relaxation on the subsequent measurement or calculation. The experimental and computed speed was chosen to be 1 mm/s in all cases and the maximum lateral displacement for traverse movement was 5 mm.

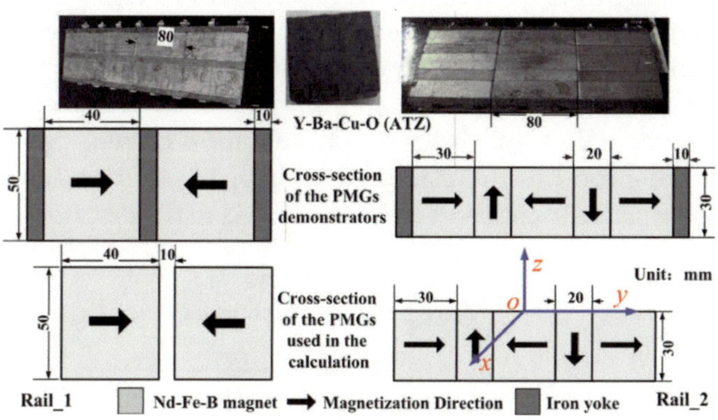

Fig. 4. Photos of the Y-Ba-Cu-O sample and magnetic rail demonstrators and associated schematic drawings of magnetic rail's cross sections.

4.2 Experimental validation of the computed results

In the calculation, the magnetic rail were calculated by a 3-D analytical model in which the finite geometry of each magnet used in the magnetic rail is taken into account (Ma, et al., 2009). With the restriction of the available experimental tools, the material parameters involved in the E-J relation and the angular-dependence of the critical current density formulation can not be directly measured. In the following calculation, the necessary material parameters were determined according to the published literatures. The reported values of the pinning potential U_0 and flow resistivity ρ_f are very scattered. Here, pinning potential and flow resistivity were chosen to be 0.1 eV and 5×10^{-10} Ωm respectively. Both of them are the frequently used values in the calculation Gou, et al.,2007a; Gou, et al.,2007b; Yoshida, et al., 1994). The anisotropic ratio for the melt-processed-single-domain Y-Ba-Cu-O has been measured to be $\sim$3 at 77 K (Murakami, et al., 1991). Critical current density in the *ab*-plane J_c^{ab}, was determined by fitting one of the levitation force versus gap curve (the first curve in the following Fig. 7). In addition, Kim's model (Kim, et al., 1962) was also employed to describe the field amplitude dependence characteristic of the critical current density in the calculation. All the parameters of the Y-Ba-Cu-O were summarized in Table 3.

4.2.1 Numerical results with different E-J constitutive relations and angular-dependence of critical current density formulas

Firstly, we evaluated the levitation force versus gap behavior with different angular-dependence of the critical current density formulations and *E*–*J* constitutive relations. In the 3-D finite-element model, the three different components of the magnetic force, i.e., vertical force perpendicular to the surface of the magnetic rail (*z*-axis), transverse force along the magnetic rail's width (*y*-axis) and longitudinal force along the magnetic rail's length (*x*-axis) can be calculated. From the computed results presented in Fig. 5, we can see that, there is no noticeable discrepancy among the computed levitation force for different test cases. This indicates that there is almost no difference between the power law model and flux flow and creep model when the pining potential U_0 and operating temperature are identical (the index *n* in power law model is ~ 15 at liquid nitrogen temperature with the pinning potential presented in Table 3). Moreover, the elliptical model is also viable to describe the anisotropic behavior of the high-Tc superconductor. Furthermore, as expected, the transverse force and longitudinal force shown in the inset of Fig. 5 are almost zero all the time with the variation of the gap because the applied field is symmetrical along the longitudinal and transverse directions of the magnetic rail.

In the following calculation, we choose the elliptical model for the purpose of introducing a new way to describe the anisotropic behavior of the critical current density and power law model to describe the *E*–*J* constitutive relation of the high-Tc superconductor because it seems to be more numerically stable to converge when compared to the flux flow and creep model.

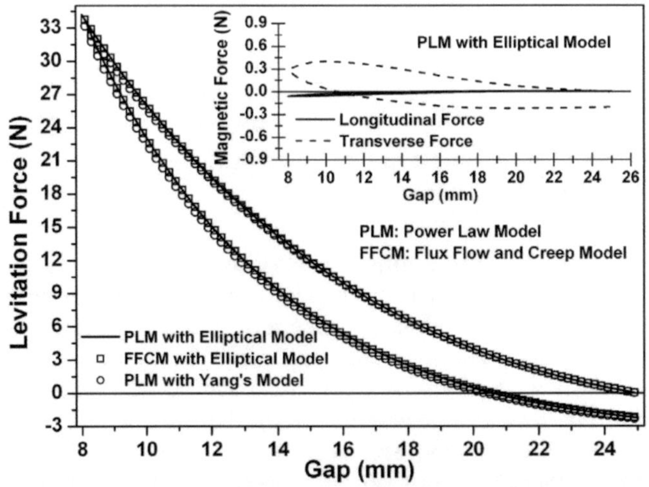

Fig. 5. Comparison of levitation force versus gap curve between different *E*–*J* constitutive relations and angular-dependent of the critical current density formulations as well as the transverse and longitudinal force versus gap behavior with power law and elliptical model (inset). The applied field was provided by Rail_1 shown in Fig. 4.

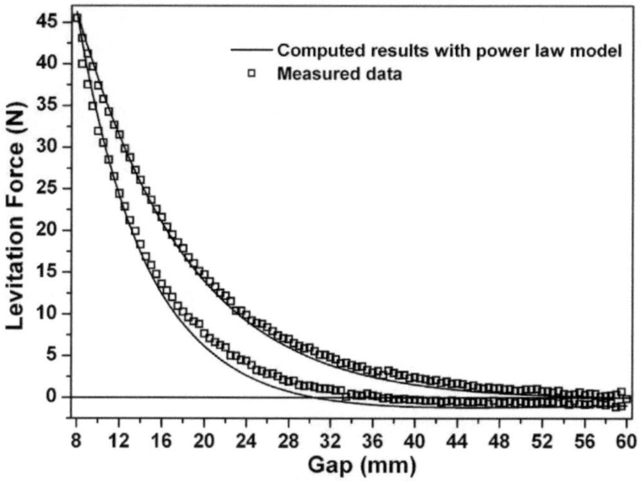

Fig. 6. Comparison of the levitation force versus gap curve between the computed results and measured results under vertical movement in the applied field generated by Rail_1 shown in Fig. 4. The Y-Ba-Cu-O was bath-cooled at a position which was 60 mm above the rail where the applied field is weak and so this case can be approximately considered as zero field-cooled condition.

4.2.2 Comparison of the levitation force under vertical movement

The vertical force of the Y-Ba-Cu-O above the center of the magnetic rail1 with different field-cooled positions was calculated and compared with the measured data. From the compared results presented in Figs. 6-8, we can see that, the computed results agree well with the measured results in both hysteresis loop as well as detailed values at a certain gap. For the levitation case shown in Figs. 6-7, we can find from both the computed and measured results that, on the downward branch of the levitation force versus gap curve, the levitation force and slope of the curve are continuously enhanced with the decrease of the gap. For the same gap, the levitation force of the downward branch is always larger than that of the upward branch, and this reveals an obvious hysteresis behavior of the levitation force. Furthermore, the levitation force versus gap behavior was calculated and measured another two times after the first one for the field-cooled test case at 25 mm above the magnetic rail1. It can be seen clearly that the computed results also compare well with the measured data for the second time and third time. This further confirms the validation of the 3-D finite-element method.

For the suspension test case shown in Fig. 8, the suspension force versus gap curve of both computed and measured cases indicates that, according to the stiffness defined in (Hull, 2000), the suspension force increases with the gap and its stiffness is always positive before the maximum absolute value of the suspension force is achieved at a certain gap. Therefore, the suspension system is always stable within this gap. The suspension force will decrease when the gap is further increased and the suspension system becomes unstable. Also, there is a clear hysteresis behavior of the suspension force by comparing the upward and downward branch of the curve. In addition, we also presented the computed results with flux flow and creep model, and the good agreement between the results of power law model and flux flow and creep model demonstrates that, the two E–J constitutive relations are also identical for the suspension case.

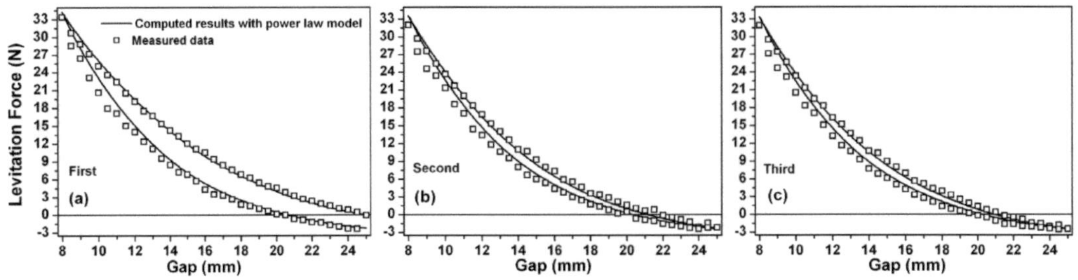

Fig. 7. Comparison of the levitation force versus gap curve between the computed results and measured results under vertical movement in the applied field generated by Rail_1 shown in Fig. 4. The Y-Ba-Cu-O was bath-cooled at a position of 25 mm above the rail which was a typical field-cooled condition. The levitation force versus gap curve was calculated and measured with continous three times here.

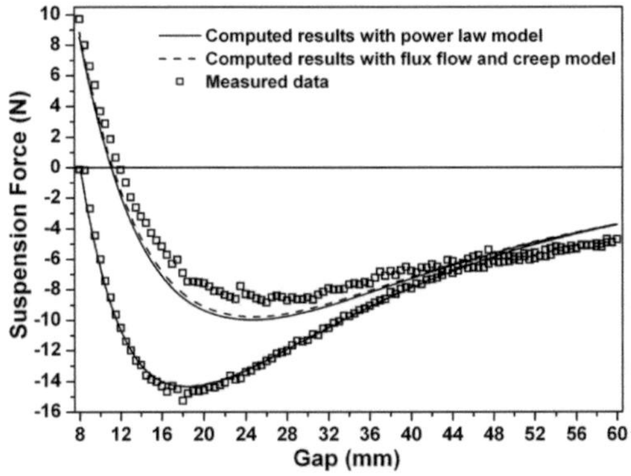

Fig. 8. Comparison of the suspension force versus gap curve between the computed results and measured results under vertical movement in the applied field generated by Rail_1 shown in Fig. 4. The Y-Ba-Cu-O is bath-cooled at a position of 8 mm above the rail.

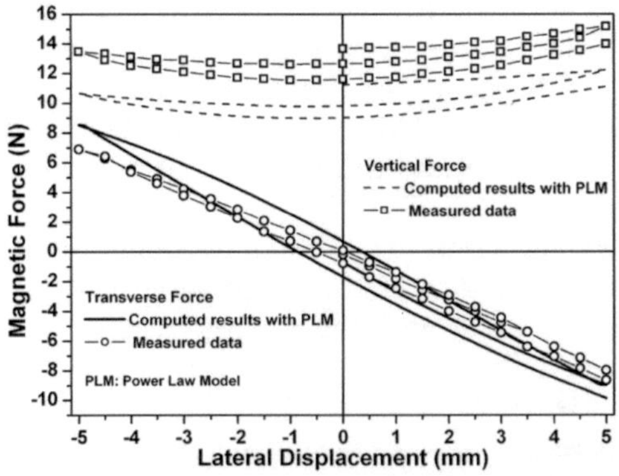

Fig. 9. Comparison of the magnetic force as a function of lateral displacement between the computed and measured results under transverse movement in the applied field generated by Rail_2 shown in Fig. 4. The Y-Ba-Cu-O sample was field-cooled at a position of 30 mm, and the levitation gap where the magnetic force was calculated or measured was 18 mm above the rail.

4.2.3 Comparisons of the magnetic force under transverse movement

In order to verify the robustness of the performance of the 3-D finite-element model, the Rail_2 shown in Fig. 4, which is a Halbach array (Halbach, 1985), was employed to produce the applied field in this section.

During the transverse movement, both the vertical force and transverse force are a function of the lateral displacement. The compared results between the numerical results and measured results for different cases, i.e., levitation case with field-cooled above the levitation position, field-cooled at the levitation position, and suspension case with field-cooled below the suspension position, are presented in Figs. 9-11, respectively.

Both the numerical results and measured results shown in Figs. 9-11 indicate that, the absolute value of the transverse force increases with the lateral displacement and the

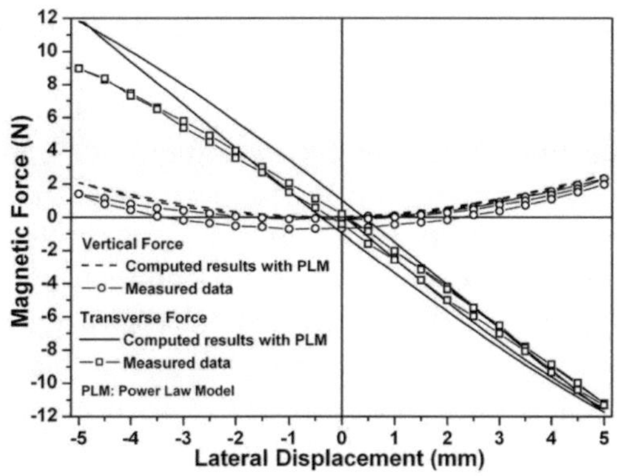

Fig. 10. Comparison of the magnetic force as a function of lateral displacement between the computed and measured results under transverse movement in the applied field generated by Rail_2 shown in Fig. 4. The Y-Ba-Cu-O sample was field-cooled at the identical height (18 mm) with the levitation height where the magnetic force was calculated or measured.

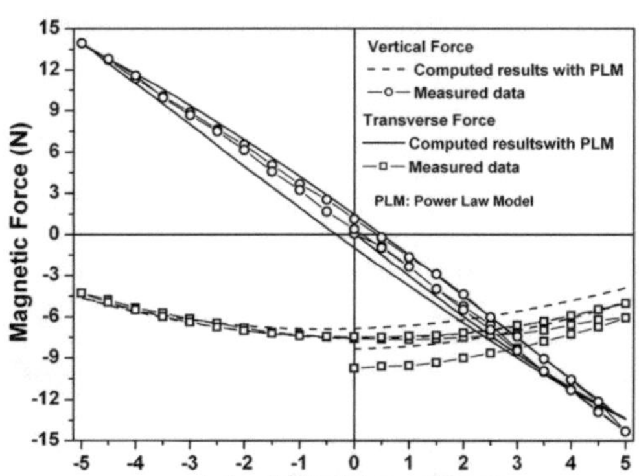

Fig. 11. Comparison of the magnetic force as a function of lateral displacement between the computed and measured results under transverse movement in the applied field generated by Rail_2 shown in Fig. 4. The Y-Ba-Cu-O sample is field-cooled at a position of 13 mm, which was below the levitation gap (18 mm) where the magnetic force was calculated or measured.

direction of the transverse force is opposite to the transverse movement, which indicates an inheret stable levitation can be acquired. Also, the magnetic force (i.e., the vertical force and the transverse force) are hysteretic as a function of the lateral displacement. Despite that the compared results of the transverse case is not as good as that of vertical case, as a whole, the numerical results are well comparable to the measured results in quality especially for the suspension case shown in Fig. 11. The fitting value of the critical current density in the *ab*-plane J_c^{ab} was derived from the case above the Rail_1 and the shift of the applied field are likely to be responsible for the discrepancy between the computed and measured results.

5. Optimization of the magnetic rail using the 3-D finite-element model

Magnetic rail is a key component to privide the applied magnetic field for the present magnet levitation system using bulk high-Tc superconductor, and the cost spend in building the magnetic rail occupies most part of the entire investment because the magnetic rail is required along the whole line. It is thereby meaningful to optimize the structure and geometric parameters of the magnetic rail in the purpose of getting a magnetic rail that holds the required levitation capability with a reduced cost. In this aspect, a lot of previous work has been reported and the dependence of the levitation force/guidance force on the parametes of the magnetic rail has been invesigated. However, all of these results were concluded from the numerical data calculated by a 2-D model that can just fit the experimental results in quality but fails in quantity with a reasonable value of critical current density (Song, et al., 2006; Dias, et al., 2010).

w_{SC} (mm)	l_{SC} (mm)	t_{SC} (mm)	J_c^{ab} (A/m^2)	E_c (V/m)	U_0 (EV)	a
42	21	9	2.5×10^8	1×10^{-4}	0.1	3

Table 4. Parameters of high-Tc superconductor used in the calculation for obtaining an optimized magnetic rail.

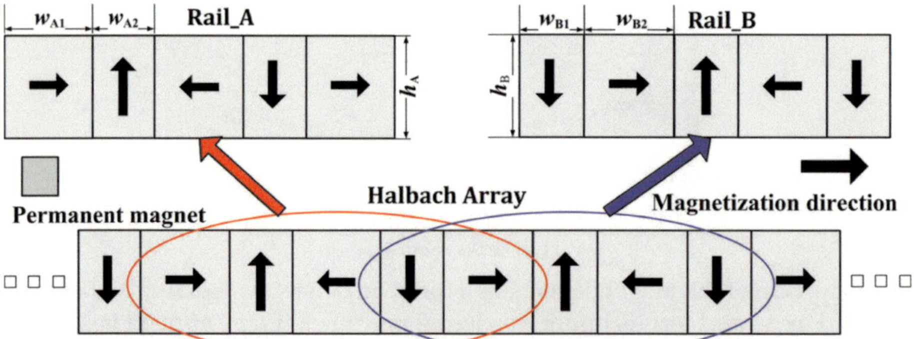

Fig. 12. Two different magnetic rails with five permanent magnets derived from the Halbach array

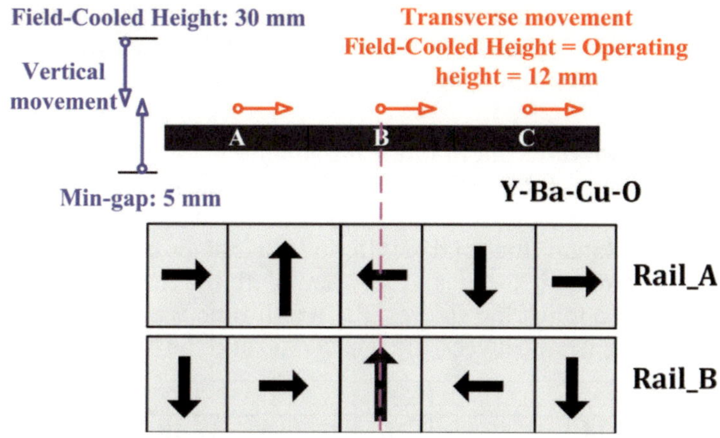

Fig. 13. Chart of the the maglev system with three bulk Y-Ba-Cu-O undergoing the vertical or transverse movement above the rail

In this section, basing on the 3-D finite-element model introduced in this chapter, we calculate the levitation force and guidance force of three Y-Ba-Cu-O bulks samples above two differnt magnetic rails deriving from the traditional Halbach array. The geometric parameters of magnetic rail such as height, width are variable in the calculation, and then the depedence of levitation capability of the bulk Y-Ba-Cu-O on those parameters was studied. Compared with the previous work, the merit of the present calculation is that, the computed levitation force/guidance force is comparable to the real system with reasonable vaule of the critical current density and thus, the computed results can be used to conduct the practical design directly.

The geometric and material parameters of the Y-Ba-Cu-O sample were shown in Table 4 The speed of the samples in both vertical and transverse direction is 1 mm/s and the magnetization M_0 of the permanent magnet employed to assemble the magnetic rail is 8.9×10^5A/m in all cases.

As for the structure of the magnetic rail, Halbach array is a better choice than the original type used in the high-Tc superconducting maglev train demonstrator because this structure can concentrate the magnetic field to its upper space where the high-Tc superconductors are placed, and thus can improve the utilization of the magnetic field (Jing, 2007). From the basic structure of the Halbach array shown in Fig. 12, we can derive two different types of magnetic rail, i.e., one has three permanent magnets magnetized in the horizontal direction and two permanent magnets magnetized in the vertical direction, the other has three permanent magnets magnetized in the vertical direction and two permanent magnets magnetized in the horizontal direction. These two magnetic rails derived from the Halbach array are presented in Fig. 12.

In the calculation, we ignore the possible interaction among the Y-Ba-Cu-O samples for simplicity, and calculate only two samples beacuse of the symmetry of the levitation system shown in Fig. 13. In the default case, the high-Tc superconductors were field-cooled at a position of 30 mm above the surface of the magnetic rail for the calculation of the levitation force, and for the calculation of the guidance force, the tranverse movement occurs at the same height as the field-cooled position, that was 12 mm above the surface of the magnetic rail. The main parameters that should be optimized in the two structure, i.e., Rail_A and

Rail_B shown in Fig. 12, are the ratio between the width of two different magnetized magnets, the width of and the height of the magnetic rail. The following section shows the computed results of the optimization by varying those parameters.

5.1 Width ratio of the two different magnetized magnets

In this section, the levitation force and guidance force on the high-Tc superconductors with the variation of the width ratio, i.e., w_{A1}/w_{A2} for Rail_A or w_{B1}/w_{B2} for Rail_B, were calculated. In this calculation, the total width and the height of both rails were assumed respectively to be 130 mm and 30 mm and invariant with the change of the width ratio. For simplicity in drawing the following figures, the width ratio was replaced by an order and the corresponding relationship between the width ratio and the order was given in Table. 5.

Order	1	2	3	4	5	6	7	8	9	10	11
	12	13	14	15	16	17	18	19	20	21	
Width ratio	0	0.01	0.1	0.16667	0.25	0.333333	0.4	0.5	0.66667	0.83	1
	1.22	1.5	2	2.5	3	4	6	10	100	∞	

Table 5. Corresponding relationship between the width ratio and the order in optimazing the width ratio of the magnetic rails.

For the case of vertical movement, twenty-one different width ratios from zero to infinity were considered and the changing curve of the levitation force along with the width ratio at position of 5, 10, 15 and 20 mm above the Rail_A and the Rail_B was shown respectively in Fig. 14(a) and Fig. 14(b). Note that the extreme case with a width ratio of zero or infinity denotes that Halbach struture disappears and permanent magnets employed in the Rail only have one magnitized direction (vertical or horizontal direction). Both Fig. 14(a) and Fig. 14(b) display that, the levitation force, at all positions we presented, increases with the growth of the width ratio and reaches to a maximum value when the width ratio was ~0.83 for Rail_A and was ~1 for Rail_B and then drops with continuous growth of the width ratio. This finding indicates, the halbach array has a better performance because the levitation

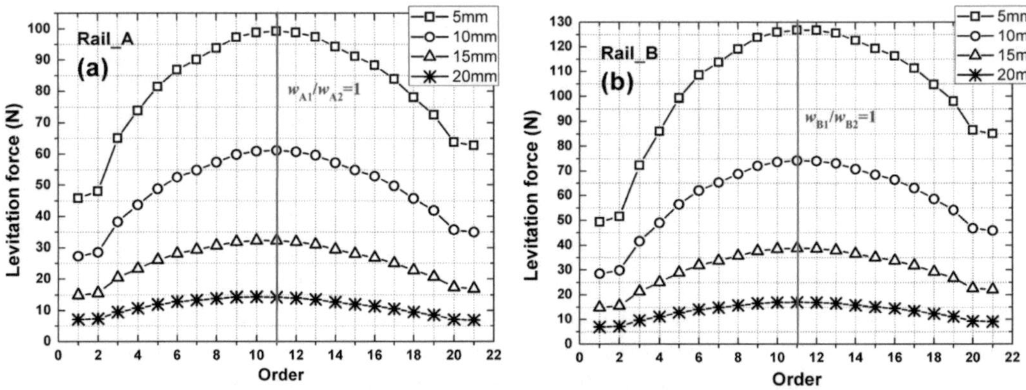

Fig. 14. The changing curve of the levitation force on the high-Tc superconductors at position of 5, 10, 15 and 20 mm above the Rail_A (a) and Rail_B (b) with the growth of the width ratio. The width ratio is replaced by an order and the corresponding relationship between the width ratio and the order (abscissa) is given in Table. 5.

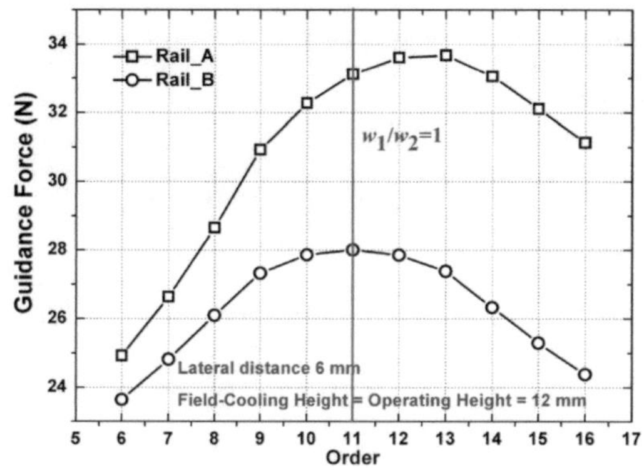

Fig. 15. The changing curve of the guidance force on high-Tc superconductors at a lateral displacement of 6 mm with the growth of the width ratio. The corresponding relationship between the width ratio and the order (abscissa) can be found in Table. 5. The total number of width ratio considered in this calculation was reduced and thus the order starts at 6 and ends at 16 .

force at the extreme case is always the smallest one as has been verified by experiment (Jing, et al., 2007; Sotelo, et al., 2010), on the other hand, the optimized struture for Rail_A is that the width of magnet magnetized in horizontal direction is slightly larger than that magnetized in vertical direction, but for Rail_B, the two type magnets have an identical width ratio.

For the case of the transverse movement, only eleven different width ratios around 1 were considered because larger levitation force can be obtained in that area according to the above computed results of levitation force. The high-Tc superconductors have a field-cooled height of 12 mm and then were drived transversely at the same height with a maximum lateral displacement of 6 mm. Fig. 15 presents the guidance force with respect to the width ratio for the Rail_A and the Rail_B. We can find from this figure that, the guidance force exhibits a same tendency with the growth of the width ratio as that found in the case of levitation force, but the width ratio where the largest guidance force occurs is found to be 1.22~1.5 for Rail_A and is still ~1 for Rail_B.

As a whole, we suggest that, for the Halbach array as the magnetic rail used in the maglev train, the optimized width ratio between the vertical and transverse magnetized magnet is ~1.

5.2 Total width of the magnetic rail

In this calculation, the total width of both rails was a variable parameter with a fixed height of 30 mm and a fixed width ratio of 1.

Fig. 16 shows the changing curve of the levitation force on the high-Tc superconductors at positions of 5, 10, 15 and 20 mm above the Rail_A and Rail_B with the growth of the total width from 60 mm to 170 mm. These figures show that, with the increase of the position, the total width where the maximum value of the levitation force appears is shift from small to large value, e.g., the levitation force increases almost linearly with the total width at the position of 20 mm for both rails and the maximum value of the levitaion force even does not appear within our calculational range. This figure also indicates that, the increase of the total width does not always bring an ehencement of the levitation force especially for the case at a

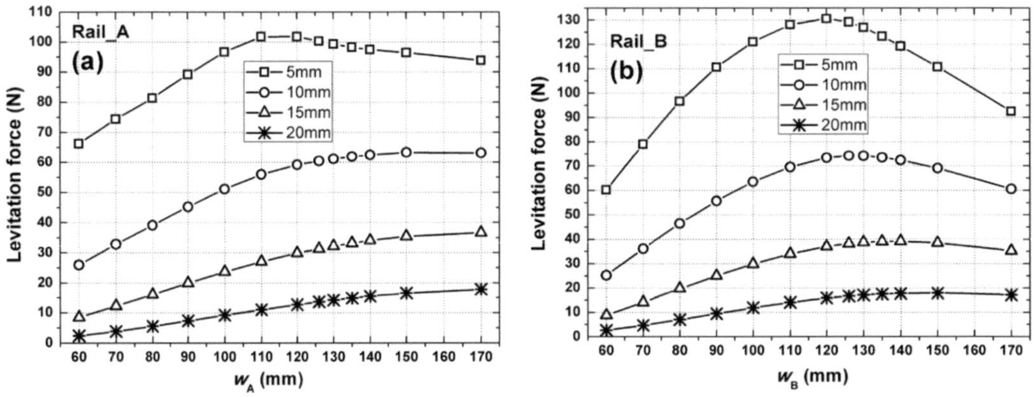

Fig. 16. The changing curve of the levitation force on the high-Tc superconductors at positions of 5, 10, 15 and 20 mm above the Rail_A (a) and Rail_B (b) with the growth of the total width. The total width was varied from 60 to 170 mm.

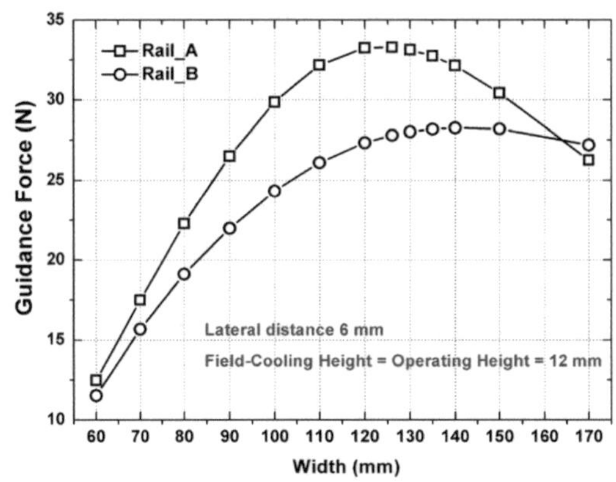

Fig. 17. The changing curve of the guidance force on the high-Tc superconductors at a lateral placement of 6 mm with the growth of the total width. The total width was varied from 60 to 170 mm.

low position. Of course, if a larger levitaion capability is required at a high position, it is a viable way to lengthen the total width of the rail. For the typical position of 15 mm in the maglev train, it seems that, the optimized value of the total width appears around 130 mm for both rails because the slope of the curve begins to dcrease at this value.

Fig. 17 shows the changing curve of the guidance force on the high-Tc superconductors at a lateral displacement of 6 mm with the growth of the total width from 60 to 170 mm. From this figure, we can find that, the guidance force also increases firstly and then decreases with the growth of the total width, and the maximum value of the guidance force occurs around 120 mm for Rail_A and around 135 mm for Rail_B. Although the total width where the maximum value of the guidance force occurs is scattered above Rail_A and Rail_B, the variation of the guidance force above both rails in the range of 120~140 mm is not evident. It is therefore this range can be considered as the optimized range which is comparable to the total width of the high-Tc superconductors (126 mm) in determining the total width of the rail for a practical design.

5.3 Height of the magnetic rail

The height of the magnetic rail is another factor influencing the levitation capability of the high-Tc superconductor. Here, we fixed the width ratio to be 1 and total width to be 130 mm, and varied the height of the magnetic rail from 15 mm to 100 mm. The levitation forces above the Rail_A and Rail_B were plotted as a function of the height and the associate curves were given in Fig. 18. This figure displays clearly that, no matter the position where the high-Tc superconductors are placed, the levitation force increases drastically at first and then goes into gradually a saturated state with heightening the rail. This phenomenon indicates that, considering the cost of the rail, it is not reasonable to get a better levitation performance by increasing the height of the rail when the height is already high enough.

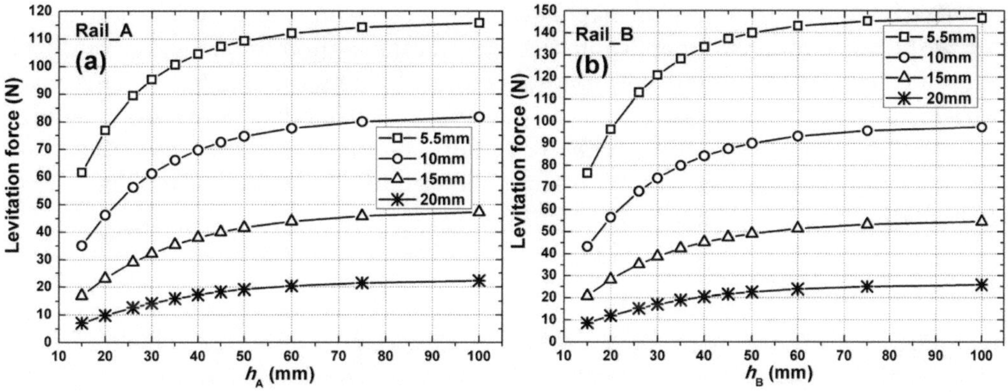

Fig. 18. The levitation force versus height of the Rail_A (a) and of the Rail_B (b). The height of both rails were varied from 15 to 100 mm.

6. Conclusion remarks

On the basis of Maxwell's equations and Helmholtz's theorem, the 3-D governing equations containing the anisotropic behavior of the HTS were deduced at length by introducing a current vector potential **T** and a simplified expression of the resistivity tensor for the high-Tc superconductor. The two common models to handle the highly nonlinear E–J characteristic, i.e., power law model and flux flow and creep model, were presented and, the critical current density J_c as a function of the angle φ between the orientation of the applied field and the c-axis of the high-Tc superconductor, was expressed by an elliptical model. The discrete matrix of the 3-D governing equations was derived by utilizing the Galerkin's finite-element method and the Crank-Nicolson-θ method to numerically treat the governing equations one by one in space and time domain respectively. The corresponding nonlinear algebraic equations of the discrete matrix were effectively resolved by incorporating the Newton-Raphson method with an extended format of the Incomplete Cholesky-Conjugate Gradientmethod.

The computed results of the levitation force of a bulk high-Tc superconductor above a magnetic rail indicate that, the levitation force and associated hysteresis behavior will approach to a saturated state with the increment of the mesh density or the number of the time steps. This result is consistent with the finite-element theory, and also, supports the approach used to simplify the coupling problem between the differential and integral term presented in the governing equations.

Subsequentely, the 3-D finite-element model was validated by comparing the computed magnetic forces of a bulk Y-Ba-Cu-O sample with the measured data under vertical and transverse movements. Basing on the 3-D finite-element model, we also found that, the computed results of magnetic force of the high-Tc superconductor are almost identical for different angular-dependence of critical current density formulations (i.e., Yang's model and elliptical model) and *E–J* constitutive relations (i.e., power law model and flux flow and creep model).

Lastly, using this 3-D finite-element model, we conduct a preliminary numerical work with aim of geting an optimized geometric parameter of the magnetic rails derived from the Halbach array. The results show that, when the width ratio of the two different magnetized magnets used in the magnetic rail is approximately identical and the total width of the rail is roughly indentical to the width of the high-Tc superconductors, the rail exhibits a better performance in considering both the levitation capability and the cost (be proportional to the voulme of the rail). Moreover, the levitation force has a saturated tendency with increasing continuely the height of the rail, so it is not reasonable to get a better levitation performance by increasing the height of the rail when the height is already high enough.

7. Acknowledgments

This work was supported by the National Natural Science Foundation of China under Grant 51007076 and 50777053, and the Innovation Foundation of Southwest Jiaotong University for Ph.D. Candidate. The authors would like to thank Dr. Frank N. Werfel of the Adelwitz Technologiezentrum GmbH (ATZ) for providing the Y-Ba-Cu-O sample employed in the experiment and to thank Dr. Hong-Hai Song, Dr. Zi-Gang Deng, Dr. Yi-Yun Lu, Dr. Min-Xian Liu, Dr. Jun Zheng, Dr. Fei Yen, Mr. Wei Liu, Mr. Qun-Xu Lin, Mr. Dong-Hui Jiang for their many fruitful discussions

8. References

Alloui, L. & Bouillault, F. & Mimoune, S. M. (2009). Numerical study of the influence of flux creep and of thermal effect on dynamic behaviour of magnetic levitation systems with a high-Tc superconductor using control volume method. *Eur. Phys. J. Appl. Phys.*, Vol. 45, pp. 20801

Alonso, D. R. & Coombs, T. A. & Campbell, A. M. (2004). Numerical analysis of high-temperature superconductors with the critical-state model. *IEEE Trans. Appl. Supercond.*, Vol. 14, No. 4, pp. 2053–2063.

Anderson, P. W. (1962). Theory of flux creep in hard superconductor. *Phys. Rev. Lett.*, Vol. 9, No. 7, pp. 309–311

Bean, C. P. (1964). Magnetization of high-field superconductors. *Rev. Mod. Phys.*, vol. 36, pp.31–39

Brandt, E. H. (1989). Levitation in physics. *Science*, vol. 243, pp. 349–355

Brandt, E. H. (1996). Superconductors of finite thickness in a perpendicular magnetic field: Strips and slabs. *Phys. Rev. B*, Vol. 54, No. 6, pp. 4246–4264

Cui, X. (1989). A new preconditional conjugate gradient algorithm. *Journal of North China Institute of Electric Power*, No. 2, pp. 1–8, (in Chinese)

Davis, L. C. & Logothetis, E. M. & Soltis, R. E. (1988). Stability of magnets levitated above superconductors. *J. Appl. Phys.*, Vol. 64, No. 8, pp. 4212–4218

Dias, D. H. N. & Motta, E. S. & Sotelo, G. G. & Andrade Jr, R. D. (2010). Experimental validation of field cooling simulations for linear superconducting magnetic bearings. *Supercond. Sci. Technol.* Vol. 23, pp. 075013

Dinger, T. R. & Worthington, T. K. & Gallagher, W. J. & and Sandstrom, R. L. (1987). Direct observation of electronic anisotropy in single-crystal $Y_1Ba_2Cu_3O_{7-x}$. *Phys. Rev. Lett.*, Vol. 58, pp. 2687–2690

Fujiwara, K. & Nakata, T. & Fusayasu, H. (1993). Acceleration of convergence characteristic of the ICCG method. *IEEE Trans. Magn.*, Vol. 29, No. 2, pp. 1958–1961

Gou, X. F. & Zheng, X. J. & Zhou, Y. H. (2007a). Drift of levitated/suspended Body in high-T_c superconducting levitation systems under vibration—part I: A criterion based on magnetic force-gap relation for gap varying with time. *IEEE Trans. Appl. Supercond.*, Vol. 17, No. 3, pp. 3795–3802

Gou, X. F. & Zheng, X. J. & Zhou, Y. H. (2007b). Drift of levitated/suspended Body in high-T_c superconducting levitation systems under vibration—Part II: drift velocity for gap varying with time. *IEEE Trans. Appl. Supercond.*, Vol. 17, No. 3, pp. 3803–3808

Grilli, F. & Stavrev, S. & Le Floch, Y. & Costa-Bouzo, M. & Vinot, E. & Klutsch, I. & Meunier, G. & Tixador, P. & Dutoit, B. (2005). Finite-element method modeling of superconductors: from 2-D to 3-D. *IEEE Trans. Appl. Supercond.*, Vol. 15, No. 1, pp. 17–25

Halbach, K. (1985). Application of permanent magnets in accelerators and electron storage rings (invited). *J. Appl. Phys.*, Vol. 57, pp. 3605–3908

Hashizume, H. & Sugiura, T. & Miya, K. & Ando, Y. & Akita, S. & Torii, S. & Kubota, Y. & T. Ogasawara. (1991). Numerical analysis of a. c. losses in superconductors. *Cryogenics*, Vol. 31, pp. 601–606

Hull, J. R. & Cansiz, A. (1999). Vertical and lateral forces between a permanent magnet and a high-temperature superconductor. *J. Appl. Phys.*, Vol. 86, No. 11, pp. 6396–6404

Hull, J. R. (2000). Superconducting bearings. *Supercond. Sci. Technol.*, vol. 13, pp. R1–R15

Jing, H. & Wang, J. & Wang, S. & Wang, L. & Liu, L. & Zheng, J. & Deng, Z. & Ma, G. & Zhang, Y. & Li. J. (2007). A two-pole Halbach permanent magnet guideway for high temperature superconducting Maglev vehicle. *Physica C*. 2007, Vol. 463–465, pp. 426–430

Kershaw, D. S. (1978). The incomplete cholesky-conjugate gradient method for the iterative solution of systems of linear equations. *Journal of Computational Physics*, Vol. 26, pp. 43–65

Kim, Y. B. & Hemptead, C. F. & Strnad, A. R. (1962). Critical persistent currents in hard superconductors. *Phys. Rev. Lett.*, Vol. 9, No. 7, pp. 306-309

Kordyuk, A. A. (1998). Magnetic levitation for hard superconductor. *J. Appl. Phys.*, Vol. 83, No. 1, pp. 610–612

Krusin-Elbaum, L. & Malozemoff, A. P. & Yeshurun, Y. & Cronemeyer, D. C. & Holtzberg, F. (1989). Temperature dependence of lower critical fields in Y-Ba-Cu-O crystals. *Phys. Rev. B*, Vol. 39, No. 4, pp. 2936–2939

Lu, Y. Y. & Wang, J. S. & Wang, S. Y. & Zheng J., (2008). 3D-Modeling numerical solutions of electromagnetic behavior of HIGH-TC SUPERCONDUCTORC bulk above permanent magnetic guideway. *J. Supercond. Nov. Magn.*, Vol. 21, pp. 467–472

Luo, Y. & Takagi, T. & Miya, K. (1999). Reduction of levitation decay in high Tc superconducting magnetic bearings. *Cryogenics*, Vol. 39, pp. 331–338

Ma, G. T. & Liu, H. F. & Wang, J. S. & Wang, S. Y. & Li, X. C. (2009). 3D modeling permanent magnet guideway for high temperature superconducting maglev vehicle application. *J. Supercond. Nov. Magn.*, Vol. 22, pp. 841–847

Ma, K. B. & Postrekhin, Y. V. & Chu, W. K. (2003). Superconductor and magnet levitation devices. *Rev. Sci. Instrum.*, Vol. 74, No. 12, pp. 4989–5017

Matsushita, T. (2007). *Flux Pinning in Superconductors.* Springer-Verlag Berlin, pp. 365-366.

Mikitik, G. P. & Brandt, E. H. (2000). Critical state in thin anisotropic superconductors of arbitrary shape. *Phys. Rev. B*, Vol.62, pp. 6800–6811

Miya, K. & Hashizume, H. (1988). Application of T-method to A.C. problem based on boundary element method. *IEEE Trans. Magn.*, Vol. 24, No. 1, pp. 134–137

Murakami, M. & Oyama, T. & Fujimoto, H. & Gotoh, S. & Yamaguchi, K. (1991). Melt processing of bulk high Tc superconductors and their application. *IEEE Trans. Magn.*, Vol. 27, No. 2, pp. 1479–1486

Napoli, A. D. & Paggi, R. (1983). A model of anisotropic grain-oriented steel. *IEEE Trans. Magn.*, Vol. 19, No. 4, pp. 1557–1561

Navau, C. & Sanchez, A. (2001). Magnetic properties of finite superconducting cylinders. II. Nonuniform applied field and levitation force. *Phys. Rev. B*, Vol. 64, pp. 214507

Pecher, R. & McCulloch, M. D. & Chapman, S. J. & Prigozhin, L. & Elliot, C. M. (2003) 3D-Modelling of bulk type-II superconductors using unconstrained H-formulation. *EUCAS 2003*, Sorrento, Italy, Sep. 14–18

Prigozhin, L. (1997). Analysis of Critical-State Problems in Type-II Superconductivity," *IEEE Trans. Appl. Supercond.*, Vol. 7, No. 4, pp. 3866–3873

Qin, M. J. & Li, G. & Liu, H. K. & Dou, S. X. & Brandt, E. H. (2002). Calculation of the hysteretic force between a superconductor and a magnet. *Phys. Rev. B*, Vol. 66, pp. 024516

Rhyner, J. (1993). Magnetic properties and AC-losses of superconductors with power law current–voltage characteristics. *Physica C*, Vol. 212, pp. 292–300

Sanchez, A. & Navau, C. (2001). Magnetic properties of finite superconducting cylinders. I. Uniform applied field. *Phys. Rev. B*, Vol. 64, pp. 214506

Sanchez, A. & Valle, N. D. & Pardo, E. & Chen, D. X. & Navau, C. (2006). Magnetic levitation of superconducting bars. *J. Appl. Phys.*, Vol. 99, pp.113904

Sawamura, M. & Tsuchimoto, M. (2000). Numerical analysis for superconductor in sheet and bulk form," *Japan J. Indust. Appl. Math.*, Vol. 17, No. 2, pp. 199–208

Schultz, L. & Haas, O. & Verges, P. & Beyer, C. & Röhlig, S. & Olsen, H. & Kühn, L. & Berger, D. & Noteboom, U. & Funk, U. (2005). Superconductively levitated transport system-the SupraTrans project. *IEEE Trans Appl. Supercond.*, Vol. 15, No.2, pp. 2301–2305

Song, H. H. & Wang, J. S. & Wang, S. Y. & Ren, Z. Y. & Wang, X. R. & Haas, O. & Fuchs, G. & Schultz, L. (2006). Studies of YBCO electromagnetic properties for high-temperature superconductor maglev technology. in *New Topics in Superconductivity Research*, Barry P. Martins Ed., New York: Nova Science Publishers, pp. 107–156

Sotelo, G. G. & Dias, D. H. N. & Andrade Jr, R. D. & Stephan, R. M. (2010). Tests on a Superconductor Linear Magnetic Bearing of a Full-Scale MagLev Vehicle. *IEEE Trans. Appl. Supercond.*, doi: 10.1109/TASC.2010.2086034, (in press)

Takagi, T. & Sugiura, T. & Miyata, K. & Norimatsu, S. & Okamura, K. & Miya, K. (1988). Iterative solution technique for 3-D eddy current analysis using T-method. *IEEE Trans. Magn.*, Vol. 24, No. 6, pp. 2682–2684

Ueda, H. & Azumaya, S. & Tsuchiya, S. & Ishiyama, A. (2006). 3D electromagnetic analysis of levitating transporter using bulk superconductor. *IEEE Trans. Appl. Supercond.*, Vol. 16, No. 2, pp. 1092–1095

Uesaka, M. & Yoshida, Y. & Takeda, N. & Miya, K. (1993). Experimental and numerical analysis of three-dimensional high-T_c superconducting levitation systems. *Int. J. Appl. Electromagn. Mater.*, Vol. 4, pp. 13–25

Wang, J. S. & Wang, S. Y. & Zeng, Y. W. & Huang, H. Y. & Luo, F. & Xu, Z. P. & Tang, Q. X. & Lin, G. B. & Zhang, C. F. & Ren, Z. Y. & Zhao, G. M. & Zhu, D. G. & Wang, S. H. & Jiang, H. & Zhu, M. & Deng, C. Y. & Hu, P. F. & Li, C. Y. & Liu, F. & Lian, J. S. & Wang, X. R. & Wan, L. H. & Shen, X. M. & Dong, X. G. (2002). The first man-loading high temperature superconducting maglev test vehicle in the world. *Physica C*, Vol. 378-381. pp.809-814

Wang, J. S. & Wang, S. Y. (2010). *High Temperature Superconducting Maglev Measurement System*. In: Milind Kr Sharma Ed., Advances in Measurement Systems, pp. 51-80

Wu, J. Z. & Hsieh, P. Y. & McGuire, A. V. & Schmidt, D. L.& Wood, L. T. & Shen, Y. & Chu, W. K. (1991). Anisotropic properties of the high-quality epitaxial $YBa_2Cu_3O_{7-\delta}$ (110) thin film. *Phys. Rev. B*, Vol. 44, No. 22, pp. 12643–12646

Yamafuji, K. & Mawatari, Y. (1992). Electromagnetic properties of high Tc superconductors: relaxation of magnetization. *Cryogenics*, Vol. 32, No. 6, pp. 569–577

Yang, W. M. & Feng, Y. & Zhou, L. & Zhang, P. X. & Wu, M. Z. & Chen, S. K. & Wu, X. Z. & Gawalek, W. (1999). The effect of the grain alignment on the levitation force in single domain $YBa_2Cu_3O_y$ bulk superconductors. *Physica C*, Vol. 319, pp. 164–168

Yang, Y. & Zheng, X. J. (2007). Method for solution of the interaction between superconductor and permanent magnet. *J. Appl. Phys.*, Vol. 101, pp. 113922

Yoshida, Y. & Uesaka, M. & K. Miya, (1994). Magnetic field and force analysis of high-Tc superconductor with flux flow and creep. *IEEE Trans. Magn.*, Vol. 30, No. 5, pp. 3503–3506

Wang Jiasu and Wang Suyu, (2005). Synthesis of Bulk Superconductors and Their Properties on Permanent Magnet Guideway. In: Anant Narlikar Ed., Frontiers in Superconducting Materials (Springer Verlag, Germany), pp.885-912

7

Epitaxial Oxide Heterostructures for Ultimate High-T_c Quantum Interferometers

Michael I. Faley
PGI-5 (Mikrostrukturforschung) Forschungszentrum Jülich GmbH
52425 Jülich
Germany

1. Introduction

The broad range of physical properties available in metal-oxide materials is now beginning to be exploited in the form of epitaxial functional oxide thin-film heterostructures for the purposes of modern nanoelectronics. The discovery of high-T_c superconductors and requirements for their structural perfection and chemical homogeneity also stimulated the rapid development of deposition and patterning technologies for thin films of conducting, insulating, ferroelectric, ferromagnetic and also multiferroic oxide materials. For example, epitaxial heterostructures with $SrTiO_3$ (STO) display high-mobility of charge carriers, two-dimensional interface conductivity, field-induced superconductivity, and the thermoelectric effect (see, e.g., Ohtomo & Hwang, 2004; Jia et al., 2009). The multifunctional metal-oxide heterostructures help to keep pace in shrinking dimensions down to the nanometer-scale and the increasing complexity of integrated electronic circuits, which are among the most distinctive features of modern times. Rapid progress in general purpose applications of the metal-oxide heterostructures has, in turn, supported further development of high-T_c superconducting devices. Up to now, the most advanced high-T_c superconducting epitaxial oxide thin-film heterostructures include thin films of $YBa_2Cu_3O_{7-x}$ (YBCO) superconductor and the technologically compatible with it metal-oxide materials whose crystal structures are mainly derived from perovskite-type crystal structures. All-oxide heterostructures based on high-quality epitaxial YBCO thin films with other metal-oxide layers are indispensable for high-T_c superconducting quantum interference devices (SQUIDs) with the highest possible magnetic field sensitivity at 77 K.

SQUIDs are used for ultrasensitive electric and magnetic measurements of all physical quantities that can be converted into magnetic flux, for example, magnetic field, magnetic field gradients, current, voltage, displacement, magnetic susceptibility etc. SQUIDs serve as extremely sensitive magnetic field sensors for biomagnetic measurements, geomagnetic survey, non-destructive evaluation, low-noise preamplifiers, picovoltmeters, etc. (see, e.g., Kleiner et al., 2004; Clarke & Braginski, 2006; Fagaly, 2006). It is important that the SQUID magnetometers measure vector components of magnetic fields. In addition, they can measure (in hardware!) spatial gradients of magnetic fields; they are able to resolve tiny changes (~1 fT) in large (~ 1 μT) signals, and have a white noise spectrum in a wide frequency range (1 Hz - 10 MHz). No other magnetic field sensor has this combination of properties, which are especially important for, e.g., biomagnetic applications. Biomagnetic

applications demand a very challenging tasks of measurements of tiny magnetic fields generated by very weak ionic currents in biological neural networks and the real time inverse calculations for the localization of these ionic currents.
High-T_c superconductors have significant potential for further developments in fundamental physics and applications (Faley, 2010c). Increased interest in the high-T_c superconducting devices is spurred by the expected up to 30-fold price increase for liquid helium due to a drastic shortage of reserves (Witchalls, 2010). This will impact, first of all, the biomagnetic applications of low-T_c DC SQUIDs, for example, for magnetoencephalography (MEG), where the implementation of cryocoolers can ruin the sensitivity of measurement systems. The high-T_c SQUIDs demonstrate low noise properties up to the temperatures of about 80 K, which can be easily reached with cheap liquid nitrogen. Most of the presently available high-T_c SQUIDs are optimized for operation at a temperature 77 K, at which the equilibrium vapour pressure of liquid nitrogen coincides with atmospheric pressure. Replacement of low-T_c SQUIDs with high-T_c SQUIDs in future MEG systems would make the MEG systems independent of helium supplies, much more user-friendly, and would save up to about €100,000.- per year in operating costs per multichannel MEG system thus stimulating general acceptance of the MEG systems for clinical use. Liquid nitrogen is about ten times cheaper than liquid ^{4}He and can be easily obtained from air. It is also sufficient to replenish the liquid nitrogen in standard cryostats for MEG only about once a month and this can be done simply through a long non-magnetic silicone tube from a stationary tank placed outside the magnetically shielded room. The optional cycle refrigeration recondensing the cryogen between measurements would be also much cheaper and simpler in the case of the high-T_c MEG system compared to the low-T_c system. Also in the case of measurements of the currents of high-energy heavy-ion beams in the radioisotope beam accelerators, the implementation of high-T_c SQUID systems can save operating costs of about $150,000 per year for each system compared to the low-T_c SQUID systems (Watanabe et al., 2010).
An important prerequisite for the application of high-T_c SQUIDs for MEG systems is magnetic field resolution better than 10 fT/√Hz down to frequencies of few Hz at 77 K. Such sensitivity of high-T_c SQUIDs has been achieved by a number of groups using superconducting multilayer flux transformers with a multiturn input coil for optimal coupling of the magnetic field to be measured to the high-T_c DC SQUID loop (Dantsker *et al.*, 1995) (Drung et al., 1996) (Faley et al., 2001, 2006b). Reduction of the excess low frequency noise of the multilayer high-T_c DC SQUID magnetometers is crucial for applications. This can be achieved only with the highest degree of crystalline perfection and stoichiometry of the high-T_c superconducting thin films in the multilayer heterostructures. The choice of the proper deposition method and deposition conditions as well as the technologically compatible materials for substrates and oxide heterostructures is important for providing the best possible epitaxial growth of all films involved (Faley et al., 2006b). Influences of lattice mismatch and difference in thermal expansion coefficients of the materials in the epitaxial oxide heterostructures should be minimized because they lead to strains inside the high-T_c superconducting films and deteriorate their properties (Faley et al., 2006a). Patterning techniques as well as proper layouts and capsulations are also essential for the total sensitivity of the high-T_c DC SQUID sensors to be optimised for particular applications.
In this chapter the technological aspects and physical properties of epitaxial oxide heterostructures are described, which are used for the preparation of high-T_c DC SQUID

flip-chip magnetometers with multilayer flux transformers having magnetic field sensitivity below 10 fT/√Hz. Deposition and structuring are outlined for the epitaxial oxide heterostructures; materials for substrates, epitaxial bilayer buffer, Josephson junctions and multilayer flux transformers; the ramp-type and bicrystal Josephson junctions; operation features, layouts, and properties of the all-oxide epitaxial multilayer high-T_c DC SQUID sensors including their encapsulation.

2. Deposition of epitaxial metal-oxide heterostructures

Significant technological efforts are required to produce high-quality samples of superconducting cuprates due to their sensitivity to the compositional and structural inhomogeneities. Accurate stoichiometry, high degree of crystallization in a single phase and proper oxidation of the film are essential. For the deposition of epitaxial YBCO films it is also important to have an appropriate substrate temperature and definite partial oxygen pressure. The required purity of c-axis orientation and 123 phase of the YBCO-films is determined by the position of the sputtering conditions in the P_{O2}-T phase diagram of YBCO. The best YBCO films are obtained along the line in the P_{O2}-T phase diagram (Hammond & Bormann, 1989) associated with initial oxygen content O_6, which corresponds to the absence of oxygen in the plane of the CuO chains and the CuO_2 planes of YBCO are undoped. Three reproducible deposition methods for the fabrication of thin-film metal-oxide heterostructures fulfilling such conditions are now mainly used: pulsed laser deposition, reactive co-evaporation, and the high oxygen pressure magnetron sputtering technique. These methods are briefly reviewed below with the emphasis on the high oxygen pressure magnetron sputtering technique, which we preferentially employ for preparation of SQUID sensors.

The reactive co-evaporation method was adopted for YBCO films by Kinder and co-workers (Prusseit et al., 2000). By rapid cycling between deposition and oxygen reaction they combined deposition in a high vacuum environment and oxygenation at a differentially high oxygen pressure enclosed in the heater. The reactive co-evaporation method is especially effective for the commercial large-scale production of epitaxial cuprate films on large wafers or on tapes intended for high current applications such as cables for transmission power lines, generators, and motors. The reactive co-evaporation method provides very high material utilization efficiency, high deposition rate, possibility of continuous deposition on km-long tapes, enabling easy switching between many elements, and fine adjustment of the composition. One of the disadvantages of the reactive co-evaporation method is the necessity of continuous rate control for each element of the compounds. An atomic absorption monitor can be used for continuous measurement of the vapour densities near to the substrates (Matias et al., 2010). The standard apparatus for reactive co-evaporation is relatively expensive with respect to initial investments and maintenance.

Pulsed laser deposition (PLD) is the most widely used method for the deposition of metal-oxide heterostructures. The material that is to be deposited is vaporized from the target by a pulsed laser beam and transported in a plasma plume to a substrate. This process can be performed in the presence of oxygen as a background gas to oxygenate the deposited metal-oxide films. The physical phenomena of laser-target interaction and film growth are quite complex. The energy of the laser pulse is first converted into electronic excitation and then into thermal, chemical and mechanical energy resulting in plasma formation, evaporation,

ablation, and, in some cases, even exfoliation. The ejected material is emitted from the target in the form of atoms, molecules, electrons, ions, clusters, and even molten globules. PLD provides a high deposition rate. A small target can be used in PLD to deposit film over large-area wafers with appropriate scanning schemes. However, this method is also relatively expensive, because a powerful laser is required. The films produced by PLD are usually relatively inhomogeneous due to ablation from a spot and contamination of the films by molten globules. There is also an angular dependence of morphology and stoichiometry of the films prepared by PLD (Sobol, 1995) (Acquaviva et al., 2005). The typical superconducting transition temperature T_c of YBCO films obtained by PLD is $\approx$ 89 K, which is significantly lower than $T_c \approx 93$ K obtained for bulk ceramic samples of YBCO.
The technique of sputtering at high oxygen pressures allows a smart and homogeneous on-axis in-situ deposition of high-quality metal-oxide thin films from stoichiometric targets (Poppe et al., 1990, 1992). Conventional sputtering is used extensively in the semiconductor industry to deposit thin films of various materials in integrated circuit processing. For the deposition of the epitaxial metal-oxide films it is necessary to heat the substrate to temperatures above 600 °C and introduce oxygen into the sputtering gas atmosphere. If conventional sputtering pressures of about 0.01 mbar are used for the on-axis deposition of cuprate superconductors, the negatively charged oxygen ions are accelerated towards the heated substrate by the bias potential and they thus resputter copper atoms from the deposited film leaving copper-deficient non-stoichiometric cuprate films (see, for example, Faley et al., 1991). With the high oxygen pressure sputtering technique, this problem is solved by multiple scattering of the oxygen ions at background gas pressures above 1 mbar with subsequent reduction of their kinetic energy down to thermal energies before they reach the substrate. This results in negligible backsputtering of the copper from the deposited films and, consequently, their good stoichiometry and electron transport properties. Typical superconducting transition temperature of the YBCO films obtained by this method is about 93 K and their critical current density is about 6 MA/cm^2 at 77.4 K.
The high oxygen pressure sputtering technique presupposes deposition at 0.5 to 5 mbar of a pure oxygen (99.999%) sputtering gas atmosphere. The main feature of the sputtering apparatus for the high oxygen pressure sputtering is the presence of a solid insulator, typically made of MACOR, between the target holder and the ground shield. The solid insulator prevents short circuit discharge at these relatively high sputtering pressures and a short mean free path ~ 0.1 mm of the accelerated electrons. If necessary, the entire range of deposition conditions from high-energy impact to low-energy thermalized quasi-condensation is accessible by changing the sputtering gas pressure in this apparatus. During deposition, the substrate typically lies unrestrained on a stainless-steel heat-resistant metal plate and is heated mainly by radiation heat transfer from a metal resistive heater. The typical substrate temperature during deposition of the films depends on the material to be deposited and for YBCO is $\approx$ 800 °C while the heater temperature is $\approx$ 920 °C.
In order to prepare multilayer heterostructures it is important that all layers should be of sufficiently homogeneous thickness. In the case of sputtering, the trivial rule is that the size of the target should significantly exceed the size of the substrate. Films deposited at an oxygen pressure $\approx$ 3.5 mbar from 50-mm magnetron targets were only about 2.5 % thinner at the corners of square 10-mm substrates and only 15 % thinner at the perimeter of round wafers of diameter 30 mm compared to the film thickness in the middle of the substrates. At the moment, the wafers up to 30 mm in diameter can be covered with heterostructures of

such homogeneous thickness by the high oxygen pressure sputtering technique. Larger area epitaxial metal-oxide heterostructures can be produced with a proper scanning apparatus or larger magnetron targets (Faley & Poppe, 2010).
Magnetron sputtering can be used at high oxygen pressure, but it has characteristic features in conditions of very short mean free path of electrons at pressures above 1 mbar. Large targets require magnetic fields in order to stabilise the sputtering plasma and the optimum distance between magnetic poles is typically in the range between 1 mm and 5 mm (Faley & Poppe, 2010). One of the magnetic poles can be replaced by a high-μ yoke made, for example, of iron (see Figure 1a). The magnetic field of the Sm_2Co_{17} magnets in such modified target holders additionally excites the sputtering plasma at positions away from the middle and perimeter of the target where otherwise the plasma tends to localize. This optimized arrangement of the Sm_2Co_{17} magnets in the magnetron target holder is mainly intended to stabilize the plasma. Figure 1b shows an example of magnetron sputtering from a 50-mm YBCO target demonstrating an approximately 3 mm wide ring of the most intensive plasma region observed at 3 mbar pressure of the pure oxygen sputtering atmosphere.

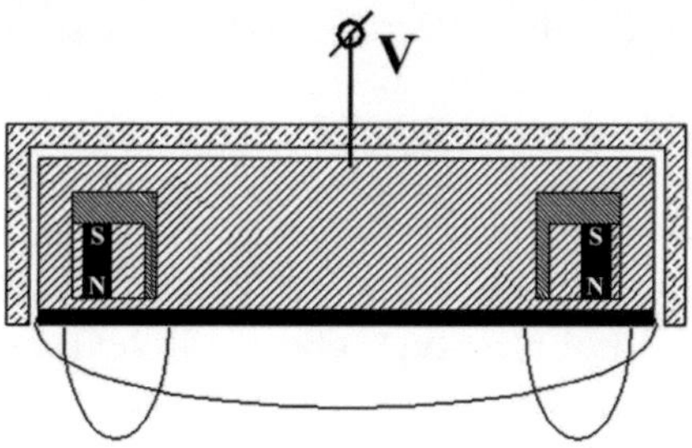

(a)

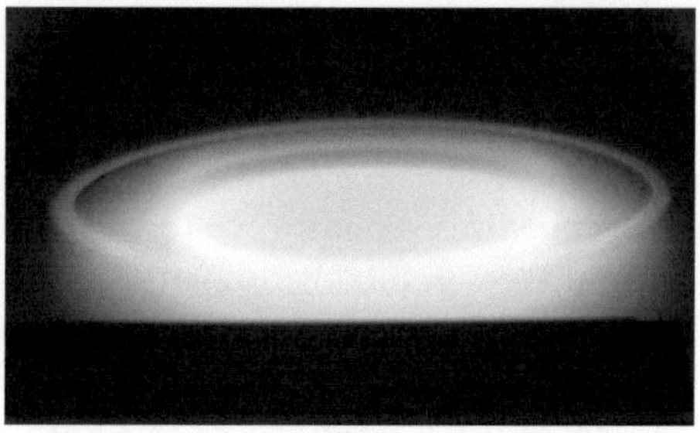

(b)

Fig. 1. High oxygen pressure magnetron sputtering: (a) sketch and (b) photograph of plasma and target holder with a YBCO target and a MACOR insulator (Faley & Poppe, 2010).

The high oxygen pressure sputtering technique is suitable for the deposition of high-quality epitaxial films of all metal-oxide materials required for the production of multilayer high-T_c DC SQUID sensors. No organic material is present in the vacuum chamber of the sputtering machine. We metallize the rear of the targets with an approximately 100 μm thick silver layer, which is partially diffused into the targets at 850 °C to a depth of about 30 μm, and we bond them to the Cu holder by soldering with AgSn solder. The diffusion coefficient of Ag into bulk YBCO ceramic samples is $D_{Ag} \approx 4.5 \times 10^{-9}$ cm²/s at 850 °C (Dogan, 2005).
The base pressure in the deposition chamber for YBCO was about $2{\cdot}10^{-7}$ mbar while sputtering of YBCO was performed at ≈ 3.5 mbar pressure of pure (99.999%) oxygen. The DC sputtering technique is usually used for deposition from sufficiently conducting targets, while in the case of more insulating targets deposition is carried out by the RF sputtering technique. The typical deposition rate obtained with the DC sputtering technique was about 90 nm/hour while in the case of RF sputtering it was about 20 nm/hour.
The surface morphology of the films is crucial for the preparation of multilayer structures. Depending on the deposited material, the epitaxial growth of metal-oxide films proceeds in the following three modes: Frank-Van der Merwe growth (layer-by-layer); Volmer-Weber growth (3-D nucleation); or Stranski-Krastanov growth (mixed mode). The YBCO films grow in the Stranski-Krastanov growth mode: initial layer-by-layer growth changes to spiral

growth for films thicker than about 20 nm (Dam et al., 2002). The growth spirals on YBCO films have an average height of about 30 nm and their in-plane size strongly depends on the deposition temperature. The optimum substrate temperature is ≈ 100 °C higher during high oxygen pressure sputtering compared to that in the case of the PLD deposition method. This explains the width of the growth spirals of up to ≈ 900 nm observed on the surface of the YBCO films deposited by the high oxygen pressure sputtering technique (Faley et al., 2006b) compared to the ≈ 200 nm wide growth spirals on the YBCO films deposited by PLD (Dam et al., 1996). The morphology of the YBCO films is one of the factors contributing to the spread of the parameters of high-T_c bicrystal Josephson junctions with misorientation angles below 24 deg and thicknesses < 60 nm as well as to the quality of the insulation layers.

3. Materials used for the high-T_c heterostructures

For the most efficient coupling of magnetic fields to a SQUID loop, a multilayer flux transformer with at least two high-T_c superconducting epitaxial, usually, YBCO layers separated by an insulator layer is required. The technological and structural compatibility of the materials involved is an important precondition for the heteroepitaxial growth of the multilayer structures of the high-T_c SQUIDs and flux transformers. The oxygenation of the bottom YBCO films is only possible if there is sufficient mobility of oxygen ions in the insulating layer. An epitaxial buffering of substrates intended for the deposition of the high-T_c heterostructures can improve further device properties.

The non-superconducting material most compatible technologically with YBCO is $PrBa_2Cu_3O_{7-x}$ (PBCO), which has thermally activated hopping-type electrical conductivity (Fisher et al., 1994) and the perovskite-derived crystal structure is isomorphic to that of YBCO. The lattice constants of PBCO are a = 3.873 Å, b = 3.915 Å, c = 11.67 Å, which are very close to those of YBCO: a = 3.823 Å, b = 3.88 Å, c = 11.68 Å. Due to the similarity of the crystal structures of PBCO and YBCO a very low charge carrier scattering and negligible contact resistance were observed for the interfaces between the films of PBCO and YBCO (Faley et al., 1993). The PBCO films were successfully used for buffer layers, tunnel barriers, and for non-superconducting insulators in the SQUID-related heterostructures with YBCO.

It was observed that the electrical insulation in the YBCO-PBCO-YBCO heterostructures could be significantly improved by passivation of the bottom YBCO layer by a brief application of ion beam etching (Faley et al., 1997a). The reason for the increased contact resistance was a cation-disordered cubic phase of YBCO that appeared after the amorphization of the surface layer of YBCO by the ion bombardment followed by the recrystallization of this surface layer at high temperatures during the deposition of the top film (Jia et al., 1995). A further improvement in insulator resistance was achieved by implementation of a PBCO-STO electrically insulating heterostructure (Faley et al., 2010). The 50 nm PBCO film served as a buffer layer followed by the 300-nm thick STO insulator film deposited *in-situ*. The PBCO film improved epitaxial growth of the STO film over the substrate and the bottom YBCO film as well as the morphology and resistance of the insulator layer in the direction normal to the substrate surface. The resistance of the PBCO film along the substrate surface contributed to dumping of microwave resonances in the input coil of the multilayer flux transformer.

The best structural and superconducting parameters of YBCO films are typically obtained on STO substrates. Epitaxial STO films have also provided an excellent template for the epitaxial growth of the top YBCO film of the top superconducting layer in the thin film superconducting flux transformers. Figure 2 shows the high-resolution transmission electron microscopy (HRTEM) image of the interface region between epitaxial STO and YBCO films produced by the high oxygen pressure sputtering technique and demonstrates the high-quality microstructure of these films.

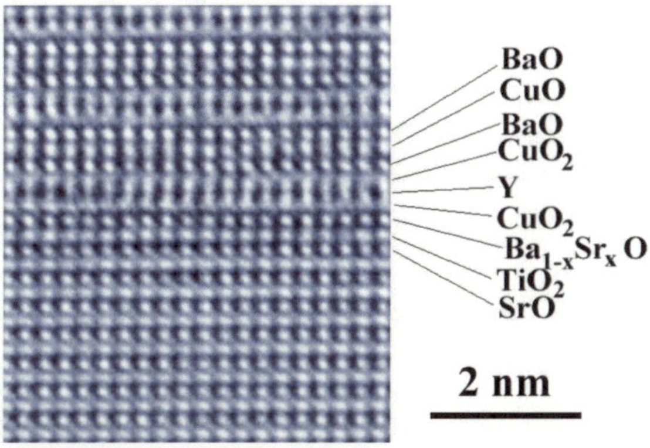

Fig. 2. Cross-sectional HRTEM image of the interface between YBCO and STO films obtained in the [110] direction (Faley et al., 2008).

Coverage of the bottom YBCO layer by the epitaxial STO films in the YBCO-STO-YBCO heterostructures does not degrade the superconducting properties of the bottom YBCO film. STO enables sufficient diffusivity of oxygen ions required for the full oxygenation of the YBCO films at, typically, about 500 °C. The diffusion coefficient of oxygen ions in single crystal STO is known to be $D_O = 5.2 \cdot 10^{-6} \cdot \exp(-11349/T)$ cm²/sec in the temperature range between 850 °C and 1500 °C (Paladino, 1965). Assuming this dependence can be extended to lower temperatures and that the diffusivity of oxygen in STO films is similar to that in single crystal STO samples, the estimated time required to oxygenate a YBCO film covered by a 0.5 μm thick STO film is about 1.5 hour at 500 °C substrate temperature. Indeed, our empirically obtained optimum oxygenation time for the YBCO-STO-YBCO heterostructures used in the high-T_c superconducting flux transformers is about 2 hours. The input coil included a 200 nm bottom YBCO film, which was covered by the approximately 400 nm PBCO-STO insulator heterostructure and 600 –1000 nm top YBCO film. A 100 nm thick silver film served to protect the top YBCO layer during structuring with AZ-photoresist.
Another useful substrate material for SQUIDs is MgO, which has a thermal expansion coefficient similar to that of YBCO (~ 14 x 10^{-6}) (see Table 1). The difference in the thermal expansion coefficients of the oxide materials such as STO, $LaAlO_3$ (LAO), $NdGaO_3$ (NGO), Al_2O_3, and YSZ often used for the substrates and films leads to a very strong tensile strain in the YBCO films degrading their superconducting properties and can even crack the films when their thickness exceeds some critical value. Much thicker multilayer high-T_c thin film structures with smaller capacitance can be produced on MgO substrates. An additional advantage of MgO is that it has a relatively low dielectric constant $\varepsilon \approx 9$ and low losses tan $\delta \approx 3.3 \cdot 10^{-7}$. It is one of the traditional materials used in microwave electronics. The low dielectric constant of MgO leads to a smaller parasitic capacitance through the substrate across the inductance of the DC SQUID loop compared to the DC SQUIDs on STO substrates. This leads to smaller voltage swings, but also lower white noise of high-T_c DC SQUIDs on MgO substrates compared to those on STO substrates (Enpuku et al, 1996).

	Linear thermal expansion (in 10^{-6}/K)	Crystal structure	Lattice constant (Å)	Dielectric constant
MgO	~ 14	cubic, rock-salt	4.21	~ 10
$BaZrO_3$	~ 7	cubic, perovskite	4.19	~ 20
$SrTiO_3$	~ 11	cubic, perovskite	3.91	~ 270
$NdGaO_3$	~ 6	orthorhombic, perovskite	3.85	~ 20
$LaAlO_3$	~ 9	rhombohedral, perovskite	3.82	~ 24
$YBa_2Cu_3O_{7-x}$	~ 13.5	orthorhombic, perovskite	3.85	~ 5

Table 1. Selected properties of materials for substrates and buffer layers used for deposition of YBCO.

Unbuffered MgO substrates demonstrate degradation of the hygroscopic surface in air and have a large lattice mismatch of ≈ 9 % with YBCO and a crystal structure that differs from YBCO. These features usually lead to appearance of in-plane 45° misoriented grains in the YBCO films deposited on MgO substrates. The average critical current density of the YBCO films is in this case usually significantly suppressed at the boundaries between the grains and the magnetic noise of the YBCO films is drastically increased. Single-layer buffers such as $BaZrO_3$ (BZO) or STO films only slightly improved this situation. At least two buffer layers are required to deposit low-noise YBCO films on MgO: the first one should provide the epitaxial growth of films with perovskite structure on the rock-salt structure of MgO, while the second buffer layer should match the lattice constants. STO and BZO films are technologically compatible with YBCO and have the required structural properties.
An epitaxial perovskite double-layer STO/BZO buffer on MgO substrates has been developed for the deposition of low-noise and crack-free YBCO films (Faley et al., 2006a). This buffer also protects the hygroscopic surface of the MgO substrates against degradation in air and/or during the lithographic procedures. Figure 3 shows a cross-sectional HRTEM image of a BZO-STO-YBCO heterostructure deposited on a MgO (100) substrate.

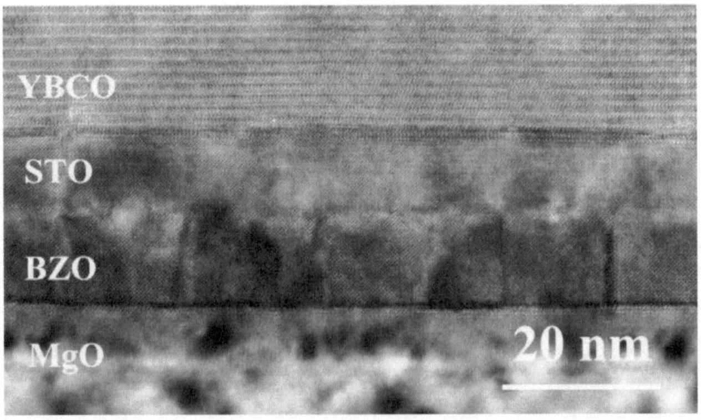

Fig. 3. Cross-sectional HRTEM image of a BZO-STO-YBCO heterostructure deposited on an MgO (100) substrate (Faley et al., 2006a).

It was observed that the antiphase boundaries (APB), which appeared at the BZO/MgO interface and spread through the BZO layer, usually disappeared at the STO/BZO interface (Mi et al., 2006). The STO layer initially grows with the lattice constant expanded to the lattice constant of BZO ≈ 4.19 Å. However, just a after few unit cells from the STO/BZO interface the lattice constant of STO already relaxed to its bulk value ≈ 3.91 Å (see Figure 4).

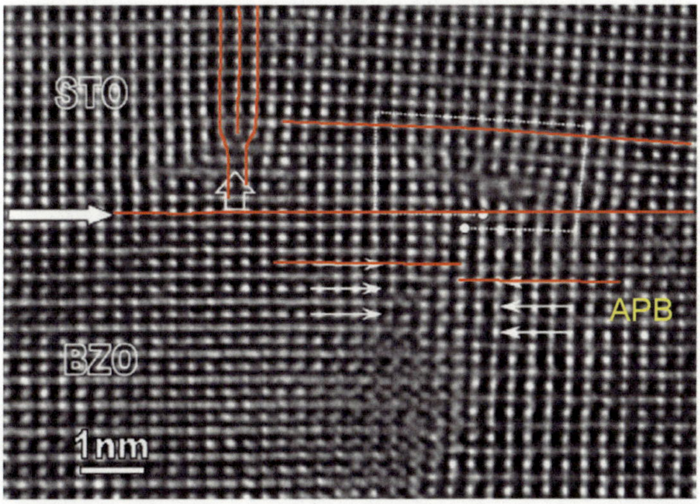

Fig. 4. Cross-sectional HRTEM image of an interface region for BZO and STO films deposited on a MgO (100) substrate (Mi et al., 2007).

Thus, at the YBCO/STO interface the lattice constant and microstructural quality of the STO layer is similar to that of the single-crystal STO substrate, but the overall thermal expansion coefficient is still determined by the 1 mm thick MgO substrate. The YBCO films deposited by high oxygen pressure sputtering technique naturally contain lattice-coherent non-superconducting Y_2O_3 nanoparticles, which are nearly spherical with a diameter of ~20 nm and are homogeneously distributed with a separation of ~30 nm (Faley et al., 2006b) and provide a strong 3D pinning of the Abrikosov vortices leading to a high critical current density J_c and a low magnetic noise in the films (Kim et al., 2007). Even 5-μm-thick YBCO films on the buffered MgO substrates do not display cracks and demonstrate a critical current density ≈ 3.5 MA/cm^2 at 77 K (Faley et al., 2008). The 1 cm wide films have an estimated total critical current of ≈ 1.7 kA at 77 K, which is about 17 times greater than the critical current of the present day 2nd-generation high-T_c superconducting tapes of similar width. Such high and homogeneous critical current densities of the high-T_c superconducting films are beneficial for production of the low-noise SQUID sensors, for high-Q microwave resonators and filters in communication technologies as well as for high-T_c superconducting tapes intended for the generation and transport of electrical power. The YBCO films deposited on the buffered MgO substrates demonstrated conductivity proportional to the film thickness for up to about 5 μm thick films (Faley et al., 2006a). The specific conductivity of YBCO films on other substrates such as STO, LAO, NGO, Al_2O_3, and YSZ was saturated or even dropped when the film thickness exceeded the critical values and cracks appeared in the YBCO films.

4. Patterning techniques for epitaxial metal-oxide multilayers

In the case of the epitaxial metal-oxide multilayers for high-T_c SQUIDs it is essential to avoid grain boundaries in the superconducting films because the thermally-activated hopping of flux vortices and fluctuations of superconducting current at the grain boundaries often act as sources of flicker noise in the SQUIDs. Patterning of bottom layers should leave chemically clean and bevelled edges of the structures for the homogeneous epitaxial growth of top superconducting layers over the edges. Such structuring can be achieved by non-aqueous chemical etching as well as by the ion beam etching methods briefly described below.

Chemical etching in a Br-ethanol solution in combination with a deep-UV photolithography of PMMA photoresist was used for the patterning of YBCO-PBCO heterostructures to prepare the high-T_c Josephson junctions, crossovers, and interconnects (Faley et al., 1993). It was observed that the chemical etching of c-axis-oriented YBCO and PBCO films through a mask of PMMA photoresist is very anisotropic: it is much faster along the ab-planes than in the c-direction of the films. This causes abnormally large undercutting, which results in very gently sloping edges of the structures (see Figures 5 and 6). The angle α of slope of the edge is about 3 degree with respect to the substrate plane. This angle can be increased by extending the etching time or in combination with ion beam etching.

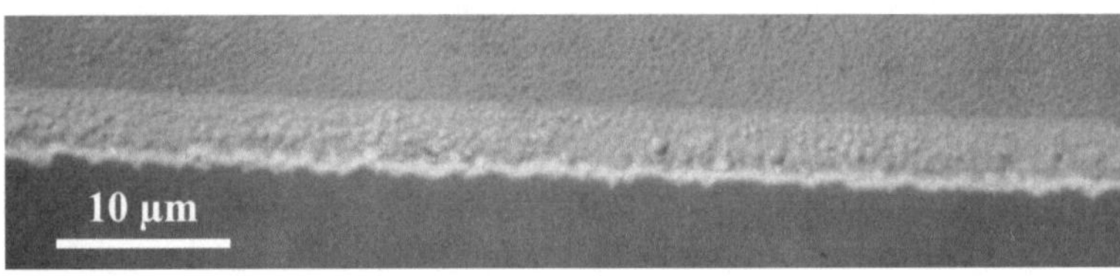

Fig. 5. Optical image of a 500-nm thick YBCO-PBCO bilayer etched through a mask of PMMA photoresist by the Br-ethanol solution. The upper part of the picture shows the film, while the lower part shows the STO substrate. The bright horizontal stripe in the middle of the picture is the chemically prepared edge.

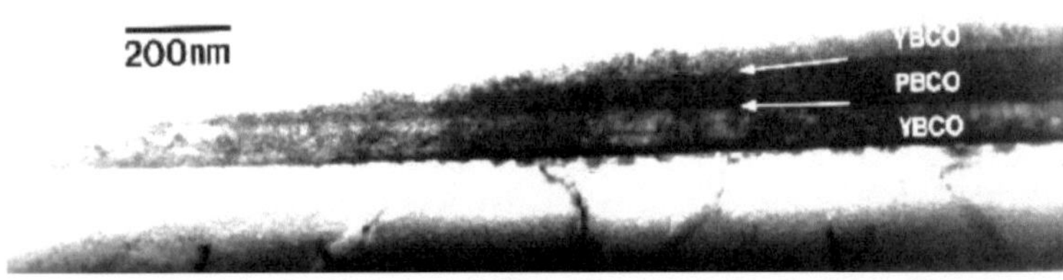

Fig. 6. A low-magnification TEM picture. This picture gives an overview of a cross-section of a YBCO-PBCO-YBCO edge structure, containing the bottom YBCO film, the insulating PBCO layer, the PBCO barrier and the top YBCO film (Faley et al., 1993).

The main advantage of the non-aqueous chemical etching in Br-ethanol solution is that the edge area is not contaminated by substrate material and shows negligible structural damage at the surface layer. Moreover, this solution does not change the local stoichiometry at the surface and, in contrast to the ion beam etching, it does not even affect the oxidation state of the copper (Vasquez et al., 1989). Bromides YBr, BaBr or CuBr are soluble in ethanol and, therefore, the surface of the edge appears to be very clean after etching followed by rinsing in ethanol. The chemical etching in Br-ethanol solution was used for the preparation of the ramp-type high-T_c Josephson junctions and the bottom layers, YBCO and PBCO, in the multilayer flux transformers with PBCO insulation layer (Faley et al., 2001). If an STO film was used for the insulation between the YBCO films, the bottom YBCO layer can also be etched by the chemical etching. The lower superconducting layer used for the return lead of the input coil and the pick-up loop does not require high precision in structuring and it was patterned with deep-UV lithography using a PMMA-photoresist and Br-ethanol chemical etching.

Ion beam etching enables sub-micrometer precision in structuring the films through masks of AZ-type (mainly AZ5214E and AZ MIR701) photoresists. Upper superconducting layers in the ramp junctions and flux transformers contain μm-size structures and required conventional patterning with AZ photoresist and ion beam etching. Ion beam etching can be also used for structuring the bottom YBCO layer and insulation layer under condition of sufficiently low-angle edges of the photoresist mask. Proper cleaning with microstructural restoration of the edge surface should follow the etching. Bevelling of the AZ-photoresist edges down to an angle below 20 degrees relative to the substrate plane can be realized by backing-out of the photoresist at 130 °C (David et al., 1994).

Cleaning and restoration of the edge surface after etching is more difficult in the case of the ion beam etching as compared to Br-ethanol chemical etching. Rinsing and mechanical polishing in acetone and methanol followed by annealing in the presence of oxygen plasma can remove the photoresist, including carbonized parts of photoresist near the edges, as well as the amorphous materials redeposited on the edges of the photoresist structures during ion beam etching. Annealing in the presence of oxygen plasma also leads to recrystallization of the surface of edges of the etched film, which partially recovers its microstructural and electron transport properties. A high quality of the crossovers and vias in the multilayer multiturn coil of the flux transformer is essential to obtain high values of the induced superconducting current. Due to the damage-free interfaces and gently sloping edges produced by Br-ethanol etching we achieved critical currents for the flux transformers of about 100 mA at 77 K. The observed 60 μT peak-to-peak dynamic range of the magnetometer having 8-mm pick-up loop ($L_{pu} \approx 20$ nH) is limited mainly by this critical current of the flux transformer.

We use both patterning techniques - non-aqueous Br-ethanol chemical etching and ion beam etching - for the preparation of sensitive high-T_c multilayer DC SQUID sensors with reduced low frequency noise, which are described in the following sections.

5. Multilayer high-T_c DC SQUID magnetometers

In this section, the review of multilayer high-T_c DC SQUID flip-chip magnetometers will include a short introduction to the principle of operation of DC SQUIDs, a description of their noise properties and basic components: high-T_c Josephson junctions, superconducting multilayer flux transformers with multiturn input coil, and capsulation. The reproducibility of the high-T_c Josephson junctions is especially important in the case of implementation of the high-T_c DC SQUID arrays. The vacuum-tight encapsulation of the sensors is a prerequisite for their long-term stability, easier handling, and for the reduction of low-frequency noise by removing the magnetic flux trapped in the superconducting films.

5.1 DC SQUIDs – principle of operation

SQUIDs consist of a loop of superconductor interrupted by one or two Josephson junctions. The operation of SQUIDs is based on the dependence of phase shift $\Delta\varphi$ of quantum wave-functions Ψ of Cooper pairs on magnetic flux Φ passing through the SQUID loop. This dependence is caused by the fundamental dependence of the canonical momentum $\vec{p} = m\vec{v} + q\vec{A}$ and, consequently, de Broglie wavelength $\lambda = h / |\vec{p}|$ and wave vector $\vec{k} = \vec{p} / \hbar$

of charged particles on magnetic vector potential $\vec{A}$. The superconducting wave function $\Psi = |\Psi|\exp(i\varphi)$ has the spatial variation of the phase $\varphi = \varphi(\vec{r},t)$ due to the presence of the vector potential $\vec{A}$ of the magnetic field threading through the SQUID loop. The phase difference $\delta\varphi_{1-2}$ of the wave function at positions x1 and x2 is $\delta\phi_{1-2} = \int_{x1}^{x2} \vec{k}d\vec{l} = \frac{q}{\hbar}\int_{x1}^{x2} \vec{A}d\vec{l} = \frac{2\pi}{\Phi_0}\int_{x1}^{x2} \vec{A}d\vec{l}$, where $\Phi_0 = h/q = h/2e \approx 2.07\ 10^{-15}$ T·m^2 is the magnetic flux quantum.

The superconducting wave function $\Psi = |\Psi|\exp(i\varphi)$ is continuous in the superconductor up to the Josephson junctions. The requirement that the superconducting wave function Ψ have a single value everywhere is an important boundary condition for SQUID operation. At the Josephson junctions, the jump of phase $\Delta\varphi$ of the wave functions in individual superconducting electrodes is detected according to the Josephson current-phase relationship $I(\varphi) = I_c\sin(\Delta\varphi)$. This quantum interference leads to a periodic dependence of the output voltage of SQUIDs on applied magnetic flux Φ threading through the SQUID loop thus enabling the SQUIDs to convert tiny changes in magnetic flux Φ into measurable voltage signals.

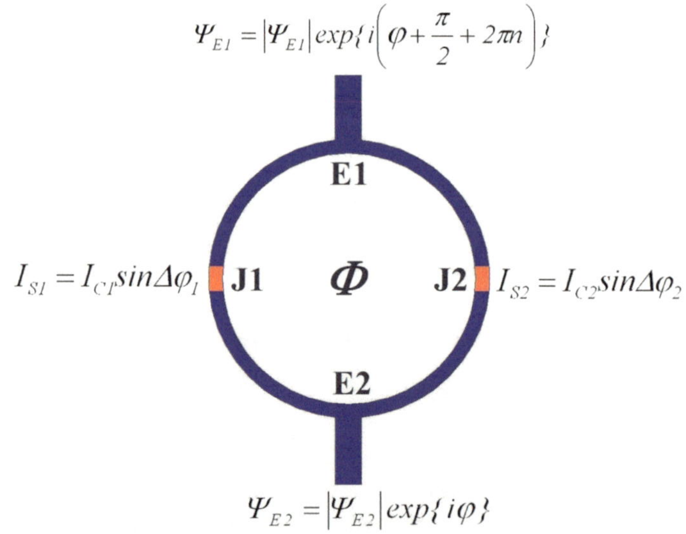

Fig. 7. Schematic representation of the DC SQUID loop with values of the superconducting wave-function Ψ, critical currents I_{S1} and I_{S2} of the Josephson junctions J1 and J2, respectively, and the magnetic flux Φ penetrating through the SQUID loop.

Direct-current SQUIDs (DC SQUIDs) consist of a loop of two superconducting electrodes E1 and E2 connected together by two Josephson junctions denoted as J1 and J2 in Figure 7. DC SQUIDs are sensitive flux-to-voltage transducers: when a flux Φ of the magnetic field penetrates the DC SQUID loop, the spatial variations of the phase of the wave function Ψ of Cooper pairs in superconducting electrodes appears. These lead to the phase shifts $\Delta\varphi_2$ and $\Delta\varphi_2$ between the wave functions in the superconducting electrodes at the Josephson junctions and, consequently, to a voltage signal on the DC SQUIDs.

The operation of DC SQUIDs can be explained most clearly in the first approximation of the zero-voltage state, for a small and symmetric DC SQUID loop. In the zero voltage state of the Josephson junctions the phase φ of the wave function of Cooper pairs does not depend on time. Without magnetic flux threading of the DC SQUID loop (Φ = 0) the maximal superconducting current $I = I_{S1}+I_{S2} = I_{C1}+I_{C2}$ is achieved at the phase difference $\Delta\varphi_1 = \Delta\varphi_2 = \pi/2+ 2\pi n$ between the phases of the wave functions in electrodes at points E1 and E2 because only in this case $I_{S1}=I_{C1}$ and $I_{S2}=I_{C2}$ (see Figure 7).

Magnetic flux $\Phi = \oint \vec{A}d\vec{l} \neq 0$ changes the phase shifts $\Delta\phi_1^* = \Delta\phi_1 + \delta\phi_1 = \frac{\pi}{2} + \delta\phi_1 + 2\pi n$ and $\Delta\phi_2^* = \Delta\phi_2 + \delta\phi_2 = \frac{\pi}{2} + \delta\phi_2 + 2\pi n$ due to the non-zero integral of the vector potential $\vec{A}$ along the superconducting parts of the SQUID loop. The phase shifts $\Delta\varphi_1$ and $\Delta\varphi_2$ are $\delta\varphi_1 = \frac{2\pi}{\Phi_0}\int_{E2}^{J1} \vec{A}d\vec{l} - \frac{2\pi}{\Phi_0}\int_{E1}^{J1} \vec{A}d\vec{l} = \frac{\pi\Phi}{\Phi_0}$ and $\delta\varphi_2 = \frac{2\pi}{\Phi_0}\int_{E2}^{J2} \vec{A}d\vec{l} - \frac{2\pi}{\Phi_0}\int_{E1}^{J2} \vec{A}d\vec{l} = -\frac{\pi\Phi}{\Phi_0}$, in the case of geometrically symmetric SQUID loops like the one shown in Figure 7. Thus, in the present of a magnetic field, the phase differences at the Josephson junctions are: $\Delta\varphi_1^* = \frac{\pi}{2} + 2\pi n + \frac{\pi\Phi}{\Phi_0}$ and $\Delta\varphi_2^* = \frac{\pi}{2} + 2\pi n - \frac{\pi\Phi}{\Phi_0}$. The total maximal superconducting current through the SQUID is thus $I = I_{S1} + I_{S2} = I_{C1}\sin(\Delta\varphi_1^*) + I_{C2}\sin(\Delta\varphi_2^*)$. In the case of Josephson junctions with similar critical currents $I_{C1} = I_{C2} = I_C$ the total current through the DC SQUID is:

$$I = I_C\left[\sin(\Delta\varphi_1^*) + \sin(\Delta\varphi_2^*)\right] = 2I_C\sin(\frac{\pi}{2} + 2\pi n)\cos\left(\frac{\pi\Phi}{\Phi_0}\right) = 2I_C\cos\left(\frac{\pi\Phi}{\Phi_0}\right) \quad (1)$$

at $\Phi < \Phi_0/2$. A further increase of flux changes the phase difference between the wave functions at points E1 and E2 from $\pi/2$ to $-\pi/2$ (in both cases I=0 at $\Phi = \Phi_0/2$) so that the maximal superconducting current through such DC SQUID I_{max} is always positive and is a periodic function of Φ with period Φ_0:

$$I_{max} = 2I_C\left|\cos\left(\frac{\pi\Phi}{\Phi_0}\right)\right| \quad (2)$$

In the dissipative regime (at bias currents $I_B > 2I_C$) there are periodic series of pulses (Josephson oscillations) of voltage U(Φ,t) across the DC SQUID. Averaging of U(Φ,t) over the period τ of the Josephson oscillations results in the dc voltage V across the DC SQUID (Tinkham, 1996):

$$V = \frac{1}{\tau}\int_0^\tau U(\Phi,t)dt \approx \frac{R_N I_B}{2}\sqrt{1-\left(\frac{2I_C}{I_B}\cos\frac{\pi\Phi}{\Phi_0}\right)^2} \quad (3)$$

where R_N is the resistance of the individual Josephson junction in the DC SQUID. The dc voltage V across the DC SQUID is a periodic function of the magnetic flux Φ through the SQUID loop.

5.2 Josephson junctions for the high-T_c DC SQUIDs

The Josephson junctions in SQUIDs transform the phase changes of the superconducting wave functions into measurable voltages owing to the Josephson current-phase relationship $I(\varphi) = I_c\sin(\Delta\varphi)$. A Josephson junction is made by sandwiching a thin layer of a non-superconducting material between two layers of superconducting material(s). With a sufficiently thin barrier, the phase of the electron wave-function in one superconductor maintains a fixed relationship with the phase of the wave-function in another superconductor. In this way, the superconductors preserve their long-range order across the insulating barrier in the Josephson junctions.

Mainly ramp-type, step edge and bicrystal grain boundary high-T_c Josephson junctions are used for high-T_c SQUIDs. The best reproducibility of the junction parameters was obtained for the ramp-type and bicrystal grain boundary high-T_c Josephson junctions shown schematically in Figure 8.

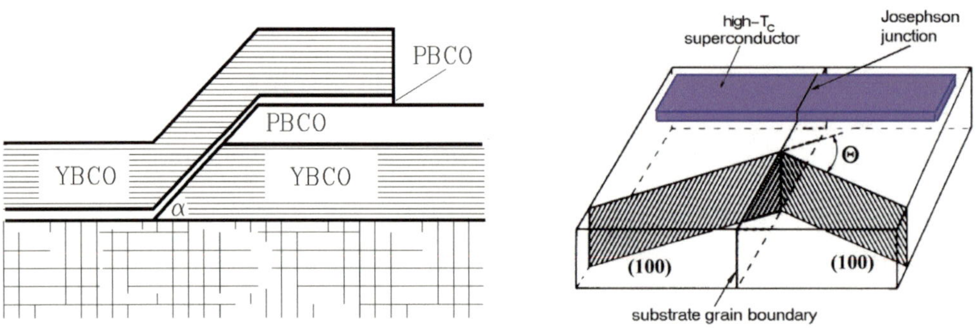

(a) (b)

Fig. 8. Schematics of the ramp-type (a) and bicrystal (b) high-T_c Josephson junctions.

Ramp-type Josephson junctions contain two superconducting layers separated by a barrier layer deposited on the edge of the bottom superconducting layer (see the Figure 8a). Electron transport on the top of the bottom layer is prevented by a thick insulation layer. The barrier material and thickness can be chosen for optimum performance in particular applications. The ramp junctions on the chemically etched edges with PBCO films as the barrier and insulation layers have proven parameters that are sufficient for many applications of DC SQUIDs (Faley et al., 1995 and 1997b). The ramp junctions have the advantage that relatively cheap single crystal substrates can be used. On the other hand, compared to the bicrystal junctions, production of SQUIDs with ramp junctions is more complicated and for a similar critical current I_c their normal state resistance, R_N, is about 3 times smaller leading to smaller voltage swings of the DC SQUID. The easier production of junctions on bicrystal substrates and the smaller contribution of the noise of SQUID control electronics to the total noise of the measurement system have led to the preferred utilization of bicrystal junctions in high-T_c DC SQUIDs.

The bicrystal junctions optimized for operation in high-T_c DC SQUID's typically have a width ≈ 1 μm, resistance $R_N \approx 6$ ohm, and a critical current $I_c \approx 25$ μA at 77 K. Dependence of the critical current density of bicrystal Josephson junctions J_c on the misorientation angle Θ can be approximated at 77 K by the following expression (Ivanov et al., 1991):

$$J_c\left(A/cm^2\right) \cong 4.16 \cdot 10^6 \cdot \exp(-\frac{\theta(\text{deg.})}{4.2}) \tag{4}$$

As was revealed by TEM studies, the effective thickness L of the distorted layer at the bicrystal grain boundary increases approximately linearly with Θ and it was suggested that the exponential decrease of $J_c(\Theta)$ is associated with this increase of the distorted layer. Typical critical current density $J_c \approx 10$ kA/cm² in the bicrystal high-T_c Josephson junctions used for DC SQUIDs operating at 77 K.

For high-T_c bicrystal Josephson junctions the resistance R_N is determined by both direct tunnelling and the resonance tunnelling components of charge carrier transport through the grain boundary (Halbritter, 1985). The total conductance of the junction with the cross-section area $\sigma = 0.2$ μm² can be approximated by the following expression (Minotani et al., 1998):

$$G = 1/R_N \cong 2150 I_c + 11 I_c^{1/2} \tag{5}$$

where the critical current $I_c = J_c\sigma$ is in amperes at 77 K while the resistance R_N is in ohms for the bicrystal junctions width ≈ 2 μm and the YBCO film thickness ≈ 100 nm. The second term in (5) representing the resonance tunnelling prevails at the bicrystal misorientation angles above about 24 degree. The I_cR_N product is ≈ 200 μV at the bicrystal in-plane misorientation angle 24 degree and temperature 77 K:

$$I_c R_N \cong 1/(2150 + 11 I_c^{-1/2}) \tag{6}$$

The I_cR_N product decreases with increase of the misorientation angle and corresponding decrease of the critical current density of the junction. This decrease is especially strong for misorientation angles above 24 degree, where it was found that $I_cR_N \propto \sqrt{J_c}$ (Gross et al., 1997). The critical current density J_c of symmetric 24 degree bicrystal Josephson junctions increases nearly linearly with the temperature dependence $J_c(T) \approx 10^4[4.1+6.4(60-T)/40]$ A/cm² in the range of operating temperatures T = (10 - 80) K. Noise parameter $\Gamma = 2\pi k_B T/I_c\Phi_0 = I_{th}/I_c$ for a particular junction depends strongly on the temperature: for the junctions having $I_c \approx 20$ μA at 77 K the noise parameter Γ is ≈ 0.16 at 77 K, Γ ≈ 0.05 at 63 K, and Γ ≈ 0.005 at 20 K. The Stewart-McCumber parameter of the Josephson junctions $\beta_C = 2\pi I_C R_N^2 C/\Phi_0$ increases with reduced temperature and increased critical current from $\beta_C \approx 0.12$ at 77 K to ≈ 0.34 at 63 K and ≈ 1 at 20 K. Below ≈ 20 K the Josephson junctions become hysteretic. The capacitance C of the bicrystal grain boundary Josephson junction $C = \sigma\varepsilon\varepsilon_0/t \approx 25$ fF for the junction width w = 2 μm and YBCO film thickness d = 100 nm (the junction area $\sigma = w\cdot d \approx 2$ 10^{-9} cm²). The grain boundary of the YBCO bicrystal junctions behaves as if it were mostly dielectric with an average barrier thickness t ≈ 2 nm (Winkler et al., 1994) and dielectric constant ε ≈ 28 for the grain boundary region in YBCO (Navacerrada, 2008).

Both the noise parameter Γ and the Stewart-McCumber parameter β_C contribute to the voltage noise of the Josephson junctions (Voss, 1981) and, consequently, to the magnetic field resolution of the DC SQUID magnetometers. Reduction of temperature from the standard operating temperature of high-T_c SQUIDs 77 K to, for example, the triple point of nitrogen 63 K leads to an increase of I_c and a reduction of Γ, but also to increase of voltage noise due to the increase of β_C. An external resistive shunting of the junctions helps to reduce β_C and, consequently, voltage noise at lower temperatures and to avoid transition of the Josephson junctions to the hysteretic mode.

5.3 Performance and noise of high-T_c DC SQUIDs

The average dc voltage V across the DC SQUID is a periodic function of magnetic flux Φ with the period equal to the magnetic flux quantum Φ_0. At final temperatures T > 0, the I(V) characteristics of Josephson junctions are rounded by thermal noise and the bias current I_B corresponding to the maximal voltage response $|\partial V/\partial \Phi|$ is smaller than the total critical current of symmetric DC SQUID $2I_c$ (Drung et al., 1996):

$$I_c \approx \frac{I_B}{2} + \frac{k_B T}{\Phi_0}\left(1 + \sqrt{1 + \frac{I_B \Phi_0}{k_B T}}\right) \tag{7}$$

The optimum bias current I_B of high-T_c DC SQUIDs is (40 ± 30) μA at 77 K. At bias currents I_B > 70 μA the voltage response $|\partial V/\partial \Phi|$ is reduced due to circulation of shielding currents in the DC SQUID loop at the typical high-T_c DC SQUID inductances L_S of about 100 pH. The inductance L_S of the DC SQUID loop reduces the voltage response to the magnetic flux in two ways: first, the induced currents shield significantly the magnetic flux through the SQUID loop at $L_S > \Phi_0/2I_c$ and, second, due to the rounding of the current-voltage characteristics by thermal flux noise $\delta\Phi = \sqrt{k_B T L_S}$. At bias currents I_B < 10 μA the critical currents I_c < 5 μA become comparable to the thermal current $I_{th} = 2\pi k_B T/\Phi_0 \sim 3.3$ μA leading to reduction of the voltage response $|\partial V/\partial \Phi|$ of the SQUIDs. Taking into account the inductance of the DC SQUID loop L_S and operating temperature T, the maximum voltage response $|\partial V/\partial \Phi|$ of DC SQUID under optimum operation conditions $\Phi \approx (2n+1)\Phi_0/4$, where n = 0, ±1..., is determined by the expression (Enpuku et al., 1995):

$$\left|\frac{\partial V}{\partial \Phi}\right| \approx \frac{4}{\Phi_0} \cdot \frac{I_c R_N}{1 + \frac{2I_c L_S}{\Phi_0}} \cdot \exp(-3.5\pi^2 \frac{k_B T L_S}{\Phi_0^2}) \tag{8}$$

Taking into account the voltage noise of the Josephson junctions, the magnetic flux noise of the high-Tc DC SQUID can be estimated according to the following expression:

$$S_\Phi = S_V\left(\frac{\partial V}{\partial \Phi}\right)^{-2} \approx 8k_B T\left[\frac{R_D^2}{R_N} + \frac{L_S^2}{4R_N}\left(\frac{\partial V}{\partial \Phi}\right)^2\right]\left[1 + \frac{1}{2}\left(\frac{2I_c}{I}\right)^2\right]\left(\frac{\partial V}{\partial \Phi}\right)^{-2} \propto \frac{L_S^2}{R_N} \tag{9}$$

for L_S > 40 pH and neglecting the noise of the preamplifier of the control electronics. The dynamic resistance of the symmetric DC SQUIDs is $R_D = \partial V/\partial I \approx R_N/\sqrt{2}$ (Ryhänen et al., 1989), where R_N is the normal state resistance of a single Josephson junction in the DC SQUID. At inductances L_S < 40 pH the R_D term in (9) prevails over the second term and the flux noise of the DC SQUID is saturated.

The white flux noise of the SQUIDs is determined mainly by the thermal fluctuations in the Josephson junctions, by the maximum voltage response to the magnetic flux $|\partial V/\partial \Phi|$ and by the noise of the preamplifier of the control electronics S_{Ve}:

$$S_\Phi = S_V / \left(\frac{\partial V}{\partial \Phi}\right)^2 \approx \left\{\frac{12k_B T}{R_N}\left[\frac{R_N^2}{2} + \frac{L_S^2}{4}\left(\frac{\partial V}{\partial \Phi}\right)^2\right] + S_{Ve}\right\} / \left(\frac{\partial V}{\partial \Phi}\right)^2 \tag{10}$$

at the bias current $I_B \approx 2I_c$.

The white flux noise of DC SQUIDs was calculated according to the expression (11) with S_{Ve} ≈ 0.2 nV. The obtained value of the flux noise ≈ 3 $\mu\Phi_0/\sqrt{Hz}$ for a 40-pH DC SQUID fits well with the measured value ≈ 4 $\mu\Phi_0/\sqrt{Hz}$. The inductances of the SQUID loop and other multi-layered superconducting circuits were estimated with the help of software package 3D-MLSI (Khapaev et al., 2003).

The high dynamic range of the DC SQUID measurement system is achieved by linearization of the DC SQUID output signal with the help of DC SQUID control electronics, which compensated most of the applied magnetic flux by a flux-lock-loop circuit. Operation of the DC SQUID control electronics in bias reversal mode led to an approximately 3-fold reduction in the intrinsic low-frequency noise originating from fluctuations of critical currents of the Josephson junctions in the high-T_c DC SQUIDs. As was verified, the high-T_c DC SQUID sensors are compatible with the commercially available bias-reversal DC SQUID control electronics from Tristan Technologies Inc., Cryoton Ltd., Supracon AG, Magnicon GmbH, and STL Systemtechnik Ludwig GmbH. The hybrid digital electronics provide especially high slew rate up to 5 $M\Phi_0$/sec, dynamic range up to about 130 dB and frequency range up to about 10 MHz for the measurement system with multilayer high-T_c DC SQUIDs operating at 77 K (Ludwig et al., 2001). The above-mentioned DC SQUID electronics have the noise of preamplifier $S_{Ve} \approx (0.2\ nV)^2/Hz$. For applications in the frequency range up to about 20 kHz, the modulation electronics with a raised transformer between the SQUID and the DC SQUID control electronics has the advantage of a convenient and stable operation of the measurement system.

5.4 High-T_c DC SQUID magnetometers with inductively coupled multiturn input coil

High sensitivity of SQUIDs magnetic fields can be provided by different superconducting flux transformers, which concentrate or convert the weak magnetic fields to be measured into the significant magnetic flux threading through the SQUID loop. The concentrating types of superconducting flux transformers such as large SQUID washer, additional single-layer thin-film concentrator or bulk flux concentrator, or direct coupled superconducting flux antenna, have relatively inefficient flux transfer and a low effective area A_{eff} caused by the large difference between the inductance of the SQUID loop L_S ~ 80 pH and the inductance of the concentrator pick-up loop L_{pu} ~ 40 nH.

Figure 9 shows the flux sensitivity $S(nT/\Phi_0) \approx 2.068/A_{eff}(mm^2)$ of the flip-chip magnetometers with multiturn input coil in comparison to one of the direct-coupled magnetometers. The effective area A_{eff} of a direct-coupled magnetometers is proportional to $A_{pu}L_S/L_{pu} \propto a$, where a is the size of the pickup coil; $A_{pu} \propto a^2$ is its area; and $L_{pu} \propto a$ is its inductance. For the single-layer directly coupled magnetometers with a pickup loop 20 mm x 20 mm, inductance of the SQUID loop ≈ 50 pH, and flux sensitivity $S \approx 4.6\ nT/\Phi_0$, the best magnetic field resolution ≈ 24 $fT/\sqrt{Hz}$ at 1 Hz and ≈ 14 $fT/\sqrt{Hz}$ at 1 kHz was obtained at 77 K (Cantor et al., 1995).

Much better coupling can be achieved with a multiturn input coil, which is connected in series to the pick-up coil and inductively coupled to the SQUID loop. The effective area $A_{eff} \propto A_{pu}\cdot\sqrt{(L_S/L_{pu})} \propto a^{3/2}$ and exceeds the effective area of a direct-coupled magnetometer in about 3 times for 8 mm pick up coils and in about 10 times for 20 mm pick-up coils.

Field resolution of the inductively coupled magnetometers with multiturn input coil is:

$$B_N = \frac{L_{pu} + L_i}{kA_{pu}\sqrt{L_i L_S}} S_\Phi^{1/2} \propto \frac{1}{k}\sqrt{\frac{L_S}{R_N}} \tag{11}$$

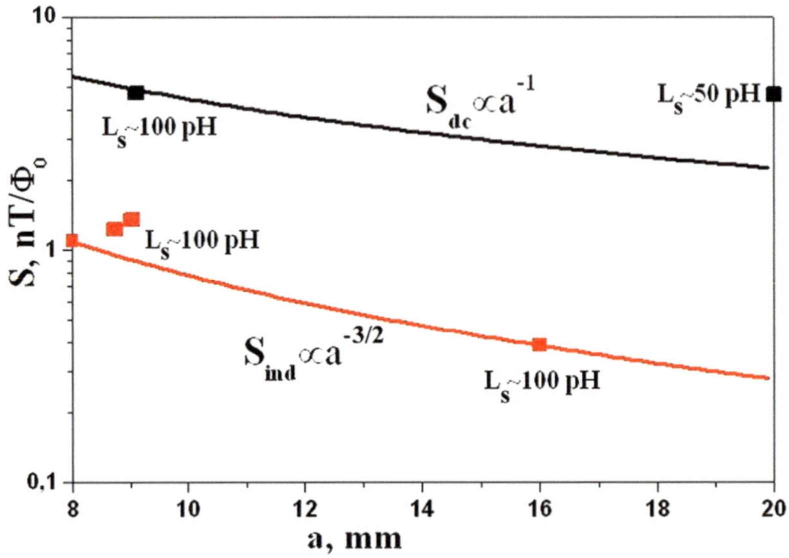

Fig. 9. Flux sensitivity S(nT/Φ_0) vs pick-up loop size a(mm) for direct coupled magnetometers (■, S_{dc}) and for the inductively coupled magnetometers with multiturn input coil (■, S_{ind}) (Faley et al., 2001).

An important prerequisite for the application of high-T_c SQUIDs for MEG is a magnetic field resolution below 10 fT/√Hz at 77 K. Such magnetic field sensitivities have only been achieved with high-T_c direct current superconducting quantum interferometers (DC SQUIDs) inductively coupled to epitaxial multilayer thin-film flux transformers with a multilayer multiturn input coil. The Berkley group (Dantsker *et al.*, 1995) has demonstrated flux sensitivity S ≈ 1.7 nT/Φ_0 magnetic field resolution ≈ 27 fT/√Hz at 1 Hz and ≈ 8.5 fT/√Hz at 1 kHz using a flip-chip magnetometer with a DC SQUID inductively coupled to a multilayer flux transformer with a 9 mm x 9 mm pickup loop. Groups from Berlin and Brondby (Drung et al., 1996) jointly reported that they achieved a magnetic field resolution ≈ 53 fT/√Hz at 1 Hz and 9.7 fT/√Hz above 1 kHz for a high-T_c DC SQUID magnetometer containing a multilayer flux transformer with a 8.3 mm x 8.6 mm pickup coil integrated on the same substrate as the SQUID.

The magnetic field resolution of the high-T_c DC SQUID magnetometers was further improved to ≈ 6 fT/√Hz above 300 Hz at 77 K by implementation of a larger pick-up loop of superconducting flux transformers made on larger wafers (Faley et al., 2001). Now this resolution can be routinely obtained, while the best resolution of the high-T_c SQUID magnetometers achieved so far is about 3.5 fT/√Hz at frequencies above 100 Hz and ≈ 6 fT/√Hz at 1 Hz and the operation temperature 77 K (Faley et al., 2006a)(see Figure 10). This magnetic field resolution is similar to the sensitivity of the currently available commercial low-T_c DC SQUID magnetometers with a 21 mm pick-up loop operating at 4.2 K (see, e.g., Elektra Neuromag®, 2006) and this is sufficient for all routine applications of SQUIDs, including biomagnetic measurements such as magnetoencephalography, which are the most demanding. Figure 11 shows a sketch and photograph of the multilayer flux transformer used for the 3 fT high-T_c DC SQUID magnetometer with 16 mm multilayer flux transformer.

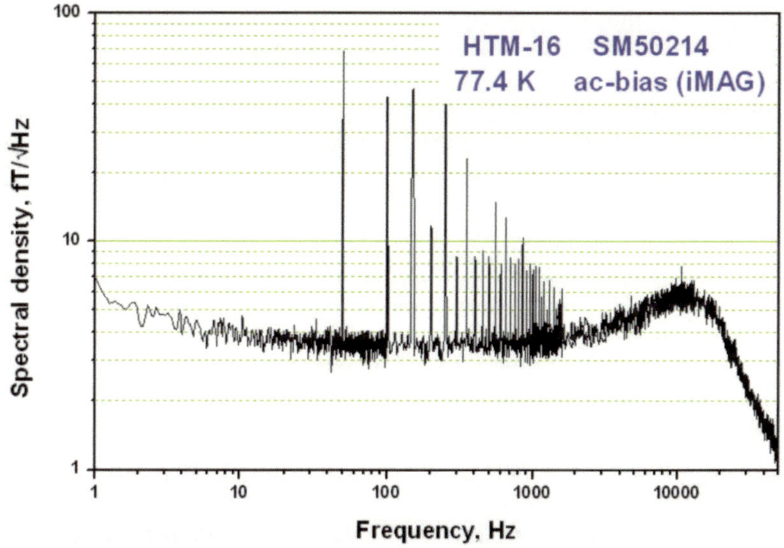

Fig. 10. Spectral density of the output signal of the measurement system based on 16-mm high-T_c DC SQUID magnetometer measured inside a 3-layer μ-metal shield and a high-T_c superconducting shield (Faley et al., 2006b).

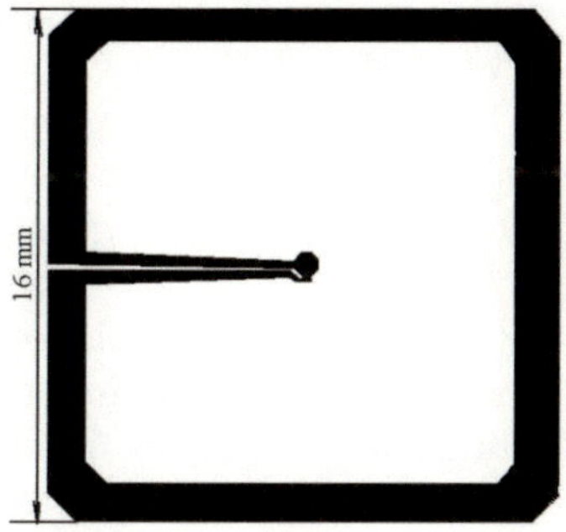

(a)

(b)

Fig. 11. (a) Sketch and (b) photograph of 16-mm high-T_c superconducting multilayer flux transformer with multiturn input coil in the middle intended for a 16-mm high-T_c DC SQUID flip-chip magnetometer.

The inductive coupling of the multiturn input coil of the multilayer flux transformer to the washer high-T_c DC SQUID is usually achieved by a flip-chip configuration with the SQUID and flux transformer made on separate substrates and subsequently placed together face-to-face. Single-crystal, 24° or 30° bicrystal 1 cm x 1 cm $SrTiO_3$ substrates were used to prepare the DC SQUIDs. For the preparation of the flux transformers, single crystal 10 mm x 10 mm $SrTiO_3$ substrates and ∅ 30 mm $LaAlO_3$ or $SrTiO_3$ wafers were used. For the flip-chip sensors, the SQUIDs and flux transformers with the lowest 1/f-noise were chosen and this procedure enabled the production of the best performing sensors so far.

For the flip-chip high-T_c DC SQUID magnetometers the optimum inductance of the pick-up loop $L_{pu} \approx 40$ nH is similar to the inductance of the input coil L_{in}. Reduction of the SQUID inductance down to about 40 pH does not appreciably degrade the field resolution, but significantly improves the voltage swings and operation stability of the DC SQUID magnetometers in the magnetically unshielded environment.

The effective inductance L_{Seff} of the SQUID loop is reduced by the screening effect in the flip-chip arrangement: $L_{seff} \approx L_S \cdot [1-k^2 L_{pu}/(L_{pu}+L_{in})] \approx L_S/2$ for the coupling coefficient $k \approx 1$ (Ryhänen et al., 1989). This effective reduction of SQUID inductance leads to an increase of the voltage swings and reduction of white noise of the SQUIDs. This effect is absent in the case of direct-coupled magnetometers.

Preparation of the multilayer high-T_c DC SQUID magnetometers is more difficult and time-consuming compared to preparation of the direct-coupled magnetometers. However, this difficulty is outweighed by much better sensitivity and reproducibility of the multilayer high-T_c DC SQUID sensors. Since 1998 high-T_c DC SQUID magnetometers having the magnetic field resolution better than 10 fT/√Hz at 77 K have already been commercially available from Forschungszentrum Jülich GmbH and Tristan Technologies Inc. (as distributor), while the commercially available direct-coupled high-T_c DC SQUID magnetometers still have the best magnetic field resolution of about 100 fT/√Hz at 77 K.

Conventional fibre-glass cryostats typically contain aluminized Mylar foil superinsulation, which is used for thermal insulation as a shield against radiation heat transfer. Due to the Nyquist noise currents in the normally conducting Al film this foil acts as a source of magnetic field noise ≈ 2 fT/√Hz. This noise often limits the total resolution of low-T_c SQUID systems and can influence the resolution of the most sensitive high-T_c systems. Reduction of the cryostat noise will reduce further the overall noise of the SQUID measurement systems.

Nyquist noise of the integrated resistance used for damping resonances in the flux transformer is one of the possible sources of the additional flux noise. We used a PBCO-STO multilayer to construct an insulation layer between the superconducting layers of the flux transformer. This has allowed us to provide sufficient insulation and resistive properties of the insulation layer also serving as an integrated resonance-damping resistor. As a result, the $V(\Phi)$ characteristics of the magnetometers were nearly sinusoidal and the estimated contribution of the Nyquist noise of the resistor to the flux noise was below $\approx 2\ \mu\Phi_0/\sqrt{Hz}$ at 77 K. Simulations of the superconducting circuits can be performed using a personal superconductor circuit analyser (PSCAN) (Polonsky et al., 1991). Possible normal-conducting micro-shorts in the insulation layer, e.g. due to CuO precipitates or defects in epitaxial growth of the insulation layer at the edges of crossovers in the input coil, can also contribute to the white noise of the flux transformers and lead to their rejection if this noise is too high.

A non-monotonous dependence of the voltage swing on the coupling between the input coil of the flux transformer and the washer of the dc-SQUID was observed: the reduction of the insulation thickness first increased the voltage swings due to the effective reduction of the SQUID inductance down to $L_{Seff} \approx 25$ pH but the reduction of the insulation thickness below ≈ 1.5 µm has led to a reduction of the voltage swings, the appearance of two maxima on the voltage swing, and an increase of the flux noise. Such effects indicated the shift of LC resonance in the DC SQUID to lower frequencies. This shift originated from a parasitic capacitive shunting of the DC SQUID loop by the flux transformer with the corresponding increase of the Stewart–McCumber parameter of the Josephson junctions.

The typical Stewart–McCumber parameter β_c of the bicrystal Josephson junctions is ≈ 0.3 at 77 K and it can increase significantly due to the capacitive coupling of the junctions with the return line of the multilayer flux transformer. In our geometry the increase of the capacitance prevails over the reduction of the inductance of the SQUID loop resulting in a reduction of the frequency of the LC resonance in the SQUID. At the bias current corresponding to the LC-resonance frequency, the voltage swing can even drop to zero. However, in most cases, by keeping the insulation thickness at ≈ 1.5 μm it was possible to avoid the appearance of the LC resonance in the vicinity of the bias current of the DC SQUID while providing sufficient inductive coupling between the input coil of the flux transformer and the washer of the DC SQUID. The estimated coupling coefficient was about $k \approx \sqrt{[2(1-L_{Seff}/L_S)]} \approx 0.87$. For further improvement of the sensors a simultaneous optimization of L_S and insulator thickness is required: a further decrease in the SQUID inductance L_S and a simultaneous decrease in the insulator thickness would keep the resonance frequency unchanged.

Further improvement of sensitivity and expanding the functionalities of high-T_c sensors are possible with, for example, larger size pick-up loops in the multilayer flux transformers and implementation of serial arrays of high-T_c DC SQUIDs. For optimum field-to-flux transformation, the increase in the inductance of the pick-up loop L_{pu} should be followed by a corresponding increase in the inductance of the input coil L_{in} of the multilayer flux transformer so that $L_{pu} \approx L_{in}$. In the case of N serial input coils each with inductance L_{in} this criterion transforms into the following: $L_{pu} \approx N \cdot L_{in}$.

Theoretically, the voltage swings of the sensors increase with implementation of serial SQUID arrays proportional to the number of SQUIDs N and this also reduces the white noise of the sensors in about √N times. At sufficiently large N, the magnetic field resolution of the high-T_c DC SQUID magnetometers with sufficiently large input coils can potentially reach values below 1 fT/√Hz at 77 K. The crucial point for the application of high-T_c DC SQUID arrays is the reproducibility of the high-T_c Josephson junctions. With high-quality substrates and photolithography the both junction types, ramp-type junctions and bicrystal junctions, have demonstrated a spread of critical currents on a chip below 10 %, which is sufficient for the preparation of arrays of high-T_c junctions (Song et al., 2010).

Serial connection of two DC SQUIDs (dual-SQUID) is the first step in the application of high-T_c DC SQUID arrays (Chen et al., 2010). Dual-SQUIDs with bicrystal Josephson junctions demonstrate a duplication of SQUID voltage swings and a reduction of noise compared to a single SQUID sensor with similar SQUID loop inductance and parameters of the Josephson junction.

Arrays of washer-SQUIDs can be inductively coupled to a single multilayer flux transformer having corresponding number of input coils and a sufficiently large inductance of the transformer pick-up loop. We suggest to couple inductively dual-SQUIDs with large area multilayer flux transformers (see Figure 12) for further improvement of the operation parameters of the high-T_c DC SQUID magnetometers and gradiometers.

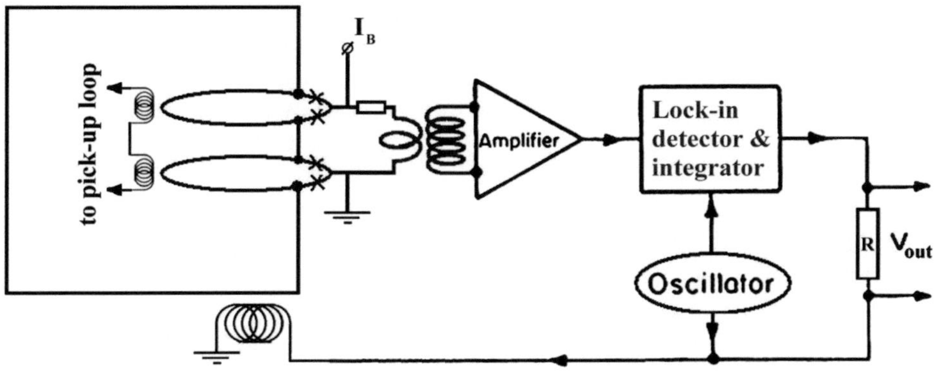

Fig. 12. Schematics of the dual-SQUID flip-chip magnetometer with control electronics.

In the suggested dual-SQUID circuit, shown schematically in Figure 12, both SQUIDs are directly coupled to a common pick-up loop, which, in turn, is inductively coupled to a feedback and modulation coil of the control electronics. The two SQUID washers are also inductively coupled to two multiturn input coils of the large-area multilayer flux transformer providing sensitivity of the sensor to the magnetic field to be measured. Application of the modulation signal to a directly coupled loop results in lower noise of the sensor. In this case, the feedback modulation signals are essentially decoupled from the pick-up loop of the multilayer flux transformer. This results in less cross-talk between the sensors compared to the case of application of the modulation signal to the pick-up coil of the multilayer flux transformer.

Further developments of the high-T_c DC SQUID sensors will include, for example, large multilayer flux transformers for more sensitive magnetometers and large-base-length gradiometers; further reduction of low-frequency noise especially for SQUIDs exposed to large permanent magnetic fields; Q-spoiler (Hilbert et al., 1985) or different types of switches (Enpuku et al., 2001, 2002), which can be used for the dissipation of parasitic circulating currents in the flux transformer during large changes of the external magnetic fields. The switches allow sensitive measurements with the SQUID magnetometers shortly after their movement in the Earth's field or after the application of strong excitation fields for low-field magnetic resonance measurements. This can be potentially used in low field magnetic resonance measurement systems for spectroscopy, biology, or security applications (Liao et al., 2010) (Espy et al., 2005, 2010).

5.5 Encapsulation of the high-T_c DC SQUIDs

The high-T_c SQUID sensors need to be encapsulated to ensure a long and reliable service life. This is especially important for the flip-chip sensors with multilayer flux transformers. The encapsulations provide mechanical and chemical protection of the sensors thus significantly simplifying handling of the sensors during characterization, applications, and for the end-users. The vacuum-tight sealing prevents the high-T_c SQUID sensors from degradation by ambient atmosphere and humidity.

The best results are obtained with enclosure of the sensors in fibreglass epoxy encapsulations. Such packaging includes the SQUID with the multilayer flux transformer as well as heater, RF filters, and a feedback coil. The thin-film Pt resistor PT-100 serves both as a thermometer and as a heater. The heater allows easy removal of trapped magnetic flux to improve the low frequency noise properties of the sensor. Good passivation was also obtained by pouring high-T_c DC SQUIDs into non-corrosive one- or two-component silicone elastomers. However, the fibreglass capsule has the advantage that, if necessary, it can be easily opened for correction or repair of the sensor followed by recapsulation.

The size and shape of the encapsulation are usually adapted to the sensor and to the specific measuring systems intended for different applications. As examples, Figure 13 shows a magnetometer and a gradiometer encapsulated into the fibreglass encapsulations. Typically, the magnetometers and gradiometers are enclosed vacuum-tight in the button-shaped encapsulations like the one shown in Figure 13a.

In the case of the flux transformers with sufficiently large-area pick-up loop, the fibre-glass capsulation can have one or two holes (Fig.13b) with vacuum-tight walls inside to permit a ferromagnetic antenna to freely tread through the pick-up coil of the thin-film flux transformer to provide better coupling to an external magnetic flux source. Such cryogenic

(a)

(b)

Fig. 13. Photograph of an encapsulated high-T_c DC SQUID magnetometer (a) and an encapsulated gradiometer with a ferromagnetic flux antenna (b).

gradiometer with ferromagnetic flux antenna can be used, for example, to measure the current of a beam of high-energy heavy-ion beams (Watanabe et al., 2004, 2010) and, potentially, it can be used in SQUID read-outs for a hot-electron microbolometer (Tarasov et al., 2002). The ferromagnetic antenna can be made from insulated Permalloy wires to suppress the circulation of macroscopic thermal (Nyquist) currents in the antenna associated with magnetic field noise.

6. Applications of the high-T_c DC SQUIDs with multilayer flux transformers

The high-T_c DC SQUIDs with superconducting thin-film multilayer flux transformers have found many applications thanks to their sensitivity, reproducibility, and relatively high operating temperature. The measurement systems equipped with the high-T_c DC SQUID sensors are used mainly for biomagnetic measurements, geomagnetic surveys, non-destructive evaluations, electronics technology, fundamental physics, and in materials science. We have developed, produced, and supplied for integration in different measurement systems worldwide more than one hundred of high-T_c SQUID sensors. These requests, in turn, supported the further development of high-T_c sensors: more than 20 types of high-T_c DC SQUID magnetometers and gradiometers prepared by the high oxygen pressure sputtering technique are now available from Forschungszentrum Jülich GmbH.

The epitaxial oxide heterostructures were used in different types of SQUID microscopes (Faley et al., 2004) (Poppe et al., 2004); in a SQUID monitor for measuring the beam current of accelerator radioisotope ions (Watanabe et al., 2004, 2010); for geomagnetic surveys (Chwala et al., 1999, Clem et al., 2001, Fagaly, 2006); for non-contact testing of semiconductor structures with a SQUID laser microscope (Daibo et al., 2002, 2005); in the NDE systems for eddy current testing of aircraft wheels and rivets (Grüneklee et al., 1997); for magnetic inspection of prestressed concrete bridges (Krause et al., 2002); for picovoltmeters (Faley et al., 1997b); and for the localization and identification of deep-seated artificial defects such as holes, slots and cracks in multilayer reinforced carbon fibre polymer panels by eddy current SQUID NDE (Valentino et al., 2002) and for magnetocardiography (MCG) measurements (Drung et al., 1995; Faley et al., 2002). Biomagnetic measurements are among those general-purpose applications for which the SQUID measurement systems are preferred due to their sensitivity and ability to measure vector components of magnetic fields. The high-T_c DC SQUID magnetometers with multilayer flux transformers arranged into an axial electronic gradiometer with ≈ 1 fT/cm·√Hz at 77 K gradient sensitivity were successfully tested in a clinical environment for MCG measurements (Faley et al., 2002).

The diversity of the applications of the multilayer high-T_c SQUID sensors is astonishing. They have already proved that it is worthwhile to further develop the technology of these sensors. Other very promising applications can be potentially added but need to be tested first. The sensitivity of the high-T_c DC SQUID sensors already obtained is also sufficient for MEG measurements, but an integration of the high-T_c MEG system in an MEG laboratory is still required. Low-field magnetic resonance imaging and nuclear quadrupole resonance with multilayer high-T_c DC SQUID sensors have also many potential applications in, for example, spectroscopy, biology, and security. For example, high-T_c SQUID preamplifiers operating at intermediate temperatures ~ 20 K can be useful for readout circuits for quantum computers. Further development of specific SQUID layouts optimized for each of these and other applications will follow.

7. Summary and outlook

The technology of multilayer high-T_c DC SQUID sensors has made significant progress: their sensitivity and yield have been further improved; the sensitive sensors can be now fabricated in batch production and have been implemented on a large scale. High-T_c DC SQUID magnetometers have achieved a magnetic field resolution of about 3 fT/√Hz at 77 K, while the planar gradiometers have achieved a gradient resolution of about 10 fT/cm·√Hz at 77 K. The mature multilayer technology of the epitaxial metal-oxide heterostructures is indispensable for reaching the ultimate sensitivity high-T_c DC SQUID sensors in white noise region and can also provide high sensitivity at low frequencies. The multilayer technology of the epitaxial metal-oxide heterostructures can be also used for many other superconducting devices and for general purpose metal-oxide heterostructures.

Bilayer epitaxial buffer helps to grow thicker YBCO heterostructures with less strain. Much thicker superconducting and insulating films can be deposited. The reproducibility of the high-T_c Josephson junctions and SQUIDs achieved so far is sufficient for the effective implementation of arrays of DC SQUIDs. This improves reproducibility, increases critical current and reduces low frequency noise of the multilayer flux transformers. The final encapsulation of the sensors with integrated electronic parts such as LP filters, heater, and feedback coil additionally improves the operation, handling, and noise properties of the sensors.
The achieved magnetic field resolution, yield, and the long-term stability of the multilayer high-T_c DC SQUID sensors enable them to be integrated into multichannel MEG measurement systems. This requires installation in a proper magnetically shielded room with an MEG infrastructure and this still remains to be demonstrated. Another prospective area of application is the low-field magnetic resonance imaging (LFMRI) and combined systems MEG-LFMRI systems, both based on high-T_c multilayer DC SQUID sensors.

8. Acknowledgments

The author gratefully acknowledges U. Poppe for fruitful discussions and R. Speen for technical assistance.

9. References

Acquaviva, S., D'Anna, E., De Giorgi, M.L., Fernandez, M., Luches, A., Majni, G., Luby, S., & Majkovacet, E. (2005). Transfer of stoichiometry during pulsed laser ablation of multicomponent magnetic targets, *Appl. Surf. Sci.*, Vol. 248, Issues 1-4, pp.286-290.

Beyer, J., Drung, D., Ludwig, F., Minotani, T., & Enpuku, K. (1998). Low-noise $YBa_2Cu_3O_{7-x}$ single layer dc superconducting quantum interference device (SQUID) magnetometer based on bicrystal junctions with 30° misorientation angle, *Appl. Phys. Lett.*, Vol. 72, No. 2, pp. 203-205.

Cantor, R., Lee, L. P., Teepe, M., Vinetskiy, V., & Longo, J. (1995). Low-noise single-layer $YBa_2Cu_3O_7$ DC-SQUID magnetometers at 77 K, *IEEE Trans. Appl. Supercond.*, Vol.5, No. 2, pp. 2927-2930.

Chen, K.-L., Yang, H.-C., Ko, P. C., & Horng H. E. (2010). Characterization of dual high transition temperature superconducting quantum interference device first-order planar gradiometers on a chip, *J. Appl. Phys.*, Vol.108, pp.064503(4).

Chwala, A., Stolz, R., Ramos, J., Schultze, V., Meyer, H.-G., & Kretzschmar, D. (1999). An HTS dc SQUID system for geomagnetic prospection, *Supercond. Sci. Technol.*, Vol.12, pp.1036–1038.

Clarke, J., & Braginski, A. I., (Editors) (2006). *The SQUID Handbook Vol.2: Applications of SQUIDs and SQUID systems*, (Weinheim: WILEY-VCH Verlag GmbH&Co. KGaA), ISBN 3-52740408-2.

Clem, T. R., Overway, D. J., Purpura. J. W., Bono, J. T., Koch, R. H., Rozen, J. R., Keefe, G. A., Willen, S., & Mohling. R. A. (2001). High-T_c SQUID gradiometer for mobile magnetic anomaly detection, *IEEE Trans. Appl. Supercond.*, Vol.11, pp.871-875.

Daibo, M., Shikoda, A., & Yoshizawa, M. (2002). Non-contact evaluation of semiconductors using a laser SQUID microscope, *Physica C*, Vol.372–376, No.1, pp.263–266.

Daibo, M., & Kamiwano, D. (2005). Examination of relationship between resistivity and photocurrent induced magnetic field in silicon wafers using laser SQUID, *IEEE Trans. Appl. Supercond.*, Vol.15, No. 2, pp. 684-687.

Dam, B., Koeman, N. J., Rector, J. H., Stäuble-Pümpin, B., Poppe, U., & Griessen, R. (1996). Growth and etching phenomena observed by STM/AFM on pulsed-laser deposited $YBa_2Cu_3O_{7-\delta}$ films, *Physica C*, Vol.261, pp.1-11.

Dam, B., Huijbregtse, J. M., & Rector, J. H. (2002). Strong pinning linear defects formed at the coherent growth transition of pulsed-laser-deposited $YBa_2Cu_3O_{7-\delta}$ films, *Phys. Rev. B*, Vol. 65, pp.064528(8).

Dantsker, E., Ludwig, F., Kleiner, R., Clarke, J., Teepe, M., Lee, L. P., Alford, N. McN., & Button, T. (1995). Addendum: "Low noise $YBa_2Cu_3O_{7-x}$ - $SrTiO_3$ - $YBa_2Cu_3O_{7-x}$ multilayers for improved superconducting magnetometers", *Appl. Phys. Lett.*, Vol.67, No. 5, pp. 725-726.

David, B. R., Grundler, D., Eckart, R., Fanghänel, K., Krumme, J.P., Doormann, V., & Dössel, O. (1994). A multi-layer process for the fabrication of HTSC flux transformers and SQUIDs, *Supercond. Sci. Technol.*, Vol.7, pp.287–289.

Dogan, O. (2005). The effect of duration of diffusion on Ag diffusion coefficients in $YBa_2Cu_3O_7$, *Journal of Quantitative Spectroscopy & Radiative Transfer*, Vol.95, pp.263-269.

Drung, D., Ludwig, F., Müller, W., Steinhoff, U., Trahms, L., Koch, H., Shen, Y. Q., Jensen, M. B., Vase, P., Holst, T., Freltoft, T., & Curio, G. (1996). Integrated $YBa_2Cu_3O_{7-x}$ magnetometer for biomagnetic measurements, *Appl. Phys. Lett.*, Vol.68, 1421.

Elekta Neuromag® (2006). System description - magnetoencephalographic and electroencephalographic System, NM21761B-A, pp.1-32.

Enpuku, K., Tokita, G., Maruo, T., & Minotani, T. (1995). Parameter dependencies of characteristics of a high-T_c dc superconducting quantum interference device, *J. Appl. Phys.*, Vol.78, pp.3498 - 3503

Enpuku, K., Maruo, T., & Minotani, T. (1996). Effect of large dielectric constant of SrTiO3 substrate on the characteristics of high T_c dc superconducting quantum interference device, *J. Appl. Phys.*, Vol.80, No.2, pp.1207 - 1213.

Enpuku, K., Minotani, T., Shiraishi, F., Kandori, A., & Kawakami, S. (1999). High T_c DC SQUID utilizing bicrystal junctions with 30 degree misorientation angle, *IEEE Trans. Appl. Supercond.*, Vol. 9, No. 2, pp. 3109-3112.

Enpuku, K., Tokimizu, D., Kuroda, D., & Hijiya, S. (2001). A four-junction switch for controlling the opening and closing of a pickup coil in high-T_c superconducting quantum interference device magnetometer, *Jpn. J. Appl. Phys.* Vol. 40 pp.L869–L 871.

Enpuku, K., Kuroda, D., Tokimizu, D., & Yang., T. Q. (2002). Suppression of thermally activated flux entry through a flux dam in high-Tc superconducting quantum interference device magnetometer, *J. App. Phys.*, Vol.92, No.8, pp. 4751- 4757.

Espy, M. A., Matlachov, A. N., Volegov, P. L., Mosher, J. C., & Kraus, Jr. R. H. (2005). SQUID-based simultaneous detection of NMR and biomagnetic signals at ultra-low magnetic fields, *IEEE Trans. Appl. Supercond.*, Vol.15, No. 2, pp. 635-639.

Espy, M., Flynn, M., Gomez, J., Hanson, C., Kraus, R., Magnelind, P., Maskaly, K., Matlashov, A., Newman, S., Owens, T., Peters, M., Sandin, H., Savukov, I., Schultz,

L., Urbaitis, A., Volegov P., & Zotev V. (2010). Ultra-low-field MRI for the detection of liquid explosives, *Supercond. Sci. Technol.*, Vol.23, No.3, pp. 034023 (8).

Fagaly, R. L. (2006). Superconducting quantum interference device instruments and applications, *Rev. Sci. Instr.*, Vol.77, pp.101101(45).

Faley, M. I., Gershenson, M.E., Kuchta, N.P. & Salun, V.S. (1991). The "in-situ" preparation and properties of Y-Ba-CuO thin films on the $SrTiO_3$, Al_2O_3 and Si substrates, *IEEE Trans. Mag.* Vol.27, No.2, pp.1475-1478.

Faley, M. I., Poppe, U., Soltner, H., Jia, C.L., Siegel, M., & Urban, K. (1993). Josephson junctions, interconnects and crossovers on chemically etched edges of $YBa_2Cu_3O_7$, *Appl. Phys. Lett.*, Vol.63, pp.2138-2140.

Faley, M. I., Poppe, U., Urban, K., Hilgenkamp, H., Hemmes, H., Aarnink, W., Flokstra, J., & Rogalla, H. (1995). Noise properties of dc-SQUIDs with quasiplanar YBa2Cu3O7 Josephson junctions, *Appl. Phys. Lett.*, Vol.67, No.14, pp.2087-2089.

Faley, M. I., Poppe, U., Jia, C.L., & Urban, K. (1997a). Size, Order and Interface Effects in $YBa_2Cu_3O_7$-$PrBa_2Cu_3O_7$-$YBa_2Cu_3O_7$ Josephson Junctions, *IEEE Trans. Appl. Supercond.*, Vol.7, No.2, pp.2514-2517.

Faley, M. I., Poppe, U., Urban, K., Krause, H.-J., Soltner, H., Hohmann, R., Lomparski, D., Kutzner, R., Wördenweber, R., Bousack, H., Braginski, A. I., Slobodchikov, V. Yu., Gapelyuk, A. V., Khanin, V. V., & Maslennikov, Yu. V. (1997b). DC-SQUID magnetometers and gradiometers on the basis of quasiplanar ramp-type Josephson junctions, *IEEE Trans. Appl. Supercond.*, Vol.7, No.2, pp.3702-3705.

Faley, M. I., Poppe, U., Urban, K., Paulson, D. N., Starr, T., & Fagaly, R. L. (2001). Low noise HTS dc-SQUID flip-chip magnetometers and gradiometers, *IEEE Trans. Appl. Supercond.*, Vol.11, No.1, pp.1383-1386.

Faley, M. I., Poppe, U., Urban, K., Slobodchikov, V. Yu., Maslennikov, Yu. V., Gapelyuk, A., Sawitzki, B., & Schirdewan, A. (2002). Operation of high-temperature superconductor magnetometer with submicrometer bicrystal junctions, *Appl. Phys. Lett.*, Vol.81, No.13, pp.2406-2408.

Faley, M. I., Pratt, K., Reineman, R., Schurig, D., Gott, S., Sarwinski, R. E., Paulson, D. N., Starr, T. N., & Fagaly R. L. (2004). HTS dc-SQUID micro-susceptometer for room temperature objects, *Supercond. Sci. Technol.*, Vol.17, pp.S324-S327.

Faley, M. I., Mi, S. B., Petraru, A., Jia, C. L., Poppe, U. & Urban, K. (2006a). Multilayer buffer for high-temperature superconductor devices on MgO, *Appl. Phys. Lett.*, Vol.89, No.8, pp.082507(3).

Faley, M. I., Jia, C. L., Houben, L., Meertens, D., Poppe, U., & Urban, K. (2006b). Meandering of the grain boundary and d-wave effects in high-T_c bicrystal Josephson junctions, *Supercond. Sci. Technol.*, Vol.19, pp.S195–S199.

Faley, M. I., Mi, S. B., Jia, C. L., Poppe, U., Urban, K., & Fagaly, R. L. (2008). Epitaxial thick film high-T_c SQUIDs, *Journal of Physics: Conf. Ser.*, Vol.97, pp.012164(6).

Faley, M. I., Poppe, U., Urban, K. & Fagaly, R. L. (2010). Noise analysis of dc-SQUIDs with damped superconducting flux transformers, *Journal of Physics: Conf. Ser.*, Vol.234, pp.042009(15).

Faley, M. I., & Poppe, U. (2010a). Sputterquellen für Hochdrucksputtern mit großen Targets und Sputterverfahren, *Patent pending*, DE 102010049329, 22 pages.

Faley, M. I., Poppe, U., Urban, K., & Fagaly, R.L. (2010b). Noise analysis of dc-SQUIDs with damped superconducting flux transformers, *Journal of Physics: Conf. Ser.*, Vol.234, pp.042009(15).

Faley, M. I. (2010c). Cuprate high-T_c superconductors, Chapter in the lecture notes of the 41st IFF Spring School "Electronic Oxides: Correlation Phenomena, Exotic Phases, and Novel Functionalities", 08 - 19 March 2010, Forschungszentrums Jülich, Germany Schriften des Forschungszentrums Jülich, Vol.37, C4 (36 pages) ISSN 1866-1807, ISSN 978-3-89336-609-5.

Fisher, B., Genossar, J., Patlagan, L., Reisner, G. M., Subramaniam, C. K., & Kaiser, A. B. (1994). Hopping conductivity in fully oxygenated $PrBa_2Cu_3O_y$, $YBa_2Cu_2CoO_y$, and $PrBa_2Cu_2CoO_y$, *Phys. Rev. B*, Vol.50, pp.4118–4124.

Hammond, R. H. & Bormann, R. (1989). Correlation between the in situ growth conditions of YBCO thin films and the thermodynamic stability criteria, *Physica C*, Vol.162-164, pp.703-704

Hilbert, C., Clarke, J., Sleator, T., & Hahn, E. L. (1985). Nuclear quadrupole resonance detected at 30 MHz with a DC SQUID, *Appl. Phys. Lett.*, Vol. 47, pp. 637–639.

Hilgenkamp, H., & Mannhart, J. (2002). Grain boundaries in high-T_c superconductors, *Rev. Mod. Phys.*, Vol.74, pp.485-549.

Gross, R., Alff, L., Heck, A., Froehlich, O. M., Koelle, D., & Marx, A. (1997). Physics and technology of high temperature superconducting Josephson junctions, *IEEE Trans. Appl. Supercond.*, Vol.7, No.2, pp.2929-2935.

Grüneklee, M., Krause, H.-J., Hohmann, R., Maus, M., Lomparski, D., Banzet, M., Schubert, J., Zander, W., Zhang, Y., Wolf, W., Bousack, H., Braginski, A. I., & Faley, M. I. (1997). HTS SQUID system for eddy current testing of airplane wheels and rivets, *Review of Progress in Quantitative NDE*, Plenum, N.Y., Vol.17.

Jia, C. L., Faley, M. I., Poppe, U., & Urban, K. (1995). Effect of chemical and ion-beam etching on the atomic structure of interfaces in $YBa_2Cu_3O_7/PrBa_2Cu_3O_7$ Josephson junctions, *Appl. Phys. Lett.*, Vol.67, No.24, pp.3635-3637.

Jia C. L., Mi, S.B., Faley, M. I., Poppe, U., Schubert, J., & Urban, K. (2009). Oxygen octahedron reconstruction in the $SrTiO_3/LaAlO_3$ heterointerfaces investigated using aberration-corrected ultrahigh-resolution transmission electron microscopy, *Phys. Rev. B*, Vol.79, pp.081405(R)

Khapaev, M. M., Kupriyanov, M. Yu., Goldobin E., & and Siegel M., (2003). Current distribution simulation for superconducting multi-layered structures, *Supercond. Sci. Technol.*, Vol.16, pp.24.

Kim, S. I., Kametani, F., Chen, Z., Gurevich, A., Larbalestier, D. C., Haugan, T., & Barnes, P. (2007). On the through-thickness critical current density of an $YBa_2Cu_3O_{7-x}$ film containing a high density of insulating, vortex-pinning nanoprecipitates, *Appl. Phys. Lett.* Vol.90, pp.252502(3)

Kleiner, R., Koelle, D., Ludwig, F., & Clarke, J. (2004). Superconducting quantum interference devices: state of the art and applications, *Proc. IEEE*, Vol.92, No.10, pp.1534-1548.

Krause, H.-J., Wolf, W., Glaas, W., Zimmermann, E., Faley, M. I., Sawade, G., Mattheus, R., Neudert, G., Gampe, U., & Krieger, J. (2002). SQUID array for magnetic inspection of prestressed concrete bridges, *Physica C*, Vol. 368, Issues 1-4, pp.91-95.

Liao, S.-H., Huang, K.-W., Yang, H.-C., Yen, C.-T., Chen, M. J., Chen, H.-H., Horng, H.-E., & Yang, S. Y. (2010). Characterization of tumors using high-T_c superconducting quantum interference device-detected nuclear magnetic resonance and imaging, *Appl. Phys. Lett.*, Vol. 97, pp.263701(3).

Ludwig, C., Kessler, C., Steinfort, A.J., & Ludwig W. (2001). Versatile High Performance Digital SQUID Electronics, *IEEE Trans. Appl. Supercond.*, Vol.11, No.1, pp.1122-1125.

Matias, V., Rowley, E. J., Coulter, Y., Maiorov, B., Holesinger, T., Yung, C., Glyantsev, V. & Moeckly, B. (2010). YBCO films grown by reactive co-evaporation on simplified IBAD-MgO coated conductor templates, *Supercond. Sci. Technol.*, Vol.23, pp.014018(5).

Mi, S. B., Jia, C. L., Faley, M. I., Poppe, U., & Urban K. (2007). High resolution electron microscopy of microstructure of $SrTiO_3/BaZrO_3$ bilayer thin films on MgO substrates, *J. Crystal Growth*, Vol.300, pp.478-482.

Minotani, T., Kawakami, S., Kuroki, Y., & Enpuku, K. (1998). Properties of Josephson junction fabricated on bicrystal substrate with different misorientation angles, *Jpn. J. Appl. Phys.*, Vol.37, pp.L718-L721.

Navacerrada, M. A., Lucía, M. L., Sánchez-Quesada, F., & Sarnelli, E. (2008). Fiske steps and hysteresis in $YBa_2Cu_3O_7$ grain boundary Josephson junctions: Structural information of the barrier by means of a non-destructive approach, *J. Appl. Phys.*, Vol.104, pp.113915(6).

Ohtomo, A., & Hwang, H. Y. (2004). A high-mobility electron gas at the $LaAlO_3/SrTiO_3$ heterointerface, *Nature*, Vol.427, pp.423-426.

Paladino, A. E. (1965). Oxidation kinetics of single-crystal $SrTiO_3$, *Journal of American Ceramic Society*, Vol.48, No.9, pp.476-478.

Polonsky, S., Semenov, V., & Shevchenko, P., (1991). PSCAN: personal superconductor circuit analyser, *Supercond. Sci. Technol.*, Vol.4, pp.667.

Poppe, U., Schubert, J., & Evers, W. (1990). Method of fabricating thin layers from high temperature oxide superconductors, Patent US4965248.

Poppe, U., Klein, N., Dähne, U., Soltner, H., Jia, C. L., Kabius, B., Urban. K., Lubig, A., Schmidt, K., Hensen, S., Orbach, S., Müller, G., & Piel, H. (1992). Low resistivity epitaxial $YBa_2Cu_3O_7$ thin films with improved microstructure and reduced microwave losses, *J. Appl. Phys.*, Vol.71, pp.5572-5578.

Poppe, U., Faley, M. I., Breunig, I., Speen, R., Urban, K., Zimmermann, E., Glaas, W., & Halling, H. (2004). HTS dc-SQUID microscope with soft-magnetic flux guide, *Supercond. Sci. Technol.*, Vol.17, pp.S191-S195.

Prusseit, W., Furtner, S. & Nemetschek, R. (2000). Series production of large area $YBa_2Cu_3O_7$ films for microwave and electrical power applications, *Supercond. Sci. Technol.*, Vol.13, pp.519-521.

Ryhänen, T., Seppä, H., Ilmoniemi, R., & Knuutila, J. (1989). SQUID magnetometers for low-frequency applications, *J. Low Temp. Phys.*, Vol.76, Nos.5/6, pp.287-386.

Sobol, E. (1995). *Phase transformations and ablation in laser-treated solids*, Wiley, New York.

Song, F., He, M., Faley, M. I., Fang, L., & Klushin, A. M. (2010). Improved coupling of Josephson junction arrays to the open space, *J. Appl. Phys.*, Vol.108, 063903(5).

Tarasov, M., Gudoshnikov, S., Kalabukhov, A., Seppa, H., Kiviranta, M., & Kuzmin, L., (2002). Towards a dc SQUID read-out for the normal metal hot-electron microbolometer, *Physica C*, Vol.368, Issues 1-4, pp.161-165.

Tinkham, M. (1996). *Introduction to Superconductivity*, 2nd edition, McGraw-Hill Inc., New York, ISBN 0-07-064878-6, p.225.

Valentino, M., Ruosi, A., Peluso, G., & Pepe, G. P. (2002). Structural health monitoring of materials by high critical temperature SQUID, *Physica C*, Vol.372-376, No.1, pp.201-208.

Voss, R.F. (1981). Noise characteristics of an ideal shunted Josephson junction, *J. Low Temp. Phys.*, Vol.42, Nos.1/2, pp.151-163.

Watanabe, T., Watanabe, S., Ikeda, T., Kase, M., Sasaki, Y., Kawaguchi, T., & Katayama, T. (2004). A prototype of a highly sensitive cryogenic current comparator with a HTS SQUID and HTS magnetic shield, *Supercond. Sci. Technol.*, Vol.17, pp.S450–S455.

Watanabe, T., Fukunishi, N., Sasaki, Y., Kase, M., Goto, A., & Kamigaito, O. (2010). Development of beam current monitor with high-T_c SQUID at RIBF, *Proceedings of Beam Instrumentation Workshop (BIW10), La Fonda on the Plaza Santa Fe, New Mexico, USA, May 2-6, 2010*, Invited talk, 10 pages.

Winkler, D., Zhang, Y. M., Nilsson, P. A., Stepantsov, E. A., & Claeson, T. (1994). Electromagnetic properties at the grain boundary interface of $YBa_2Cu_3O_{7-\delta}$ bicrystal Josephson junctions, *Phys. Rev. Lett.*, Vol.72, No.8, pp.1260-1263.

Witchalls C. (2010). Nobel prizewinner: We are running out of helium. *New Scientist*. 18 August 2010, Witchalls C. (2010). One minute with... Robert Richardson, *The New Scientist*, Vol.207, Issue 2773, 14 August 2010, p.29.

8

Thermophysical Properties of Bi-based High-Tc Superconductors

Asghari Maqsood[1] and M. Anis-ur-Rehman[2]
[1]*Thermal Transport Laboratory, School of Chemical and Materials Engineering, National University of Sciences and Technology (NUST),*
[2]*Applied Thermal Physics Laboratory, Department of Physics, COMSATS Institute of Information Technology, Islamabad, Pakistan*

1. Introduction

Since the discovery of 90 K superconductivity in the Ba-Y-Cu oxide system (Wu, et.al., 1987) a number of studies have been published. A true superconductor not only shows zero resistance but also excludes a magnetic field completely (the Meissner effect). A visual demonstration of the Meissner effect was carried out by placing a small magnet on a pellet of $Dy_1Ba_2Cu_3O_{7-x}$ and cooling the system to liquid-nitrogen temperature. The levitation of the magnet due to ejection of magnetic lines of flux from the superconductor is shown in Figure 1 (Maqsood, et.al., 1989).
Dissipation phenomena in high temperature superconductors are governed by the microstructure that develops during the preparation process. Therefore, detailed investigations of the electrical and thermal transport and ac magnetic susceptibilities in

Fig. 1. The $Dy_1Ba_2Cu_30_7$-x specimen, showing the Meissner effect at liquid-nitrogen temperature.

superconductors prepared either in the form of single crystals, thin films or polycrystalline are important for understanding superconductivity as well as for practical applications (Rehman & Maqsood, 2005).

Among high-T_c superconductors, (Bi, Pb)-2223 appears to be the most promising candidate for the application of power transmission cables at liquid nitrogen temperature. Unlike other high-Tc superconductors (HTS), such as $YBa_2Cu_3O_{7-\delta}$ (Y-123), it is still a problem to control and increase its critical temperature and current density. The Bi based superconductors offer potential advantages in comparison to the Y-based superconductors. The studies of transport properties, such as electrical resistivity, thermoelectric power (S) and thermal conductivity, are important for exploring the conduction mechanisms. The transport properties are very sensitive to the sample preparation methods.

The BISCCO samples substituted with Fe, Cr, Co, Gd, Er, Nd, Sm, Ag, V, Ga, Zn, Cd, etc. have been widely prepared using conventional solid state reaction and glass–ceramics techniques (Aksan & Yakyncy, 2004; Chatterjee, et.al. 1998; Cloots, et.al., 1994; Coskun, et.al. 2005; Dorbolo, et.al. 1999; Ekicibil, et.al., 2004, 2005; Mandal, et.al., 1992; Munakata, et.al., 1992; Nanda, et.al., 1995; Ozhanli,et.al., 2002; Rao,et.al. 1990; Sera,et.al. 1992; Varoy, et.al. 1992). Investigation of thermal conductivity, $\lambda(T)$, also gives important information about the scattering mechanism of charge carriers, electron–phonon interaction and other physical properties, such as carrier density and phonon mean free path (Aksan, et.al. 1999; Houssa & Ausloos, 1996; Knizek, et.al. 1998; Natividad,et.al. 2002; Uher,et.al. 1994; Yankyncy 1997). In the last decade, many investigations have been made on $\lambda(T)$ of high-T_c materials (Aksan, et.al. 1999; Castellazzi, et. al. 1997; Houssa, et.al. 1996; Hui, et.al. 1999; Knizek, et.al. 1998; Natividad,et.al. 2002; Uher,et.al. 1994; Yankyncy 1997; Wermbter, 1991) and almost similar results are reported. In general, for $\lambda(T)$ investigation of high-T_c materials, three important approaches can be considered to the total $\lambda(T)$ calculations: (i) phonon contribution; (ii) electron contribution; and (iii) both electron and phonon contributions. Many research groups have investigated these valuable approaches for high-T_c materials and results are published (Castellazzi, et.al., 1997; Peacor, et.al., 1991; Tewordt & Wolkhausen, 1989,1990; Wermbter, et.al., 1996; Yu, et.al. 1992). However, there exists a difficulty in the $\lambda(T)$ properties of the high-T_c materials. In particular, compared with conventional metallic structures, the high-T_c superconductors show unusual behavior just below their T_c. At that point, thermal conductivity rises and reaches to the maximum and then drops sharply. The explanation of the rapid rise and the maximum point seen in a wide range just below T_c, is summed up through two main points (Uher, et.al., 1994). Firstly, decrease on the scattering mechanism, because of the superconducting state ($T < T_c$ ($R = 0$), and secondly, an increase in the electron mean free path due to decrease in the phonon scattering. In many investigations, the maximal value was also found to depend on the preparation method and chemical composition (Cohn, et.al. 1992; Jezowski,et.al. 1987; Morelli,et.al. 1987; Peacor, et.al. 1991; Uher 1992;). However, it is important to see the effect of the quasi-particle contribution on the rapid rise of $\lambda(T)$ below the T_c, as explained by many groups (Castellazzi, et.al. 1997; Yu, et.al. 1992). There exist some other models that have been widely accepted for materials in solid state. Particularly, for the graded materials, effective medium approximation (EMA) (Hirai, 1996; Hui, et.al. 1999) and another model developed for the conventional low-Tc superconductors by Bardeen et al (Bardeen, et.al., 1959) that describes the phonon thermal conductivity in the superconducting state. This model then was generalized by Tewordt and Wolkhausen in order to describe the phonon thermal

conductivity of high-T_c superconductors in a wide range of temperatures (Tewordt & Wolkhuasen, 1989). However, still many efforts have to be made both experimentally and theoretically to understand the $\lambda(T)$ mechanism of high-T_c materials.
Thermoelectric power being sensitive to the energy dependence of the electron lifetime and the density of states near the Fermi level energy, provides valuable information regarding many fundamental aspects of charge carrier transport in the materials. The thermoelectric power (S) of high-temperature superconductors has been widely studied and reported a positive 'S', while Khim et al. (Khim et. al. 1987) and others have reported a negative 'S' for the same compositions. Later studies proved that the sign of the thermoelectric power is sensitive to the oxygen content present in the compound (Lee, et. al. 1988). This behavior was also observed in other materials like Cheverly phase compounds as was seen by Vasudeva Rao et al. (Rao, et.al. 1984). In the BISCCO compounds, the TEP studies earlier reported on the (2201), (2212) and (2223) phases. Sera et al. (Sera et. al. 1992) have studied the *S* behavior of La doped (2201) cuprates. The 'S' behavior of the $Bi_2Sr_2Ca_{1-x}Y_xCu_2O_y$ compounds was investigated as a function of temperature (Mandal, et.al. 1992; Munakata, et.al. 1992; Varoy, et. al. 1992) and it was found that with increasing temperature 'S' increases, exhibits a maximum value and decreases thereafter. From the 'S' data they concluded that the substitution of Y for Ca decreases the hole concentration from the optimum value. Varoy et. al. have found a systematic fall to more negative values of thermoelectric power as holes are added with the introduction of more lead or oxygen into BISCCO compounds. In this chapter, we report the structural, electrical, magnetic and thermal transport properties of vanadium-substituted BISCCO (2223) systems.

2. Experimental details

In the Bi-based high-T_c superconductors the Bi-2223 phase is stable within a narrow temperature range and exhibits phase equilibria with only a few of the compounds existing in the system. Precise control over the processing parameters is required to obtain the phase-pure material (Rehman & Maqsood, 2005). Bismuth-based superconducting powder with chemical formula $Bi_{1.3}V_{0.3}Pb_{0.4}Sr_2Ca_2Cu_3O_\delta$ was prepared by the solid state reaction method. The starting powders corresponding to stoichiometric quantities of high purity (99.9%) Bi_2O_3, PbO, $SrCO_3$, $CaCO_3$, V_2O_5, CuO were weighed on digital balance within 1 mg accuracy. The powders were mixed and ground in an agate mortar and pestle for 2 h and then the mixed powders were calcined in air at 810^{0}C for 24 h. The porcelain crucibles were used for calcinations and sintering. The porcelain crucibles have a melting point of 1400^{0}C, are not contaminated with the materials being used and are economical. The aim of the calcination was to eliminate the carbonates and to produce an oxide with a nominal composition $Bi_{1.3}V_{0.3}Pb_{0.4}Sr_2Ca_2Cu_3O_\delta$. The polyvinyl alcohol (PVA) was used as a binder for pellet formation. A pressure of 50 kN was applied to all the pellets by a hydraulic press. The pellets with nominal composition $Bi_{1.3}V_{0.3}Pb_{0.4}Sr_2Ca_2Cu_3O_\delta$ were finally sintered in air at 840 ±5^{0}C for 216 h in three steps (first step 96 h, second step 72 h and the third 48 h) and gradually cooled down to room temperature. Phase purity and lattice parameters were determined by X-ray diffraction (XRD). XRD data were taken using CuK*a* (1.5406 Å) radiation. Grain size was analyzed from scanning electron microscopic (SEM) image taken on the surface on several cross-sectional cuts. No appreciable variations of size and aspect ratio of grains were observed for the different zones. DC electrical resistivity was used to characterize superconducting properties of the samples using standard four-probe DC technique, in the temperature range from 77 to 300 K. The critical current density of the sample was measured by the four-probe method in such a way that the current may pass through the bulk of the specimen. For the calculation of Jc the following equation is used

$$J_c = I_c / A = I_c / (WT) \tag{1}$$

where A is the area of cross section, W is the width, I_c is the critical current and T is the thickness of the sample. Considerable care was taken to ensure that results were not affected by heating at the current contacts. The mutual inductance bridge method was used for AC magnetic susceptibility measurements. For these measurements, a very low AC field (0.5 Oe) was applied parallel to the axis of the rectangular bar-shaped specimen. The temperature range was again 77–300 K. The thermal conductivity as a function of temperature was measured using the advantageous transient plane source techniques (ATPS). The detail of the ATPS technique is described in (Rehman & Maqsood, 2005; Gusstafsson, 1991; Rehman, 2009) at length. Two identical sizes of the sample (diameter 25 mm × thickness 10 mm) were used for measurements. Thermoelectric power S(T) was obtained by taking the ratio of the voltage difference to the temperature gradient as

$$S = \Delta V / \Delta T \tag{2}$$

where ΔV is the voltage arising (in the absence of an applied field) across the sample with a temperature difference of ΔT between its ends, given by the relation

$$S = E / \Delta T \tag{3}$$

where E is the electric field in the sample. The detail of the apparatus used is already published (Rehman & Maqsood, 2005).

3. Results and discussion

X-ray diffraction (XRD) pattern of the sample with nominal composition $Bi_{1.3}V_{0.3}Pb_{0.4}Sr_2Ca_2Cu_3O_\delta$ after the final sintering of 216 h is shown in Figure 2. XRD pattern was taken using CuKa radiation (λ = 1.5046 Å) and measurements were taken at room temperature. All the peaks in the XRD pattern are indexed. XRD pattern showed peaks belonging to the high-T_c (2223) phase as a major component along with peaks due to the low-T_c (2212) phase. The dominant is the high-T_c (2223) phase having orthorhombic structure with lattice constants: a = 5.417(6) Å, b = 5.392(7) Å and c = 37.164(3) Å. The SEM result of the sample is shown in Figure 3.

It is clear from the figure that the superconducting grains are connected with each other, but with the unfilled spaces between them. This type of granular morphology has been rarely discovered in conventional high temperature crystalline superconducting samples. Furthermore, the average grain size was calculated at different spots of the sample and found to lie between 3 and 4 µm. The electrical property of the sample was examined by DC electrical resistivity measurements. The DC electrical resistivity as a function of temperature is also shown in Figure 4, after the final sintering step of 216 h. The DC electrical resistivity measurements show a well defined metallic behavior and the superconducting transitions. The resistivity versus temperature plot shows that the resistivity decreases linearly with temperature in the normal state. The onset temperature T_c (onset) and zero resistivity critical temperature T_c $(R = 0)$ were found to be 112 ± 1 K and 106 ± 1 K, respectively. The measured

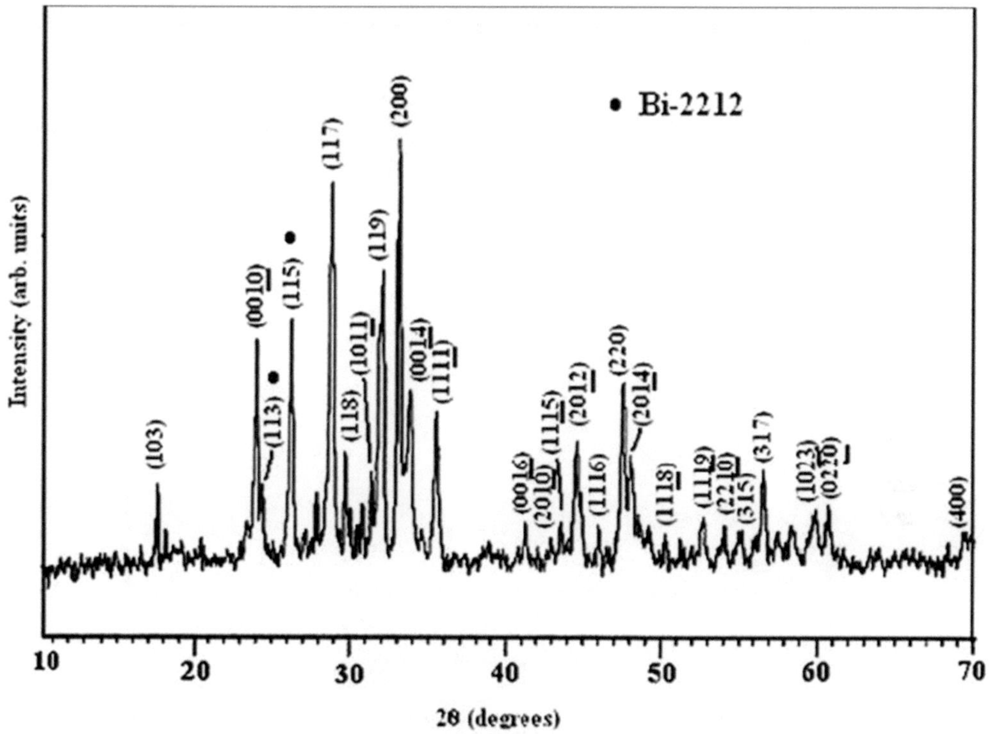

Fig. 2. Indexed X-ray pattern of $Bi_{1.3}V_{0.3}Pb_{0.4}Sr_2Ca_2Cu_3O_\delta$ superconductor

Fig. 3. SEM micrograph of the sample with composition $Bi_{1.3}V_{0.3}Pb_{0.4}Sr_2Ca_2Cu_3O_\delta$

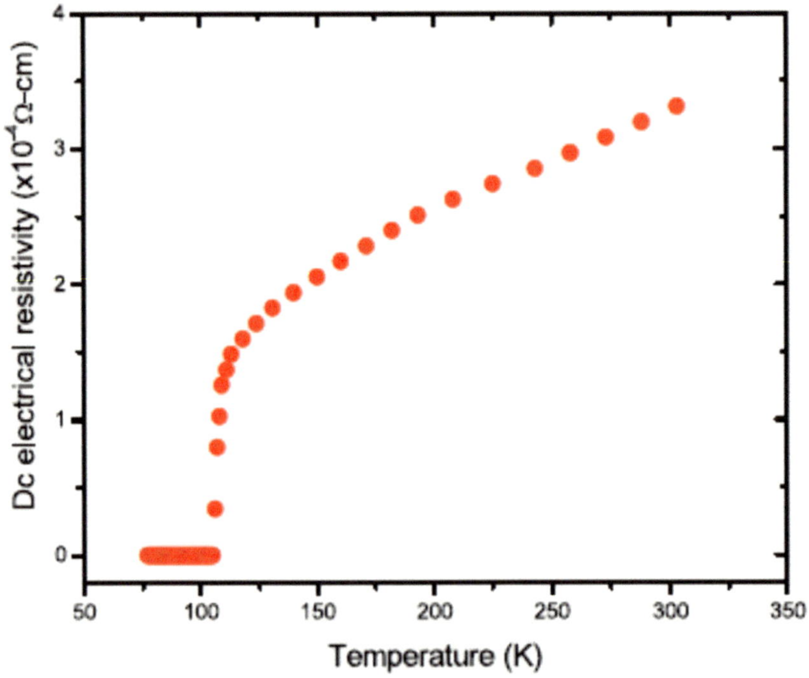

Fig. 4. Temperature dependence of DC electrical resistivity of the sample with nominal composition $Bi_{1.3}V_{0.3}Pb_{0.4}Sr_2Ca_2Cu_3O_\delta$ after final sintering time of 216 h

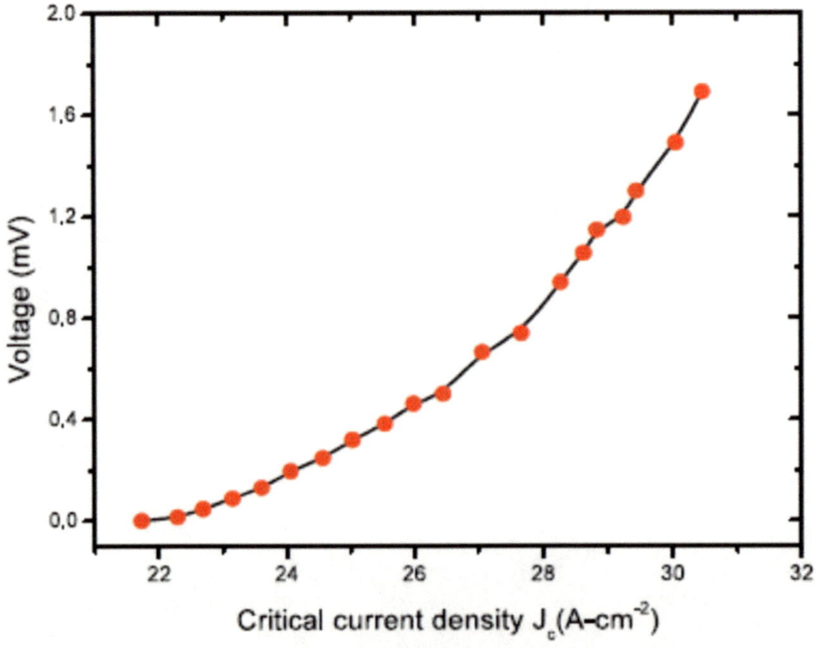

Fig. 5. Critical current density of the sample with composition $Bi_{1.3}V_{0.3}Pb_{0.4}Sr_2Ca_2Cu_3O_\delta$ at 77 K

mass density of the sample was 3.58±0.01 g/cm³. The critical current density of the samples was measured at a temperature of 77K in zero magnetic field as shown in Figure 5. It is clear that the value of the critical current density increases as the dopant concentration of vanadium increases, as compared to the undoped samples (Gul et. al., 2005). The increase in the critical current density is due to the fact that the increase in the concentration of vanadium increases the flux pinning, and strong coupling between the grain results in high values of critical current density (J_c).

The measurement of the real part χ' of the AC magnetic susceptibility on the sample $Bi_{1.3}V_{0.3}Pb_{0.4}Sr_2Ca_2Cu_3O_\delta$ is clearly showing that there are two phases. The curve of Figure 6 displays a two-step process which reflects the flux penetration between and into the grains, as temperature decreases.

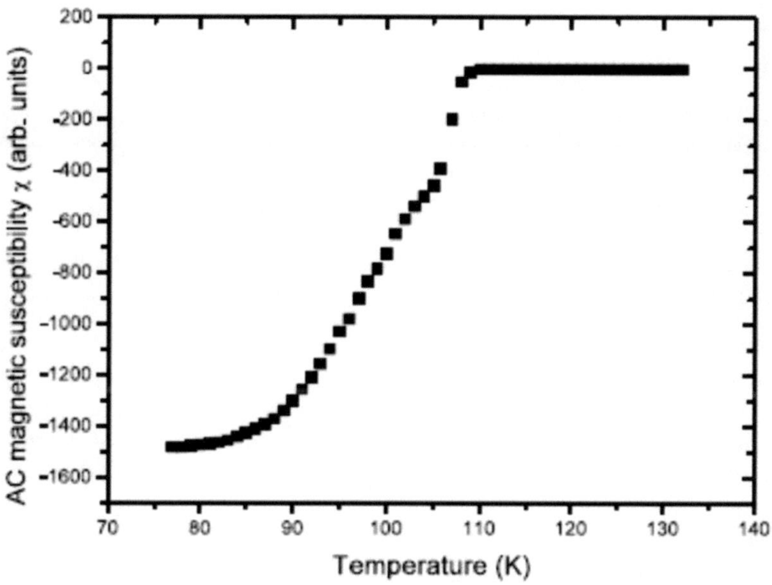

Fig. 6. Temperature dependence of the real part of AC susceptibility of the sample with composition $Bi_{1.3}V_{0.3}Pb_{0.4}Sr_2Ca_2Cu_3O_\delta$

The diamagnetic transition in χ' seems to occur in the first phase with onset temperature T_c (onset) around 113 ± 1 K. Another drop in diamagnetic signal is observed around 106 ± 1 K. This second phase has comparatively slow transition and it reflects the presence of low-*Tc* (2212) component, which is also confirmed by X-ray analysis. It is well known that the demagnetizing correction would cause $\chi' = -1$ for a sufficiently low temperature. It is evident from Figure 2 that X-ray diffraction measurement supports the observation of AC magnetic susceptibility. When the sample is at a temperature just below T_c (R = 0), the superconducting grains shield the applied magnetic field. As the temperature decreases further, the inter-granular component χ' appears and rapidly grows proportionally to the quality of interconnectivity of the grains, and volume fraction of the sample is expected to be shielded by the super current circulates in the samples, and hence the curve of χ' versus temperature saturates. Thermal conductivity measurement as a function of temperature ranging from 77 to 300 K is done using advantageous transient plane source (ATPS) technique. Thermal conductivity results are shown in Figure 7.

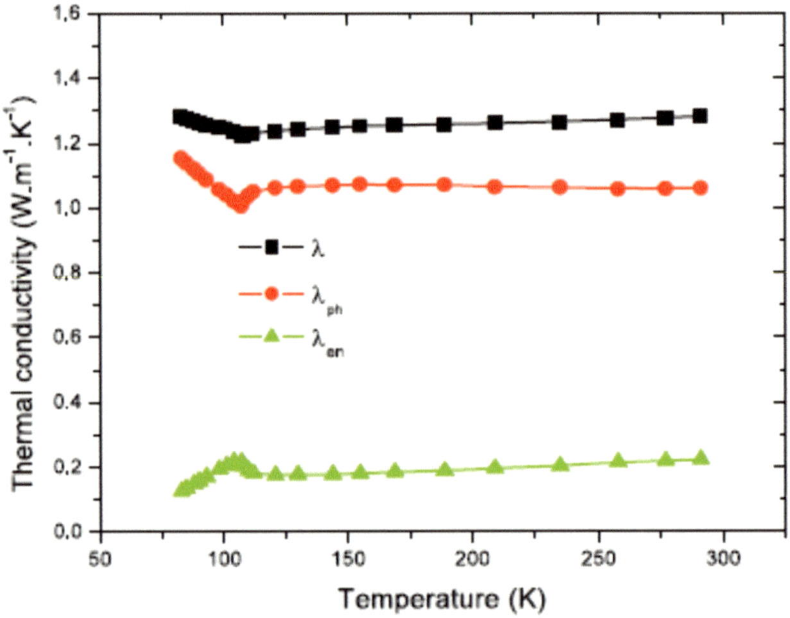

Fig. 7. Thermal conductivity (λ) as a function of temperature along with electron (λ_e) and phonon (λ_{ph}) contributions

This plot shows that there is a change in thermal conductivity at 105±1 K corresponding to T_c (R = 0). There is an increase in slope close to T_c (R = 0), which is in good agreement with the literature (Aliev, et.al. 1992; Dey, et.al. 1991; Jezowski, 1992) and also agrees with the DC electrical resistivity and AC magnetic susceptibility measurements. In superconducting state the electrons, which become Cooper pairs, no longer are allowed to exchange energy with phonon; therefore, the lattice conduction rises as more and more Cooper pairs are formed. To find the electron-phonon scattering time (τ_{e-ph}), the relation reported by Hook and Hall (Hook & Hall, 1991) is used as

$$1 / \rho_{e-ph} = ne^2\tau_{e-ph} / m_e \tag{4}$$

where n, e and m_e are the number density, charge and mass of the electron, respectively, ρ_{e-ph} being the resistivity at a particular temperature. This gave τ_{e-ph} = 2 × 10^{-15} s at 118 K, the temperature above which the resistivity is linear with temperature. Also, the phonon-limited mobility can be calculated using the relation (Hook & Hall, 1991)

$$\mu_{e-ph} = e\tau_{e-ph}/ m_e \tag{5}$$

The calculated value of the phonon-limited mobility is 3 × 10^{-4} C-S-Kg^{-1}. To estimate the size of the electron–phonon coupling constant, the lowest order variational solution of the Bloch–Boltzmann transport equation (Pinksi, et.al. 1981) is

$$\hbar / \tau_{e-ph} = 2\pi d_{e-ph} k_B T \tag{6}$$

here *de−ph* is the electron–phonon coupling constant which is 6. These results are in good agreement with the results reported by other authors (Heremans, et.al. 1988; Suleiman, et.al. 1993). The heat flux in solids consists of two contributions: energy transport associated with the flow of the charge carriers that are referred to as the electronic thermal conductivity, and thermal energy carried by lattice vibrations that is known as the lattice or phonon thermal conductivity. As is well known, the thermal conductivity (λ) of metals is given by these two contributions:

$$\lambda = \lambda_e + \lambda_{ph} \tag{7}$$

here λ_e is the thermal conductivity due to electrons and λ_{ph} is the thermal conductivity due to phonons. In simple metals, the separation of the two components of the thermal conductivity is made by use of the Wiedemann–Franz (WF) law (Ziman, 1963),

$$\lambda_{en}\,\rho / T = (1/3)\,\pi^2\,(k^2_B/e^2) = 2.45\times10^{-8}\ \mathrm{W.\Omega.K^{-2}} \tag{8}$$

where, λ_{en} is the electronic thermal conductivity in the normal state, ρ is the electrical resistivity, k_B is the Boltzmann constant. In ordinary metals, the WF law fails at intermediate temperatures where the electrical resistivity ρ deviates from the T linear dependence. The electrical resistivity of oxide superconductors shows the characteristic T linear dependence over quite a wide temperature range, as can be seen in Figure 4. Accordingly, the WF law is expected to hold over the entire temperature range above T_c (R = 0) for the oxide superconductors and to result in constant and small λ_{en}, making a marked contrast to ordinary metals. Below T_c (R = 0) the charge carriers that have condensed in the ground state do not contribute to the heat conduction, and the electronic component λ_{es} is expected to decrease with lowering temperature. Among several theories (Bardeen, et.al. 1962; Tewordt, 1963) that treat λ_{es}, we refer to revised Kadanoff's formulation adapted by Ikebe et al. (Ikebe, et.al. 1994) to observe the linear dependence of ρ,

$$\lambda_{es}/\lambda_{en} = (3/2\pi^2)\int_0^{\infty} d\varepsilon\,\varepsilon^2\,\sec h^2\,\{1/2[\varepsilon^2+(\beta\Delta)^2]^{1/2}\}\ \mathrm{x}\ [1+a(T/T_c)/(\varepsilon/[\varepsilon^2+(\beta\Delta)^2)]^{1/2}+a(T/T_c)] \tag{9}$$

here, a represents the ratio of the T -linear electrical resistance at T_c(R = 0) to the residual resistance, $\beta = 1/k_BT$, and Δ is the BCS energy gap in the Buckingham (Bardeen, et.al. 1957; Buckingham, 1956) form,

$$\Delta = 3.2k_BT_c\,(1- T/T_c)^{1/2} \tag{10}$$

The electronic and phonon contribution to thermal conductivity is estimated by (7), (8) in normal and superconducting state, respectively, as is shown in Figure 7, along with λ. The thermoelectric power is used as a medium to investigate the physical properties, in particular the properties that can be correlated with the transport mechanism that occurs in these superconductors. Thermoelectric power is a sensitive property in high-temperature superconductors, particularly to determine the sign of the charge carriers. The apparatus for the measurement of thermoelectric power was being assembled. In the aim to test the apparatus, it was calibrated with copper. Figure 8 shows that the behavior of the thermoelectric power of copper is almost linear with temperature. The magnitude of thermoelectric power of copper is approximately what one expects in noble metals such as silver and gold.

The thermoelectric power of these metals is positive except at very low temperatures. The positive sign of these metals is due to that of the Fermi surfaces touching the Brillouin zone boundaries. The Fermi surface areas could decrease with an increase of the electron energy (Barnard, 1972). Thermoelectric power (thermopower or Seebeck coefficient) determines the interaction between electrical and thermal currents in a conductor. The phonon–phonon

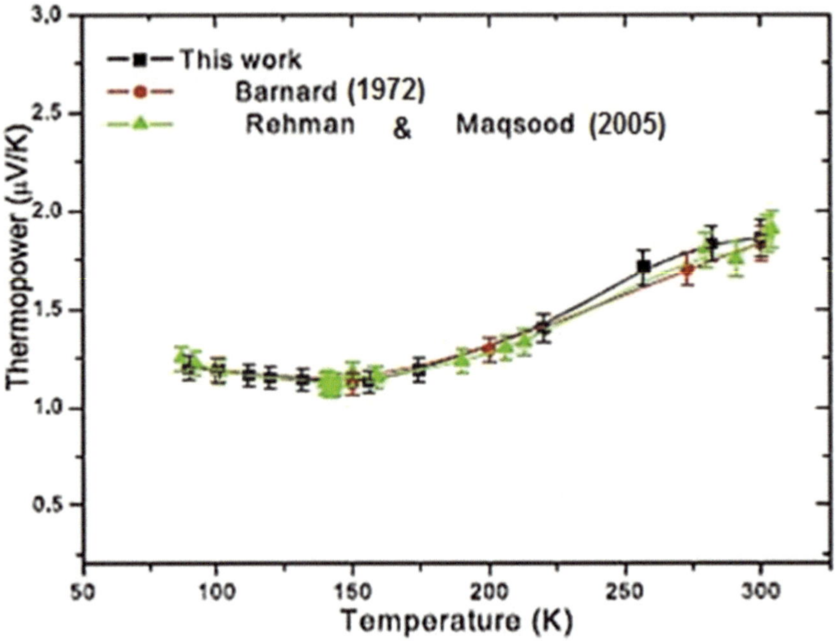

Fig. 8. Absolute thermoelectric power of copper as a function of temperature

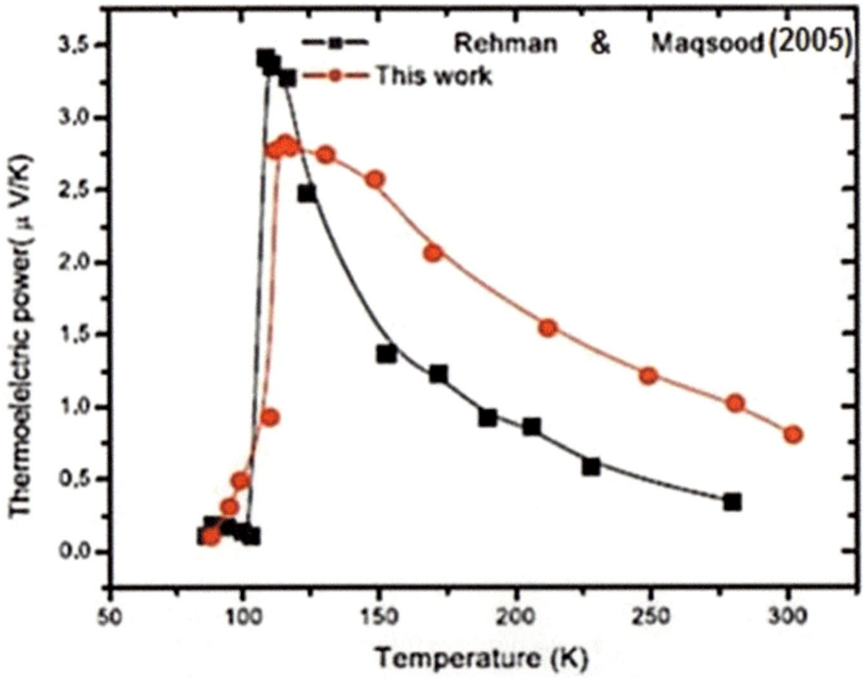

Fig. 9. Thermoelectric power of the sample with composition $Bi_{1.3}Pb_{0.4}V_{0.3}Sr_2Ca_2Cu_3O_\delta$

interaction is dominant at higher temperature, arises from anharmonicity in potential and this increase with temperature. At low temperature phonon-phonon interaction is less important. The precise form of the curve at very low temperatures is uncertain because the thermopower increases rapidly associated with the metal impurities, notably iron. Thus minute quantities of iron were the cause of this increase in thermopower even in the ultra-pure copper at very low temperature (Gul, et. al., 2005). Thermoelectric power of the superconducting sample measured in the temperature range 85-300 K is shown in Figure 9. From the figure, it is clear that thermoelectric power increases with decreasing temperature and drops rapidly to zero at superconducting transition phase. From transition to room temperature, thermoelectric power is closely linear with temperature and its profile is similar to that of other high-*Tc* superconductors (Mitra & Trefny, 1988). The critical temperature in thermoelectric power is observed at 116 ± 1 K, after that thermoelectric power begins to decrease and becomes almost zero at 88 ± 1 K. The critical temperature T_c already determined from electrical resistivity was 106±1 K. To explain these results, we assume that two kinds of carriers existed simultaneously in our sample: electrons and holes. The values of thermoelectric power in our sample are very small so we can say that a very week Seebeck coefficient shows that the behavior of the sample is typically metallic. In any superconductor, two types of contributions are important: one from phonons and the other from electrons. But contribution due to phonons is not important above the liquid nitrogen temperature (Yan, et.al., 1988).

It is also noted that the onset temperature drops of thermoelectric power are higher than those of the resistance drops. A possible explanation might be as follows. Since these materials are granular, one expects high electrical resistance between grains. On the other hand, the temperature drops between grains are expected to be small and, consequently, the granular nature would have less effect on thermoelectric power than the electrical resistivity (Lim, et,al. 1989). Polycrystalline HTSC*S* are generally viewed as an agglomeration of grains compacted together. On deoxygenation, the surface of the grains loses more and more oxygen than the grain itself. For T <T_c (R = 0), the grains become good superconductors where they are coupled together by weak links formed within the intergranular region. The weak links behave as superconducting region with very small critical current. These links become weaker and weaker with increasing deoxygenation. In such a situation, a very small current used in the critical resistivity measurement can exceed the *Jc* of the weak links, resulting in the breakdown of the percolation chain. Consequently, the resistance can persist. The temperature T_c (R = 0)S is lower than the T_c (R = 0) of the bulk material as long as *Jc* of the weak links is smaller than the transport current density. However, contrary to the resistivity measurement, the thermopower experiment is performed in the absence of any current. Hence the percolation chain of the superconducting grains can persist up to temperatures higher than T_c (R = 0) RES even in the presence of a small temperature gradient. This shifts T_c (R = 0)S to relatively higher temperatures (Dey, et. al., 1998).

4. Conclusions

A superconducting sample with nominal composition $Bi_{1.3}V_{0.3}Pb_{0.4}Sr_2Ca_2Cu_3O_\delta$ was prepared by the solid state reaction method. By resistivity measurement, T_c (R = 0)RES obtained for this composition was 106± 1 K. Thermal conductivity variation with temperature showed initially a slight decrease and then a pronounced increase around T_c (R = 0). Although the expected theoretical trend is similar, the peak near *T/2* was not observed due to temperature limitations of the temperature controller. A similar behavior is observed in all hole-type CuO_2 plane superconductors and in all their structural forms. This effect is due to phonon or quasiparticle scattering. The Wiedemann–Franz law is applied to predict the magnitude of electron and phonon contributions to the total thermal conductivity of the samples, whereas T_c (onset) observed was at 112 ± 1 K. Thermoelectric power of the sample was positive in the temperature range 85– 300 K. The behavior of thermoelectric power in this superconducting sample was approximately linear with temperature as observed in other Bismuth-based high-T_c superconductors. The transition temperature of this superconductor T_c (R = 0)S was measured to be 116 ± 1 K, whereas the transition temperature of this sample measured with resistivity, T_c (R = 0)RES, method was 106±1 K. The difference between T_c (R = 0)RES and T_c (R = 0)S is 10 K. This difference increases as more and more oxygen is being taken out of the compound by deoxygenation.

5. Acknowledgements

The authors would like to acknowledge the Higher Education Commission (HEC) Islamabad, Pakistan for the financial support. Mr. A. Abdullah is acknowledged for useful support in preparation of the manuscript.

6. References

Aksan, M.A., Yakinci, M.E., Balci, Y. & Ates, H. (1999). Thermal conductivity properties of $Bi_{2-\alpha}Tl_{\alpha}Sr_2Ca_2Cu_3O_{10+z}$ glass-ceramic HT_c superconductor rods. *J. Low Temp. Phys.* Vol. 117, No. 3-4, 957-961

Aksan, M.A. & Yakyncy, M.E. (2004). Synthesis and characterization of Er-substituted Bi-2223 H-T_c glass–ceramic superconductors. *J. Alloy. Compd.* Vol. 385, No. 1-2, 33-43

Aliev, F.G., Moshchalkov, V.V., Pryadun, V.V. & Monge, M.A.A. (1992). Thermal conductivity of $TmBa_2Cu_3O_x$ and $Bi_2Sr_2CaCu_2O_x$ single crystals in the vicinity of the superconducting transition. *Solid State Commun.* Vol. 82, No. 4, 241-244

Bardeen, J., Cooper, L.N. & Schrieffer, J.R. (1957). Theory of superconductivity *Phys. Rev.* Vol. No. 5, 108, 1175-1204

Bardeen, J., Cooper, L.N., Schrieffer, J.R. (1962). Theory of the thermal conductivity of superconductors. *Phys. Rev.* Vol. 113, No. 4, 982-994

Barnard, R.D.: *Thermoelectricity in Metals and Alloys.* Taylor and Francis Ltd., London (1972)

Buckingham, M.J. (1956). Very high frequency absorption in superconductors *Phys. Rev. Vol.* 101, No. 4, 1431-1432

Chanda, B. & Dey, T.K. (1994). Thermal conductivity of bismuth-lead (2223) superconducting cuprates with vanadium/nickel substitutions and evidence for strong electron-phonon coupling. *Solid State Commun.* Vol. 89, No. 4, 353-361

Castellazzi, S., Cimberle, M.R., Ferdeghini, C., Giannini, E., Grasso, G., Marre, D., Putti, M. & Siri, A.S. (1997). Thermal conductivity of a BSCCO(2223) *c*-oriented tape: a discussion on the origin of the peak. *Physica C* Vol. 273, No. 3-4, 314-322

Chatterjee, S., Pal, P.K., Bhattacharya, S. & Chaudhuri, B.K. (1998). Depression of superconductivity and the phonon-drag effect in glass-ceramic superconductors obtained by annealing the Fe-doped (Bi, Pb)$_4$Sr$_3$Ca$_3$Cu$_{4-m}$Fe$_m$O$_x$glassy precursor. *Phys. Rev. B*. Vol. 58, No. 18, 12427–12432

Cloots, R., Bougrine, H., Houssa,M., Stassen, S., D'Urzo, L., Rulmont, A. & Ausloss, M. (1994). Bi-based 2223 superconducting polycrystalline materials prepared by either a solid state route or a glassy 'matrix' precursor method: Chemical analysis as well as electrical and thermal transport properties. *Physica C*. Vol. 231, No. 3-4, 259-270

Cohn, J.L., Skelton, E.F., Wolf, S.A., Liu, J.Z. & Shelton, R.N. (1992). Cyclotron resonance of both magnetopolaron branches for polar and neutral optical phonon coupling in the layer compound InSe.*Phys. Rev. B* vol. 45, No. 20, 12144-12147

Coskun, A., Ekicibil, A., Ozcelik, B. & Kyymaç, K. (2005). Effects of annealing time on the magnetic properties of a $Bi_{1.7}Pb_{0.29}Gd_{0.01}Sr_2Ca_3Cu_4O_{12+y}$ superconductor prepared by the melt-quenching method. *Chin. J. Phys.* Vol. 43, No. 2, 372

Dey, T.K. (1998). Thermoelectric power of deoxygenated $Bi_{1.6}Pb_{0.4}Sr_2Ca_2Cu_3O_{10+\delta}$ sintered superconducting pellets. J. Supercond. Vol.11, 367-372

Dorbolo, S., Ausloss, M., Bougrine, H., Robertz, B., Cloots, R., Mucha, J. & Durczewski, K. (1999). Effect of synthesis process and substrate on electrical and thermal transport properties of Bi-2212. *J. Supercond.* Vol. 12, Number 5, 623-629

Ekicibil, A., Coskun, A., Ozcelik, B. & Kyymacç, K. (2005). The magnetic and electrical properties of rare earth Sm^{3+} substituted $Bi_{1.7}Pb_{0.3}Sr_2Ca_{2-x}Sm_xCu_3O_{12}$ system. *Mod. Phys. Lett. B* Vol. 331, No. 6, 331-340

Gul, I.H., Rehman, M.A., Ali, M. & Maqsood, A. (2005). Effect of vanadium and barium on the Bi-based (2223) superconductors. *Physica C* Vol. 432, No. 1-2, 71-80

Gusstafsson, S.E.(1991). Transient plane source techniques for thermal conductivity and thermal diffusivity measurements of solid materials. *Rev. Sci. Instrum.* Vol. 62, No. 3, 797-804

Heremans, J., Morelli, D.T., Smith, G.W., Strite, S.C. (1988). Thermal and electronic properties of rare-earth $Ba_2Cu_3O_x$ superconductors. *Phys. Rev. B* Vol. 37, 1604-1610

Hirai, T.: In: Brook, R.J. (ed.) (1996). *Processing of Ceramics*, Part 2. VCH Verlagsgesellschaft mbH, Weinheim

Hook, J.R., Hall, H.E. (1991). *Solid State Phys*, 2nd edn. Wiley, Chichester

Houssa, M. & Ausloos, M. (1995). Thermal conductivity of superconducting $Bi_2Sr_2CaCu_2O_8$ and $YBa_2Cu_3O_{7-y}$. *Phys. Rev. B* Vol. 51, No. 14, 9372-9374

Houssa, M. & Ausloos, M. (1996). n-plane electronic thermal conductivity of layered d-wave high-T_c superconductors. *Physica C* Vol. 257, No. 3-4, 321-331

Houssa,M., Ausloos,M. & Sergeenkov, S. (1996). The electronic contribution to the thermal conductivity of layered high-T_c materials. *J. Phys.: Condens. Phys.* Vol. 8, No. 12, 2043-2052

Hui, P.M., Zhang, X. ,Markworth, J. & Stroud, D. (1999). Thermal conductivity of graded composites: Numerical simulations and an effective medium approximation. *J. Mater. Sci.* Vol. 34, Number 22, 5497-5503

Ikebe, M. Fujisshiro, H. Naito, T., Noto, K. (1994). Simultaneous measurement of thermal diffusivity and conductivity applied to Bi-2223ceramic superconductor. *J. Phys. Soc. Jpn*. Vol. 63, 3107-3114

Jezowski, A., Mucha, J., Rogaci, K., Horyn, R., Bukovski, Z. & Horobiowski, M. (1987). Thermal conductivity and electrical resistivity of the high-T_c superconductor $YBa_2Cu_3O_{9-\Delta}$. *Phys. Lett. A* Vol. 122, No. 8, 431-433

Khim, Z.G., Lee, S.C., Lee, J.H., Suh, B.J., Park, Y.W., Park, C., Yu, I.S. & Park, J.C. (1987). Superconductivity in single-phase $Y_1Ba_2Cu_3O_{9-x}$ and thermoelectric power measurement. *Phys. Rev. B* Vol. 36, No. 4, 2305-2307

Knizek, K., Veverka, M., Hadova, E., Hejtmanek, J., Sedmidubsky, D. & Pollert, E. (1998). Synthesis of $HgBa_2CuO_{4+\delta}$ by sol–gel method under controlled oxygen pressure; electron and thermal transport properties. *Physica C* Vol. 302, No. 4, 290-298

Lee, S.C., Lee, J.H., Suh, B.J., Moon, S.H., Lim, C.J. & Khim, Z.G. (1988). Thermoelectric power and superconducting properties of $Y_1Ba_2Cu_3O_{7-\delta}$ and$R_1Ba_2Cu_3O_{7-\delta}$. *Phys. Rev. B* Vol. 37, No. 4, 2285-2288

Lim, Z.S., Han, K.H., Salk, S.H. & Song, Y.S. (1989). Charged-particle interaction with liquids: ripplon excitations. *Phys. Rev. B* Vol. 40, No. 1, 10-19

Mandal, J.B., Keshri, S., Mandal, P., Poddar, A., Das, A.N. & Ghosh, B. (1992). Thermoelectric power of the $Bi_2Sr_2Ca_{1-x}Y_xCu_2O_{8+y}$ (x=0–1.0) system. *Phys. Rev. B* Vol. 46, No. 18, 11840-11846

Maqsood A., Mahmood M.S., Suleman B. & Tasneem A. (1989). Preparation and characterization of superconducting $Dy_1Ba_2Cu_3O_{7-x}$ material. *J. Mat. Sci. Lett.* Vol. 8, No. 7, 757-758

Mitra, N. & Trefny, J. (1988). Thermoelectric power of the Tl-Ca-Ba-Cu-O superconductor. *Phys. Rev. B* Vol. 38, No. 10, 7064-7066

Morelli, D.T., Heremans, J. & Swets, D.E. (1987). Thermal conductivity of superconductive Y-Ba-Cu-O. *Phys. Rev. B* Vol. 36, No. 7, 3917-3919

Munakata, F., Matsuura, K., Kubo, K., Kawano, T. & Yamauchi, H. (1992). Thermoelectric power of $Bi_2Sr_2Ca_{1-x}Y_xCu_2O_{8+y}$. *Phys. Rev. B* Vol. 45, No. 18, 10604-10608

Nanda Kishore, K., Satyavathi, S., Muralidhar, A., Pena, O. & Hari Babu, V. (1995). Thermoelectric power studies on the Sm substituted BPSCCO (2223) superconductors. *Physica C* Vol. 252, No. 1-2, 49-53

Natividad, E., Castro, M., Burriel, R., Angurel, L.A., Diez, J.C. & Navarro, R. (2002). Correlation of normal and superconducting transport properties on textured Bi-2212 ceramic thin rods. *Supercond. Sci. Technol.* Vol. 15, No. 7, 1022

Ozhanli, Z., Yakýncý, M.E., Balcý, Y. & Aksan, M.A. (2002). Crystallization activation energy and hole concentration properties of the $Bi_2Sr_2Ca_{1-x}Cd_xCu_2O_{8+y}$ glass-ceramic superconductor Rods. *J. Supercond.* Vol. 15, No. 6, 543-547

Peacor, S.D., Cohn, J.L. & Uher, C. (1991). Effect of magnetic field on thermal conductivity of $YBa_2Cu_3O_{7-\delta}$ single crystals. *Phys. Rev. B* Vol. 43, No. 10, 8721-8724

Pinksi, F.J., Allen, P.B. & Butler, W.H. (1981). Calculated electrical and thermal resistivities of Nb and Pd. *Phys. Rev. B* Vol. No. 10,23, 5080-5096

Rao, C.N.R., Ramakrishnan, T.V. & Kumar, N. (1990). Systematics in the thermopower behaviour of several series of bismuth and thallium cuprate superconductors: An interpretation of the temperature variation and the sign of the thermopower. *Physica C* Vol. 165, No. 2, 183-*188*

Rehman, M.A. & Maqsood, A. (2005). Study of the thermal behaviour of Bi(Pb)Sr(Ba)-2223 high-T_c granular superconductors. *Physica C*. Vol. 418, No. 3-4, 121-130

Rehman M.A. (2009). Thermal and electrothermal characterization of bismuth based high-T_c superconductors *J. Alloys Comp.* Vol. 469, No. 1-2, 66-72

Sera,M., Tanaka, S., Sato,M. & Fujishita, H. (1992). Anomalous thermoelectric power of $Bi_{1.6}Pb_{0.5}Sr_{1.9-y}La_yCu_{1.05}O_z$. *Solid State Commun.* Vol. 81, No. 5, 415-417

Suleiman, B.M., Haq, I., Karawacki, E.,Maqsood, A. & Gustafsson, S.E. (1993). Thermal conductivity and electrical resistivity of the Y- and Er-substituted 1:2:3 superconducting compounds in the vicinity of the transition temperature. *Phys. Rev. B* Vol. 48, No. 6, 4095-4102

Tewordt, L. (1963). Theory of the intrinsic electronic thermal conductivity of superconductors *Phys. Rev.* Vol. 129, No.2, 657-663

Tewordt, L. & Wolkhausen, T. (1989). Theory of thermal conductivity of the lattice for high-T_c superconductors. *Solid State Commun.* Vol. 70, No. 8, 839-844

Tewordt, L. & Wolkhausen, T. (1990). Theory of phonon thermal conductivity for strong-coupling s- and d-wave pairing in high T_c superconductors. *Solid State Commun.* Vol. 75, No. 6, 515-519

Uher, C. & Kaiser, A.B. (1987). Thermal transport properties of $YBa_2Cu_3O_7$ superconductors. *Phys. Rev. B* Vol. 36, No. 10, 5680

Uher, C. & Wang, W.N. (1989). Thermoelectric power and thermal conductivity of neutron-irradiated $YBa_2Cu_3O_{7-\delta}$, *Phys. Rev. B* Vol. 40, No. 4, 2694-2697

Uher, C.: In: Ginsberg, D.M. (ed.) *Physical Properties of HTSC*,vol. 3. World Scientific, Singapore (1992)

Vasudeva R. V., Rangarajan, G. & Srinivasan, R. (1984). Thermoelectric power in the normal state of Chevrel-phase superconductors of the type $Cu_{1.8}Mo_6S_{8-y}Se_y$, $0 \leq y \leq 8$, and $Cu_{1.8}Mo_6S_{8-y}Te_y$ $0 \leq y \leq 4$. *J Phys.F* Vol.14, 973-980

Varoy, C.R., Trodahl, H.J., Buckley, R.G. & Kaiser, A.B. (1992). Thermopower of $Bi_{2-x}Pb_xSr_2CaCu_2O_{8+\delta}$. *Phys. Rev. B* Vol. 46, No. 1, 463-468

Yan, S., Lu, P. & Li, Q. (1988). Thermoelectric power of single phase_$YBa_2Cu_3O_{7-x}$ superconductors. *Solid State Commun.* Vol. 65, No. 5, 355-358

Yu, R.C., Salamon, M.B., Lu, J.P. & Lee, W.C. (1992). Thermal conductivity of an untwinned $YBa_2Cu_3O_{7-\delta}$ single crystal and a new interpretation of the superconducting state thermal transport. *Phys. Rev. Lett.* Vol. 69, No. 9, 1431-1434

Varoy, C.R., Trodahl, H.J., Buckley, R.G. & Kaiser, A.B. (1992). Thermopower of $Bi_{2-x}Pb_xSr_2CaCu_2O_{8+\delta}$. *Phys. Rev. B* Vol. 46, No. 1, 463-468

Wang, J., Wakata, M., Kaneko, T., Takano, S. & Ya-mauchi, H.(1993). Enhancement of T_c in (Bi,Pb)-2223 superconductor by vacuum encapsulation and post-annealing. *Physica C* vol. 208, No. 3-4, 323-327

Wermbter, S. & Tewordt, L. (1991). Theory of thermal conductivity and nuclear relaxation rate for high temperature superconductors. *Physica* C Vol. 183, No. 4-6, 365-371

Wu M.K., Ashburn J.R., Torng C.J. & Horetal P.H. (1987). Superconductivity at 93 K in a new mixed-phase Y-Ba-Cu-O compound system at ambient pressure. *Phys. Rev. Lett.* Vol. 58, No. 9, 908-910

Ziman, J.M.: *Electrons and Phonons*. Clarendon, Oxford (1963)

9

Chemical Solution Deposition Based Oxide Buffers and YBCO Coated Conductors

M. Parans Paranthaman
Chemical Sciences Division
Oak Ridge National Laboratory
USA

1. Introduction

The main objective of this work is to conduct fundamental research in the broad areas of chemical solution based buffer and high temperature superconductor, namely Yttrium Barium Copper Oxide (YBCO) development. The results of this research provide new insights in buffer/superconductor areas and suggest methods to improve buffer/superconductor multi-layer thin film fabrication. The overall purpose is to develop a potentially lower-cost, high throughput, high yield, manufacturing processes for buffer/superconductor thin multi-layer film fabrication, and to gain fundamental understanding of the growth of solution buffer/superconductor layers for Rolling Assisted Biaxially Textured Substrate (RABiTS) templates. This understanding is critical to the development of a reliable, robust, long-length manufacturing process of second-generation (2G) wires for electric-power applications. In order to reduce the cost of superconductor wires, it is necessary to replace the existing physical vapor deposited three buffer layer RABiTS architecture of Yttrium Oxide, Y_2O_3 seed/Yttria Stabilized Zirconia, YSZ barrier/Cerium Oxide, CeO_2 cap with buffers deposited by industrially scalable methods, such as slot-die coating of chemical solution deposition (CSD) precursors [1-11]. Spin-coating is typically used to deposit short samples for optimizing the CSD film growth conditions. In a typical chemical solution process, metal organic precursors in suitable solvents are spin/dip/slot-die coated on either single crystal or biaxially textured substrates followed by heat-treating in a tube furnace under controlled conditions. Chemical Solution Deposition (CSD) process offers significant cost advantages compared to physical vapor deposition (PVD) processes [5-11]. Solution coating is amenable to complex oxides, and the materials utilization (yield) is almost 100%. The high-temperature superconductors (HTS) such as $(Bi,Pb)_2Sr_2Ca_2Cu_3O_{10}$ (BSCCO or 2223 with a critical temperature, T_c of 110 K) and $YBa_2Cu_3O_{7-\delta}$ (YBCO or 123 with a T_c of 91 K) have emerged as the leading candidate materials for the first generation (1G) and second generation (2G) high temperature superconductor wires or tapes that will carry high critical current density in liquid nitrogen temperatures [1,2]. Here, we report the growth of buffer/YBCO superconductor film growth using a chemical solution method towards fabrication of second generation superconductor wires.

2. Chemical solution deposition of oxide buffers

The schematic of the standard RABiTS architecture developed by Oak Ridge National Laboratory and American Superconductor Corporation [3,4] is shown in Figure 1. The main goal is to replace the most commonly used RABiTS architectures with a starting template of biaxially textured Ni-5 at.% W substrate with a physical vapor deposited (PVD) seed layer

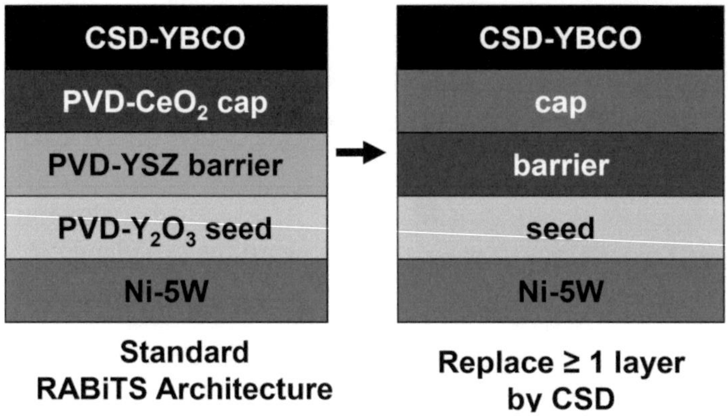

Fig. 1. The schematic of the standard RABiTS architecture.

Buffers	Structure type		% lattice mismatch		Ref.
	Cubic lattice parameter	Pseudocubic a/2 2 or	vs. YBCO	vs. Ni	
$BaCeO_3$	4.377		13.55	21.59	12
$BaZrO_3$	4.193		9.27	17.34	13
$Pb_{0.56}La_{0.3}TiO_3$	3.916		2.44	10.54	14
La_3TaO_7	11.054	3.908	2.37	10.34	15-17
$SrTiO_3$	3.905		2.16	10.26	18,19
Eu_2O_3	10.868	3.843	0.54	8.64	20,21
$NdGaO_3$	5.431	3.841	0.51	8.61	22
CeO_2	5.411	3.826	0.12	8.22	23,39
Gd_2O_3	10.813	3.824	0.07	8.17	24
$La_2Zr_2O_7$	10.786	3.814	- 0.20	7.90	20,25-27
$LaAlO_3$	5.364	3.793	- 0.75	7.35	28,29
Gd_3NbO_7	10.659	3.769	- 1.42	6.95	30
Y_2O_3	10.604	3.750	- 1.89	6.22	31
$Gd_2Zr_2O_7$	5.264	3.722	- 2.64	5.47	32,33
Y_3NbO_7	5.250	3.713	- 2.88	5.23	30
Yb_2O_3	10.436	3.690	- 3.50	4.61	34
Ni	3.524				

Table 1. Structure, lattice misfit data and chemical solution deposition (CSD) methods for various buffer layers. The lattice parameters were obtained from the International Center for Diffraction Data, Powder Diffraction Files. * Rhombohedral; ♦ Orthorhombic

of Y_2O_3, a barrier layer of YSZ, and a CeO_2 cap layer by a chemical solution deposition method. To develop an all solution buffer/YBCO, it is necessary to either replace all three layers or reduce the number of buffer layers to one. The role of the Y_2O_3 seed layer is to improve the out-of-plane texture of buffer layer compared to the underlying Ni-5W substrate and Y_2O_3 is also an excellent W diffusion and good oxygen barrier [4]. The role of YSZ barrier layer is to contain the diffusion of Ni from the substrate into superconductor. In order to grow YBCO superconductor films with critical current densities, it is necessary to contain the poisoning of Ni into YBCO. Finally, the CeO_2 cap layer is compatible with CSD based REBCO films and has enabled high critical current density REBCO films. The optimized film thickness for each buffer layer is 75 nm and the typical YBCO layer thickness is ~ 1 μm carrying a critical current of 250-300 A/cm-width at 77 K and self-field. Researchers all over the world have developed several chemical solution deposited oxide buffer layers that are suitable for YBCO film growth. A partial list of several epitaxial oxide buffers grown using a CSD method have been reported in Table 1 [4]. It is possible for us to select a buffer layer to lattice match with either the substrate Ni/Ni-W or with YBCO. The list of chemical solution deposited buffer layers with YBCO superconductor films deposited on such buffers is reported in Table 2.

CSD Buffer Layers	Stacking for YBCO	J_c (MA/cm^2)	Reference
CeO_2	YBCO (CSD)/CeO_2 (Sputtered)/YSZ (Sputtered)/CeO_2 (CSD)/Ni-W	3.3	39
YSZ	YBCO (CSD)/CeO_2 (CSD)/YSZ (CSD)/ CeO_2 (CSD)/Ni	0.5	35
Y_2O_3	YBCO (PLD)/CeO_2 (Sputtered)/YSZ (Sputtered)/Y_2O_3 (CSD)/Ni-W	1.2	31
Eu_2O_3	YBCO (ex-situ BaF_2)/CeO_2 (Sputtered)/ YSZ (Sputtered)/Eu_2O_3 (CSD)/Ni	1.1	20
Gd_2O_3	YBCO (PLD)/CeO_2 (Sputtered)/YSZ (Sputtered)/Gd_2O_3 (CSD)/Ni-W-Fe	1	36
Ce-Gd-O	YBCO (CSD)/CeO_2 (CSD)/CGO (CSD)/ Gd_2O_3 (CSD)/Ni	0.1	37
$SrTiO_3$	YBCO (CSD)/STO (CSD)/Ni	1.3	38
$La_2Zr_2O_7$	YBCO (e-beam)/CeO_2 (Sputtered)/YSZ (Sputtered)/LZO (CSD)/Ni	0.48	26
$La_{1/4}Zr_{3/4}O_y$	YBCO (PLD)/$La_{1/4}Zr_{3/4}O_y$ (CSD)/Ni-W	0.55	42
$Gd_2Zr_2O_7$	YBCO (MOCVD)/GZO (CSD)/Ni	1.3	33
Gd_3NbO_7	YBCO (PLD)/GNO (CSD)/Ni-W	1.1	30

Table 2. List of chemical solution deposited oxide buffer layers with J_c of the high temperature superconducting YBCO films deposited on such buffers.

3. Chemical solution deposition of REBCO

Currently, chemical solution based synthesis of YBCO uses a trifluoroacetate (TFA) based precursor approach [5]. In this approach, the precursor solution is prepared by dissolving Yttrium, Barium and Copper trifluoroacetates in methanol. Then the precursor solution is spin/slot-die coated on RABiTS templates followed a two-stage heat-treatment to convert the precursor films to high quality YBCO. In the first stage (pyrolysis), there is a significant bottle neck to processing rates for these films because the shrinkage stresses developed in the films during pyrolysis need to be accommodated using very slow heating rates. The reactions taking place during the synthesis are illustrated below:

$$Y(OOCCF_3)_3 + 2\,Ba(OOCCF_3)_2 + 3\,Cu(OOCCF_3)_3$$

$$\rightarrow 0.5\,Y_2O_3 + 2\,BaF_2 + 3\,CuO + (nCO_2 + mC_xO_yF_2) \quad (1)$$

$$0.5\,Y_2O_3 + 2\,BaF_2 + 3\,CuO + 2\,H_2O \rightarrow YBa_2Cu_3O_{7-\delta} + 4HF \quad (2)$$

Significant efforts were made to increase the growth rate by replacing part of the metal TFA precursors with non-fluorine based precursors and also adjust the water and oxygen pressure during the growth of YBCO films. Another advantage of the TFA process is to introduce mixed rare earths and Zirconium oxides into the starting precursors to enhance the flux-pinning properties of REBCO films [5,40,41]. Chemical solution deposition method may prove to be a promising route for producing a low-cost all-CSD buffer/YBCO based coated conductors. The main challenge is to fabricate high-temperature superconductor tapes in kilometer lengths in carrying 1000 A/cm-width. Industries from US and Japan are leading in this area while industries from Europe, Korea, and China are only few years away.

4. Summary

In summary, RABiTS template with several possible architectures based on chemical solution deposition methods have been developed and superconductivity industries around the world are in the process of taking the technology to the pilot scale to produce commercially acceptable 500 meter lengths. The research in the area of second generation high temperature superconductor wire technology to increase the flux pinning properties of YBCO superconductor and to reduce the ac loss in these wires for various electric-power applications such as transmission cables, fault-current limiters and high-field magnets is continuing ahead.

5. Acknowledgements

This work was supported by the U.S. Department of Energy, Office of Electricity Delivery and Energy Reliability (OE) – Advanced Conductors and Cables Program.

6. References

[1] M. Parans Paranthaman and T. Izumi, Editors, "High-Performance YBCO-Coated Superconductor Wires," MRS Bulletin 29 (2004) 533-536.

[2] M. Parans Paranthaman, "Superconductor Wires," in McGraw-Hill 2006 Yearbook of Science and Technology, McGraw-Hill Publishers, New York (2006) pp. 319-322.
[3] A. Goyal, D.P. Norton, J.D. Budai, M. Paranthaman, E.D. Specht, D.M. Kroeger, D.K. Christen, Q. He, B. Saffian, F.A. List, D.F. Lee, P.M. Martin, C.E. Klabunde, E. Hatfield, V.K. Sikka, "Fabrication of Long Range, Biaxially Textured, High Temperature Superconducting Tape on Rolled Ni Substrates," Appl. Phys. Lett. 69 (1996) 1795.
[4] A. Goyal, M. Paranthaman, and U. Schoop, "The RABiTS Approach: Using Rolling-Assisted Biaxially Textured Substrates for High-Performance YBCO Superconductors," MRS Bulletin 29 (2004) 552-561.
[5] M.W. Rupich, D.T. Verebelyi, W. Zhang, T. Kodenkandath, and X. Li, Metalorganic Deposition of YBCO Films for Second-Generation High-Temperature Superconductor Wires," MRS Bulletin 29 (2004) 572-578.
[6] C.J. Brinker, A.J. Hurd, P.R. Schunk, G.C. Frye, C.S. Ashley, "Review of Sol-gel Thin Film Formation," J. Non-Cryst. Solids 147 (1992) 424.
[7] F.F. Lange, "Chemical Solution Routes to Single-crystal Thin Films," Science 273 (1996) 903.
[8] R.W. Schwartz, "Chemical Solution Deposition of Perovskite Thin Films," Chem. Mater. 9 (1997) 2325.
[9] S. Sathyamurthy, M. Paranthaman, Lee Heatherly, Patrick M. Martin, E.D. Specht, Amit Goyal, Thomas Kodenkandath, Xiaoping Li, Martin W. Rupich, "Solution-processed lanthanum zirconium oxide as a barrier layer for high I_c coated conductors," J. Mater. Res. 21 (2006) 910.
[10] M.P. Paranthaman, S. Sathyamurthy, M.S. Bhuiyan, P.M. Martin, T. Aytug, K. Kim, M. Fayek, K.J. Leonard, J. Li, A. Goyal, T. Kodenkandath, X. Li, W. Zhang, M.W. Rupich, "MOD buffer/YBCO approach to fabricate low-cost second generation HTS wires," IEEE Trans. On Appl. Supercond. 17 (2007) 3332.
[11] M. Coll, J. Gazquez, R. Huhne, B. Holzapfel, Y. Morilla, J. Garcia-Lopez, A. Pomar, F. Sandiumenge, T. Puig, X. Obradors, "All chemical $YBa_2Cu_3O_7$ superconducting multilayers: Critical role of CeO_2 cap layer flatness," J. Mater. Res. 24 (2009) 1446.
[12] G.N. Glavee, R.D. Hunt and M. Paranthaman, "Low Temperature Preparation of $BaCeO_3$ and $Ce_{0.75}Zr_{0.25}O_2$ Thin Films Using Sol-gel Processing Techniques," Materials Research Bulletin 34 (1999) 817-825.
[13] M. Paranthaman, S.S. Shoup, D.B. Beach, R.K. Williams and E.D. Specht, "Epitaxial Growth of $BaZrO_3$ Films on Single Crystal Oxide Substrates Using Sol-gel Alkoxide Precursors," Materials Research Bulletin 32 (1997) 1697-1704.
[14] N.J. Ali, P. Clem and S.J. Milne, "Synthesis of sols for the production of La- modified $PbTiO_3$ thin films," J. Mater. Sci. Lett. 14 (1995) 837-840.
[15] M. P. Paranthaman, M. S. Bhuiyan, S. Sathyamurthy, L. Heatherly, C. Cantoni and A. Goyal, "Improved textured $La_2Zr_2O_7$ buffer on La_3TaO_7 seed for all-MOD Buffer/YBCO coated conductors," Physica C 468 (2008) 1587.
[16] M.S. Bhuiyan, M. Paranthaman, S. Sathyamurthy, "Chemical solution-based epitaxial oxide films on biaxially textured Ni-W substrates with improved out-of-plane texture for YBCO coated conductors," J. Electronic Mater. 36 (2007) 1270.

[17] M.S. Bhuiyan, M. Paranthaman, A. Goyal, L. Heatherly and D.B. Beach, "Deposition of rare earth tantalite buffers on textured Ni-W substrates for YBCO coated conductor using chemical solution deposition approach," J. Mater. Res. 21 (2006) 767-773.

[18] J.T. Dawley, R.J. Ong and P.G. Clem, "Improving sol-gel $YBa_2Cu_3O_{7-\delta}$ film morphology using high-boiling-point solvents," J. Mater. Res. 17 (2002) 1678-1685.

[19] S. Sathyamurthy and K. Salama, "Application of metal–organic decomposition techniques for the deposition of buffer layers and Y123 for coated-conductor fabrication," Physica C 329 (2000) 58-68.

[20] M. Paranthaman, T.G. Chirayil, F.A. List, X. Cui, A. Goyal, D.F. Lee, E.D. Specht, P.M. Martin, R.K. Williams, D.M. Kroeger, J.S. Morrell, D.B. Beach, R. Feenstra and D.K. Christen, "Fabrication of Long Lengths of Epitaxial Buffer Layers on Biaxially Textured-Ni Substrates using a Continuous Reel-to-reel Dip-coating Unit," J. Amer. Ceram. Soc. 84 (2001) 273-278.

[21] Y. Akin, Z.K. Heiba, W. Sigmund and Y.S. Hascicek, "Engineered oxide thin films as 100% lattice match buffer layers for YBCO coated conductors," Solid-State Electronics 47 (2003) 2171-2175.

[22] C.Y. Yang, A. Ichinose, S.E. Babcock, J.S. Morrell, J.E. Mathis, D.T. Verebelyi, M. Paranthaman, D.B. Beach and D.K. Christen, "Microstructure of a High J_c Laser-ablated $YBa_2Cu_3O_{7-\delta}$/Sol-gel Deposited $NdGaO_3$ Buffer Layer/(001) $SrTiO_3$ Multi-layer Structure," Physica C 331 (2000) 73-78.

[23] M.S. Bhuiyan, M. Paranthaman, S. Sathyamurthy, T. Aytug, S. Kang, D. F. Lee, A. Goyal, E. A. Payzant, and K. Salama, "MOD approach for the growth of epitaxial CeO_2 buffer layers on biaxially textured Ni-W substrates for YBCO coated conductors," Superconductor Science & Technology 16 (2003) 1305-1309.

[24] J.S. Morrell, Z.B. Xue, E.D. Specht, A. Goyal, P.M. Martin, D.F. Lee, R. Feenstra, D.T. Verebelyi, D.K. Christen, T.G. Chirayil, M. Paranthaman, C.E. Vallet and D.B. Beach, "Epitaxial Growth of Gadolinium Oxide on Roll-Textured Nickel Using a Solution Growth Technique," J. Mater. Res. 15 (2000) 621-628.

[25] T.G. Chirayil, M. Paranthaman, D.B. Beach, D.F. Lee, A. Goyal, R.K. Williams, X. Cui, D.M. Kroeger, R. Feenstra, D.T. Verebelyi, and D.K. Christen, "Epitaxial growth of $La_2Zr_2O_7$ thin films on rolled Ni-substrates by sol–gel process for high T_c superconducting tapes," Physica C 336 (2000) 63-69.

[26] S. Sathyamurthy, M. Paranthaman, H.Y. Zhai, H.M. Christen, P.M. Martin and A. Goyal, "Lanthanum zirconate: A single buffer layer processed by solution deposition for coated conductor fabrication," J. Mater. Res. 17 (2002) 2181-2184.

[27] M. Parans Paranthaman, T. Aytug, S. Sathyamurthy, D.B. Beach, A. Golyal, D.F. Lee, B.W. Kang, L. Heatherly, E.D. Specht, K.J. Leonard et al., "Bulk Solution Techniques to Fabricate High Jc YBCO Coated Conductors," Physica C 378-381 (2002) 1009-1012.

[28] S. S. Shoup, M. Paranthaman, A. Goyal, E. D. Specht, D. F. Lee, D. M. Kroeger and D. B. Beach, "Epitaxial Thin Film Growth of Lanthanum and Neodymium Aluminate Films on Roll-Textured Nickel Using a Sol-Gel Method," J. Amer. Ceram. Soc. 81 (1998) 3019-3021.

[29] M.W. Rupich, W. Palm, W. Zhang, E. Siegal, S. Annavarapu, L. Fritzemeier, M.D. Teplitsky, C. Thieme and M. Paranthaman, "Growth and Characterization of Oxide Buffer Layers for YBCO Coated Conductors," IEEE Trans. on Appl. Supercond. 9 (1999) 1527-1530.
[30] M. Paranthaman, M.S. Bhuiyan, S. Sathyamurthy, H.Y. Zhai, A. Goyal and K. Salama, "Epitaxial Growth of Solution Based Rare Earth Niobate, RE_3NbO_7 Films on Biaxially Textured Ni-W Substrates," J. Mater. Res. 20 (2005) 6-9.
[31] M.S. Bhuiyan, M. Paranthaman, S. Kang, D.F. Lee, and K. Salama, "Growth of epitaxial Y_2O_3 buffer layers on biaxially textured Ni-W substrates for YBCO coated conductors by MOD approach," Physica C 422 (2005) 95-101.
[32] T. Aytug, M. Paranthaman, K.J. Leonard, H.Y. Zhai, M.S. Bhuiyan, E.A. Payzant, A. Goyal, S. Sathyamurthy, D.B. Beach, P.M. Martin, D.K. Christen, X. Li, T. Kodenkandath, U. Schoop, M.W. Rupich, H.E. Smith, T. Haugan, P.N. Barnes, "Assessment of chemical solution synthesis and properties of $Gd_2Zr_2O_7$ thin films as buffer layers for second-generation high-temperature superconductor wires" J. Mater. Res. 20 (2005) 2988-2996.
[33] Y.X. Zhou, X. Zhang, H. Fang, R.T. Putman and K. Salama, "Development of Single Solution Buffer Layers on Textured Ni Substrate for HTS Coated Conductors," IEEE Trans. On Appl. Supercond. 15 (2005) 2711-2714.
[34] T.G. Chirayil, M. Paranthaman, D.B. Beach, J.S. Morrell, E.Y. Sun, A. Goyal, R.K. Williams, D.F. Lee, P.M. Martin, D.M. Kroeger, R. Feenstra, D.T. Verebelyi and D.K. Christen, "Epitaxial Growth of Yb_2O_3 Buffer Layers on Biaxially Textured-Ni (100) substrates by Sol-gel Process," Mat. Res. Soc. Symp. Proc. (1999) 51-56.
[35] Y. Akin, Z. Aslanoglu, E. Celik, L. Arda, W. Sigmund and Y.S. Hascicek, "Textured Growth of Multi-Layered Buffer Layers on Ni tape by Sol-Gel Process," IEEE Trans. Appl. Supercond. 13 (2003) 2673.
[36] T. Aytug, M.P. Paranthaman, B.W. Kang, D.B. Beach, S. Sathyamurthy, E.D. Specht, D.F. Lee, R. Feenstra, A. Goyal, D.M. Kroeger, K.J. Leonard, P.M. Martin and D.K. Christen, "Reel-to-reel continuous chemical solution deposition of epitaxial Gd_2O_3 buffer layers on biaxially textured metal tapes for the fabrication of $YBa_2Cu_3O_{7-\delta}$ coated conductors," Journal of The American Ceramic Society 86 (2003) 257-265.
[37] Y. Takahashi, Y. Aoki, T. Hasegawa, T. Watanabe, T. Maeda, T. Honjo and Y. Shiohara, "In-plane textured buffer layer for the TFA-MOD method on <100>{001} Ni tapes using MOD process," Physica C 392-396 (2003) 887.
[38] M.P. Siegal, P.G. Clem, J.T. Dawley, R.J. Ong, M. A. Rodriguez and D.L. Overmyer, "All solution-chemistry approach for $YBa_2Cu_3O_{7-\delta}$ coated conductors," Appl. Phys. Lett. 80 (2002) 2710.
[39] M. P. Paranthaman, X. Qiu, K. Kim, Y. Zhang, X. Li, S. Sathyamurthy, C. Thieme, and M. W. Rupich, "Development of Solution Buffer Layers for RABiTS Based YBCO Coated Conductors," IEEE Trans. On Appl. Supercond. (2011) in press.
[40] Y Shiohara, M Yoshizumi, T Izumi and Y Yamada, "Present status and future prospect of coated conductor development and its application in Japan," Superconductor Science and Technology 21 (2008) 034002.

[41] J. Gutiérrez, A. Llordés, J. Gazquez, M. Gibert, N. Romá, S. Ricart, A. Pomar, F. Sandiumenge, N. Mestres, T. Puig and X. Obradors, "Strong isotropic flux pinning in solution-derived $YBa_2Cu_3O_{7-x}$ nanocomposite superconductor films," Nature Materials 6 (2007) 367-373.

[42] M. P. Paranthaman, S. Sathyamurthy, X. P. Li, E. D. Specht, S. H. Wee, C. Cantoni, A. Goyal, and M. W. Rupich, "Modified Lanthanum Zirconium Oxide buffer layers for low-cost, high performance YBCO coated conductors," Physica C 470 (2010) 352.

10

Superconducting Properties of Graphene Doped Magnesium Diboride

Xun Xu, Wenxian Li, Xiaolin Wang and Shi-Xue Dou
University of Wollongong/ Institute for Superconducting and Electronic Materials Australia

1. Introduction

Graphene, carbon in the form of monolayer sheets, has revealed astonishing and unique chemical and physical properties, which have made it an extremely active research topic in both materials science and physics (Novoselov, K. S. et al., 2004). Through chemical and materials integration, graphene is being actively exploited in a range of technological applications (Stankovich, S. et al., 2006). Superconductors can carry electrical current without any energy dissipation. The combination of both graphene and a superconductor into a composite has great potential for electrical devices and large scale applications. MgB_2, a superconductor with a simple composition and two-gap feature has great potential for large current carrying applications, as demonstrated through a series of chemical dopings (Dou, S. X. et al. 2007). In the case of graphene's, the strict two-dimensionality and its high electrical and thermal conductivities, make it an ideal candidate for integrating/doping into MgB_2 in order to improve the superconducting properties.

Substitutional chemistry can modify, in a controlled way, the electronic structures of superconductors and their superconducting properties, such as the transition temperature (T_c), critical current density (J_c), upper critical field (H_{c2}), and irreversibility field (H_{irr}). In particular, carbon containing dopants, including nano-meter sized carbon (nano-C), silicon carbide (SiC), carbon nanotubes (CNTs) and hydrocarbons/carbohydrates are effective means to enhance the J_c- field dependence and H_{c2} (Ma, Y. et al., 2006, Senlowocz, B. J. et al., 2005, Kumakura, H. et al., 2004, Sumption, M. et al., 2005, Dou, S. X. et al., 2003, Kim, J. H. et al., 2006, Wilke, R. H. T. et al., 2008). Upon graphene incorporation into MgB_2 it is expected that H_{c2} and the flux pinning properties should be improved. Recently, high-throughput solution processing of large-scale graphene has been reported by a number of groups (Tung, V. C. et al., 2009, Kim, K. S. et al., 2009, Hernandez, Y. et al., 2008, Li, D. et al., 2008, Li, X. et al., 2008, Choucair, M. et al., 2009).

Based on the works of Choucair et al., sufficient quantities of graphene were obtained for doping the bulk MgB_2 samples via a diffusion process. it was demonstrate that graphene is the most efficient among all the carbon-based dopants used to date, in terms of enhancing the flux pinning behaviour in MgB_2. Very low levels of graphene doping (e,g. 0.9 at%) have been shown to be sufficient to lead to a significant improvement in the critical current density - field performance ($J_c(B)$), with little change in the transition temperature (T_c). At 3.7 at% graphene doping of MgB_2 an optimal enhancement in $J_c(B)$ was reached by a factor of 30 at 5 K and 10 T, compared to the un-doped reference sample. This improvement is believed to be related to: the single carbon sheet two dimensional (2-D) geometry, the negative thermal expansion coefficient, high electron mobility, low resistivity, high thermal conductivity, and high mechanical strength of graphene. In contrast to many carbon-based dopants, where carbon substitution and hence scattering, is the dominant factor for enhancement of $J_c(B)$, the enhanced flux pinning is largely attributable to strong strain effects induced by the large difference between the negative coefficient of thermal expansion of graphene and the large, anisotropic coefficient of thermal expansion of MgB_2.

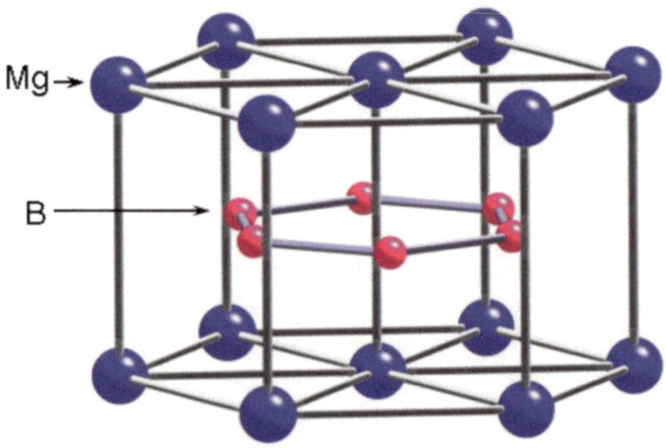

Fig. 1. Crystal structure of MgB_2, a=3.086 Å and c=3.524 Å (Nagamatsu, J. et al., 2001).

The most exciting result is the soften E_{2g} mode in graphene added samples. It is well known that the superconductivity of intermetallic compound MgB_2 is self-optimized both in electronic structure and in phonon dispersion and slim chance is left to improve its superconducting transition temperature, T_c, above ~39.4 K (Nagamatsu, J. et al., 2001). Higher T_c of $Mg^{10}B_2$ and band structure calculation indicate that MgB_2 is a phonon-mediated BCS-Eliashberg superconductor with multiple gaps and strong electron-phonon coupling (EPC). As a two-gap superconductor, MgB_2 shows two qualitatively different Fermi surfaces: σ-band with extremely strong pairing strength and π-band with weak pairing strength. The σ-band contribution is dominant and specifically the contribution from its coupling with the B-B stretch modes with E_{2g} symmetry in boron layer, the only Raman active phonon modes according to first principle calculation (Kunc, K. et al, 2001). the T_c does not show improvements in the alloys due to carbon substitution and impurity scattering, a tensile strain effects assuredly exist in the alloys judging from both the width of superconducting energy gaps and the phonon behaviors.

2. Graphene doped polycrystalline MgB_2

MgB_2 is an intermetallic, polycrystalline material that had been known since 1953 and was used in the commercial preparation of elemental boron (B). It was not known to be superconducting until this property was accidentally discovered by Jun Akimitsu and his colleagues (Nagamatsu, J. et al., 2001). MgB_2 consists of alternating layers of boron and magnesium atoms in a hexagonal structure, as shown in Figure 1. It was analyzed that the crystal structure of MgB_2 using by the Rietveld method . It was found that the compound has hexagonal symmetry (Space Group P6/mmm) with the unit cell lattice parameters: a =

0.3086 nm and c = 0.3524 nm. Similar conclusions on the crystal structure were soon made on the basis of studies using high resolution transmission electron microscopy (HRTEM), high resolution powder neutron diffraction and electron energy loss spectroscopy .

Up to now, large number of experimental and theoretical of works has been studied because of its high T_c of 39 K. Beside the high T_c, simple crystal structure, large coherence length, high critical field, transparency of grain boundaries, and low normal state resistivity are a fascinating topic to study for both large scale application and electronic devices. Moreover, the presence of two-gap superconductivity (π and σ band) has been theoretically and experimentally established besides it shows a significant isotope effect like a low temperature superconductor.

2.1 Dual reaction model and C doping effects in MgB_2

The "*pure*" MgB_2 is not so good for applications due to the poor flux pinning and low J_c, H_{c2} and H_{irr} values. A significant enhancement of critical current density (J_c) in MgB_2 can be achieved through chemical doping with various kinds of material. Chemical doping is a simple and readily scalable technique. A significant enhancement of critical current density (J_c) in MgB_2 can be achieved through chemical doping with carbon (C) containing composites or compounds, such as SiC, C, B_4C, or carbon nanotubes (CNT), hydrocarbon, carbohydrates etc. The C can enter the MgB_2 structure by substituting into boron (B) sites, and thus J_c and H_{c2} are significantly enhanced due to the increased impurity scattering in the two-band MgB_2.

In 2007, Dou et al has proposed a dual reaction model based on the following factors to explain the mechanism of enhancing the electromagnetic properties due to the SiC and C doping (Dou, S. X. et al. 2007):

- Demonstrated a dual reaction model in which the C substitution due to nano SiC doping and MgB_2 formation take place simultaneously at temperatures
- The C substitution and induced highly localised fluctuation in structure and T_c are responsible for the enhancement in J_c, Hirr and Hc_2 by SiC doping.
- Disorder created by all means has universal influence on electromagnetic properties of MgB2
- free and reactive C is the key to searching for better do pants
- Fig. 2 shows the effects of sintering temperature on the J_c(H) for different carbon based dopants. The hydrocarbon and SiC doped MgB_2 show significant enhancement in J_c for the samples sintered at lower temperature, whereas the carbon and CNT doped MgB_2 need to be sintered at higher temperature for high J_c. The low sintering temperature results in small grain size, high concentrations of impurities and defects, and large lattice distortion, which are all responsible for a strong flux pinning force (Soltanian et al., 2005, Yamamoto et al., 2005). Furthermore, the hydrocarbon and SiC can release fresh and active free carbon at very low temperature, which means that the carbon substitution effects take place simultaneously with the MgB_2 formation. A high sintering temperature will perfect the crystallization and decrease the flux pinning centers in the MgB_2 matrix. That is the reason why high sintering temperature degrades the J_c performance. Although high sintering temperature has the same shortcomings in nanosized carbon and CNT doped MgB_2, the carbon substitution effects improve their J_c values. The high sintering temperature is necessary for carbon and CNT doped MgB_2 because the carbon and CNT are quite stable at low temperature and the substitution effects are absent if the sintering temperature is not high enough.

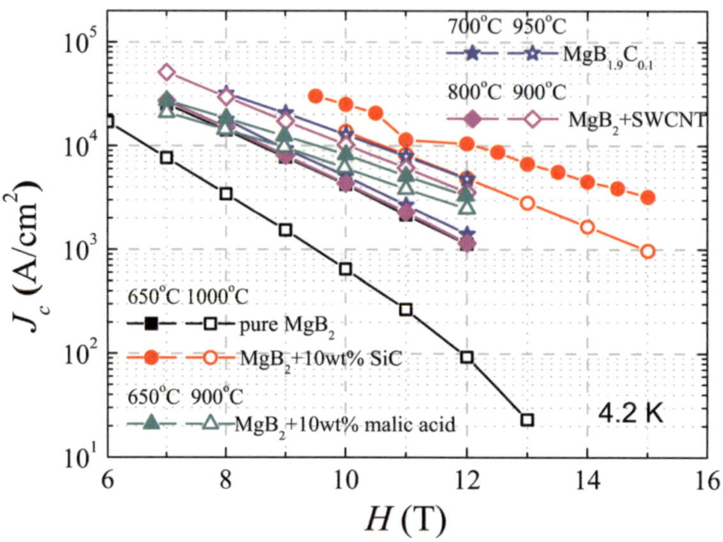

Fig. 2. The critical current density (J_c) at 4.2 K versus magnetic field for wires of pure MgB_2 and MgB_2 doped with C, SiC, SWCNTs, and malic acid that were sintered at different temperatures (Dou et al., 2002; Yeoh et al., 2006; Dou et al., 2007; Kim et al., 2008)

2.2 Gram-scale production of graphene

Graphite is the most common allotrope of carbon. The name isderived from the Greek verb graphein, 'to write', which relates very literally to the compound we now know as graphene, as single sheets of graphene were first isolated by simply tracing a sample of bulk graphite across a substrate in a process known as micromechanical cleavage (Novoselov, K. S. et al., 2004). Further attempts to synthesize isolated graphene have been based on intercalation (Li, X. et al., 2008), sonication in various solvents (Hernandez, Y. et al., 2008) and the chemical reduction of graphite to yield few-layer graphite oxide (Li, D. et al., 2008). Approaches developed for the production of carbon nanotubes have also been explored, but so far they have only been able to produce graphitic film (Dikin, D. A. et al., 2007). However, single- and few-layer graphene sheets have been grown epitaxially by the chemical vapour deposition of hydrocarbons on metal and nonmetal substrates, substrate-free deposition and by thermal decomposition of SiC. Given the lack of a reliable top-down approach for the large-scale production of graphene, attention has turned to bottom-up approaches that might be able to deliver the economies of scale that are found in the chemical and pharmaceutical industries. Choucair et al. have demonstrated that single-layer graphene can be synthesized by low-temperature flash pyrolysis of a solvothermal product of sodium and ethanol, followed by gentle sonication of the nanoporous carbon product (Choucair, M. et al., 2009).

All solvothermal reactions were performed in a Teflon-lined Parr Instrument Company 4749 reactor having a maximum volume of 23 ml. A typical synthesis consists of heating a 1:1 molar ratio of sodium (2 g) and ethanol (5 ml) in a sealed reactor vessel at 220°C for 72 h to yield the solid solvothermal product—the graphene precursor. This material is then rapidly pyrolysed, and the remaining product washed with deionized water (100 ml). The suspended solid is then vacuum filtered and dried in a vacuum oven at 100°C for 24 h. The final yield of graphene is approximately 0.1 g per 1 ml of ethanol—typically yielding ~0.5 g

Fig. 3. Example of the bulk quantity of graphene product. The image consists of approximately 2 g of sample (Choucair, M. et al., 2009).

per solvothermal reaction.The product of this reaction is then washed in water and dried to obtain pure graphene (Fig. 3). Solvothermal reactions have been reported to lead to the production of carbon nanosheets, although such materials do not appear to consist of single sheets having both high degrees of planarityand crystallinity, suggesting an amorphous or disordered graphitic structure (Choucair, M. et al., 2009).

2.3 Graphene doping effects in MgB_2

Based on the works of Choucair et al., sufficient quantities of graphene were made available for doping the bulk MgB_2 samples. The bulk MgB_2 samples were fabricated via a diffusion process. The crystalline boron powder (0.2 to 2.4 μm) 99.999% without and with graphene was prepared by ball milling with toluene medium. Then, the powders were dried in a vacuum oven to evaporate the medium. These powders were mixed and pressed into pellets. The pellets were then put into an iron tube filled with Mg powder (-325mesh 99%). The samples were sintered at 850°C for 10 hrs in a quartz tube; the heating rate was 5^oCmin^{-1} under high purity argon (Ar 99.9%) gas. The phase and crystal structure of all the samples were investigated by X-ray diffraction (XRD). T_c was defined as the onset temperature at which diamagnetic properties were observed. The magnetic J_c was derived from the width of the magnetization loop using Bean's model by a Physical Properties Measurement System (PPMS). Transport measurements for resistivity (ρ) were done using a standard AC four probe method. In addition, $H_{c2}(T)$ and $H_{irr}(T)$ were defined as the fields where the temperature dependent resistance at constant magnetic field $R(H_{c2}, T) = 0.9R_{ns}$ and $R(H_{irr}, T) = 0.1R_{ns}$ with R_{ns} being the normal state resistance near 40 K. The hysteresis loops of the MgB_2 sample every 1.5 K in the 17- 35 K range. The symmetric hysteresis loops with respect to the magnetic field indicate the dominance of bulk pinning up to temperatures near T_c.

MgB_2/Fe monofilament wires were prepared by an *in situ* reaction process and the powder-in-tube method. Magnesium (99%, 325 mesh), and the different boron powders with the nominal atomic ratio of Mg : B = 1.1 : 2 were mixed through grinding and were put into Fe tubes with a length of 140 mm, an outer diameter (O.D) of 10 mm, and an inner diameter (I.D) of 8 mm. The packing process was carried out in air. Both ends of the tubes were sealed with aluminum pieces, and then the tubes were drawn to a wire with a diameter of 1.4 mm. Short wire samples (4 cm each) were sealed with Zr foil, then sintered with a heating rate of 5 $^{o}C\ min^{-1}$ in flowing high purity Ar to 700-800 ^{o}C and held at the final temperature for 30 minutes, followed by a furnace cooling to room temperature. The transport current (I_c) at 4.2 K was measured by the standard dc four-probe resistive method with a criterion of 1 μVcm^{-1} in magnetic fields up to 12 T. T_c was determined using the standard ac four-probe method. In addition, $H_{c2}(T)$ and $H_{irr}(T)$ were defined as the fields where the temperature dependent resistance at constant magnetic field $R(H_{c2}, T) = 0.9R_{ns}$ and $R(H_{irr},T) = 0.1R_{ns}$, respectively, with R_{ns} being the normal state resistance near 40 K.

The common format $Mg(B_{1-x}C_x)_2$ with x=0, 0.037, was used. The composition of graphene doped MgB_2 were 0, 3.7 at%, and as such, the samples are designated as G000, and G037respectively.

2.3.1 XRD analysis for bulk samples

Fig. 4 shows XRD patterns for the pure and graphene doped MgB_2 samples. All the peaks are inherent to MgB_2. No impurity phase was detected in the XRD data, however a slight peak shift to higher angle is noted for the G087 samples. MgO, which is commonly present in most bulk MgB_2 samples is absent in all the samples.

The lattice parameters, *a*, *c*, the ratio of *a/c*, grain size, strain, and full width at half maximum of the representative peak (110) calculated from the XRD patterns are shown in Table 1. Both the *a*-axis and *c*-axis parameters vary little with increasing graphene doping level of 3.7%, apart from G087 sample, which shows a notable decrease in the *a*-axis parameter, suggesting that carbon (C) likely substitutes into the boron (B) sites, leading to a slight drop in T_c (36.7 K) for the G087 sample. it also observed that the full width at half maximum (FWHM) of the (110) peak increases with increasing graphene doping level. Such a peak broadening is caused by both grain size reduction and an increase in lattice strain. The calculated results on grain size and lattice strain from a Williamson-Hall plot (Williamson, G. K., and Hall, W. H., 1953) are given in Table 1. Transition temperatures (T_c) for the graphene doped samples were determined by AC susceptibility measurements. Also, the T_c onset determined from the AC susceptibility measurement is 38.9 K for the un-doped sample, dropping only slightly to 37.7 and 36.7 K for the G037 and G087 samples, respectively.

Sample	Lattice Constants			Grain Size (nm)	Strain (%)	FWHM (110) (°)	T_c (onset) (K)
	a (Å)	*c* (Å)	c/a				
G000	3.084(1)	3.525(1)	1.143(1)	216(10)	0.1198(188)	0.288	38.9
G037	3.082(1)	3.527(1)	1.144(1)	170(8)	0.1685(250)	0.400	37.7
G087	3.075(1)	3.525(1)	1.146(1)	171(11)	0.1782(330)	0.414	36.7

Table 1. The full width at half maximum (FWHM) of the (110) peak, the lattice parameters, and the transition temperature (T_c) for the MgB_2 samples, made with 0, 3.7, and 8.7 at% graphene doping via a diffusion process (Xu, X. et al., 2010).

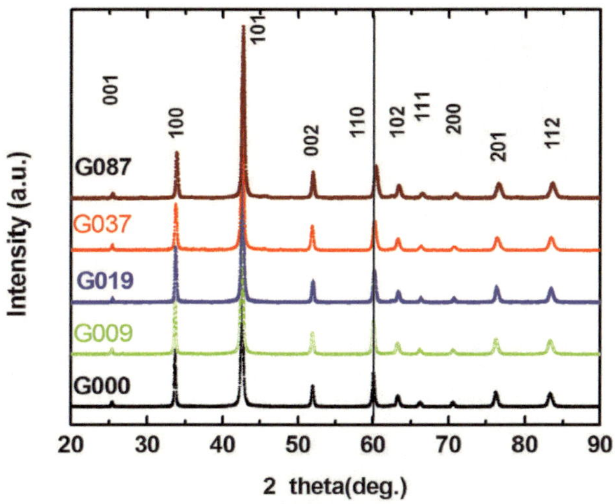

Fig. 4. XRD patterns for graphene doped MgB_2 samples: G000, G009, G019, G037, and G087, showing that all diffraction peaks can be indexed to MgB_2 with undetectable impurity phases.

2.3.2 Critical current density

Graphene is the most efficient among all the carbon-based dopants used to date, in terms of enhancing the flux pinning behaviour in MgB_2. Very low levels of graphene doping (e,g. 0.9 at%) have been shown to be sufficient to lead to a significant improvement in the critical current density - field performance ($J_c(B)$), with little change in the transition temperature (T_c). At 3.7 at% graphene doping of MgB_2 an optimal enhancement in $J_c(B)$ was reached by a factor of 30 at 5 K and 10 T, compared to the un-doped reference sample. In contrast to many carbon-based dopants, where carbon substitution and hence scattering, is the dominant factor for enhancement of $J_c(B)$, the enhanced flux pinning is largely attributable to strong strain effects induced by the large difference between the negative coefficient of thermal expansion of graphene and the large, anisotropic coefficient of thermal expansion of MgB_2.

The common format $Mg(B_{1-x}C_x)_2$ with x=0, 0.037, and 0.087 was used. The composition of graphene doped MgB_2 were 0, 3.7, and 8.7 at%, and as such, the samples are designated as G000, G037, and G087, respectively. Fig.5 shows the magnetic $J_c(B)$ curves at 5 K and 20K for all the samples, which were sintered at 850°C for 10 hours. The $J_c(B)$ values for all the doped samples are higher than the un-doped sample at high fields. The sample G037 gives the highest J_c at high fields: J_c increases by a factor of 30 at 5 K for the field of 10 T, as compared to the un-doped sample, G000. Even though the J_c in the low field regime is depressed, a higher doping level (G087), still results in the rate of J_c dropping much slower than the un-doped sample, clearly indicating strong flux pinning induced by the graphene doping. The most significant effect of graphene doping is the high effectiveness of graphene to improve flux pinning at lower doping levels, which distinguishes graphene from any other C containing dopants, for example, the J_c for G037 reached 20,000 A/cm^2 at 5 K and 8 T, exceeding or matching the best J_c resulting from dopants such as SiC, CNT, and carbohydrates at their optimal doping level of 10 at%, as well as nano-C at its optimal

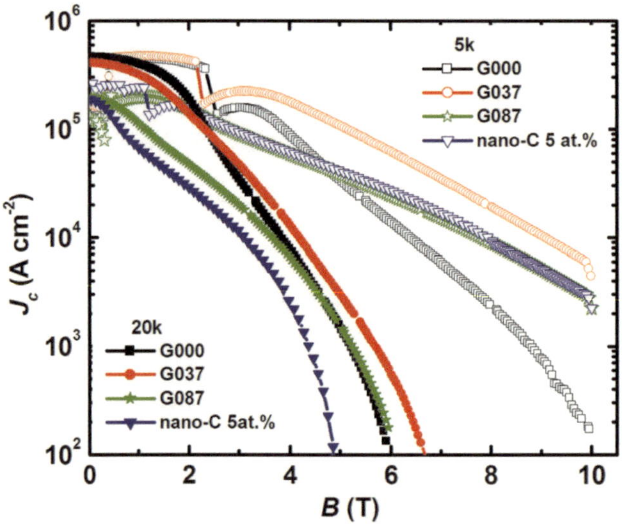

Fig. 5. Magnetic critical current density as a function of magnetic field at 5 K and 20K for with and without graphene doped bulk samples. 5 at% nano-C doped sample for a comparable result at the same sample preparation route (Xu, X. et al., 2010).

doping level of 5-6.4 at.%. In the latter case, the T_c is substantially reduced to temperatures as low as 30 K. Compared to the graphite doped MgB_2 pallets prepared through the ball-milling and HIP the J_c of graphite doped MgB_2 is better than graphene doping at 5 K (Yamamoto et al., 2005), but at 20K, the J_c for graphene doping is much better than graphite. For example, the J_c for the graphene doped MgB_2 at 20 K and 6 T is larger than that for graphite doped MgB_2 by a factor of 50 (Yamamoto et al., 2005). In comparison, low levels of graphene doping have little effect on T_c and cause only a very small increase in impurities, not compromising the significant enhancement in J_c in high fields by the degradation in low-field J_c, which is a common issue for all other C based dopants. In order to see the difference with other C based dopants, the same preparation route was applied to 5 at% nano-C doped sample and the resultant decrease in J_c at 20K can be seen in the figure 5, because the T_c for nano- C doped MgB_2 is only 34K.

The transport current J_{ct}-B performance of the two kinds of the samples sintered at temperatures from 700 °C to 800 °C, denoted by, WG37S7, WG37S8 for 3.7 at% graphene doped and WG0S7, and WG0S8 for undoped wire sample, respectively, is shown in Figure 6. It can be clearly seen that J_{ct} of samples prepared from the doped wire sample showed better performance in the field range of 4 to 15 T. This indicates that the most significant effect of graphene doping is the high effectiveness of graphene to improve flux pinning at lower doping levels. The strong enhancement in the flux pinning may be attributable to a combination of C substitution for B and thermal strain-induced defects. It is very surprised that if compared to the magnetic J_{cm} with the bulk MgB_2 samples via a diffusion process under magnetic fields of 5 to 10 T, the transport J_{ct} increment with the wire MgB_2 via the powder-in-tube (PIT) method is tremendous lower from Fig.6. The transport current capacity is the real useful J_c that flows through the whole of the sample. It can be understand, the difference between J_{cm} and J_{ct} in MgB_2 may be related to features of the microstructure of the superconducting MgB_2 core, such as porosity, agglomeration of

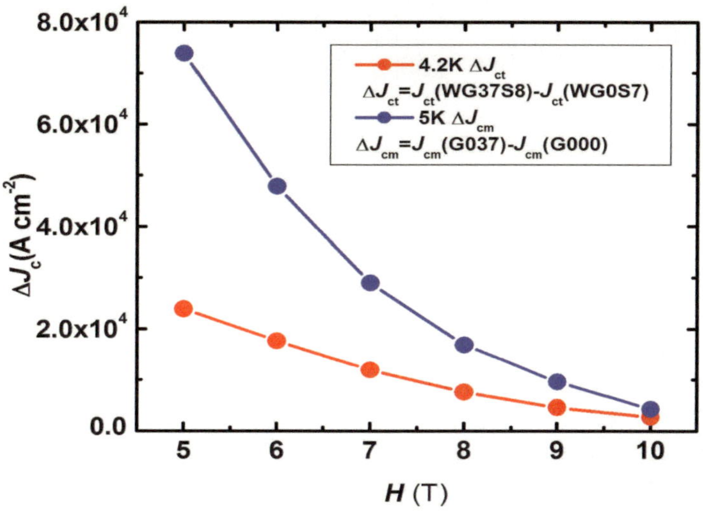

Fig. 6. Compared to the increment of magnetic J_{cm} at 5K and transport J_{ct} at 4.2K

superconducting crystals, and fraction of impurities as the main secondary phase by different fabricated processing (Horvat, J. et al., 2008). It is clearly that the graphene doped bulk sample via the diffusion process had the highest mass density, which improved the most inter-grain connectivity to improve the J_c so much. At the same time, according to the Rowell connectivity analysis, the calculated active cross-sectional area fraction (A_F) represents the connectivity factor between adjacent grains, which is estimated by comparing the measured value with that of a single crystal. (Rowell, J. M., 2003). The A_F for all wire samples via the powder-in-tube (PIT) method is almost half of the bulk sample via diffusion process. With the wire doped samples, the A_F value was increased as the sintering temperature increased. This indicates that additional grain growth occurs due to high temperature sintering. The larger grains are also accompanied by improved density and grain connectivity. So, in order to improve the J_c of the wire sample, the key point is how to improve the inter-grain connectivity.

2.3.3 Flux pinning mechanism

Regarding the flux pinning mechanism, it is established that the core interaction, which stands for the coupling of the locally distorted superconducting properties with the periodic variation of the superconducting order parameter is dominant over the magnetic interaction for MgB_2 due to its large GL coefficient κ (~26 in MgB_2) . The core interaction includes two types of mechanism: δTc and δl pinning. The δTc pinning refers to the spatial variation of the GL coefficient associated with disorder due to variation in the transition temperature T_c, while δl pinning is associated with the variation in the charge-carrier mean free path l near lattice defects . According to the collective pinning model, the disorder induced spatial fluctuations in the vortex lattice can be clearly divided into different regimes depending on the strength of the applied field: single-vortex, small-bundle, large-bundle, and charge-density-wave (CDW)-type relaxation of the vortex lattice. The crossover field, B_{sb} is defined as a field separating single vortex regime into small bundles of vortices. Below B_{sb}, J_c is almost field independent. The B_{sb} as a function of reduced temperature (t=T/T_c) is described by the equation (Qin, M. J. et al, 2002):

$$B_{sb} = B_{sb}(0)\left(\frac{1-t^2}{1+t^2}\right)^{2/3} \tag{1}$$

for δT_c pinning,

$$B_{sb} = B_{sb}(0)\left(\frac{1-t^2}{1+t^2}\right)^{2} \tag{2}$$

for δl pinning.

To define the pinning mechanism in our grapheme doped the samples, the crossover field, B_{sb}, as a function of temperature with graphene doped sample (G037) is plotted in Figure 7 as red squares. B_{sb} is defined as a field where J_c drops by 5% only compared to J_c at zero field. It can be seen that the curve for δT_c pinning calculated from q. (1) is in a good agreement with the experimental data, whereas, the curve for δl pinning according to Eq. (2) does not fit to the experimental data. For polycrystalline, thin film, and single crystalline MgB_2 samples, it has been found that the dominant pinning mechanism is δT_c pinning, which is related to spatial fluctuation of the transition temperature while most C-doped MgB_2 samples displayed δl pinning mechanism (Wang, J. L. et al., 2008) as a result of strong scattering and hence the shortening of the mean free path l owing to the presence of large amount of impurities in the doped samples. This is reflected by the significant increase in the residual resistivity. The local strain was suggested to be one of potential pinning centres.

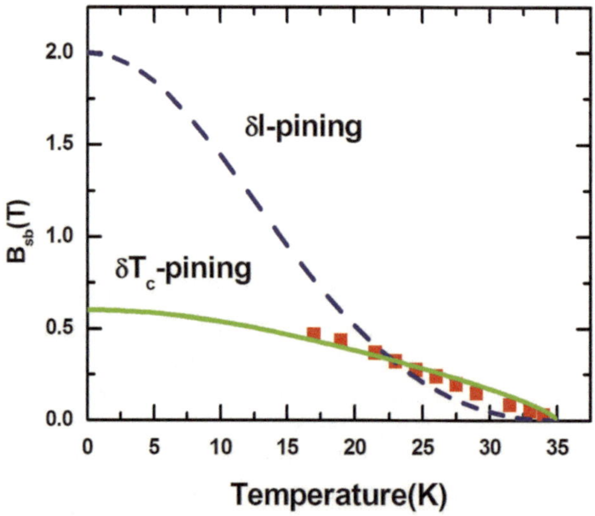

Fig. 7. The crossover field B_{sb} as a function of temperature with graphene doped sample (G037) (Xu, X. et al., 2010)

However, we do not have strong evidence that the dominant pinning in the graphene doped MgB2 is due to the local strain effect alone. In contrast, the graphene doping sets an exceptional example, following the δT_c pinning rather than δl pinning mechanism. This demonstrates the unique feature of the graphene doping. The amorphous phases can also act pinning centres, which is in favour for δT_c pinning. Although the graphene doped samples have a lot of defects these samples contain low concentration of impurities compared to the samples by other forms of carbon dopants. One of major differences of graphene doping from other dopants is that the samples are relatively pure as evidenced by the low resistivity (20 μΩ cm) in the grapheme doped samples. Normally, the resistivity in carbon doped MgB_2 ranges from 60 μΩ cm to as high as 300 μΩ cm. The high electrical connectivity is beneficial for J_c in low magnetic fields and high field performance; however we can not find any correlation between electrical co*n*nectivity with the J_c in the case here. The graphene doped samples have higher resistivity than the un-doped MgB_2 sample (3 μΩ cm), indicating electron scattering caused by graphene doping levels. But, it should be pointed out that the increase in resistivity is much smaller than for any other forms of carbon doped MgB_2, Which is shown in Figure 8.

2.3.3 E_{2g} mode and Raman peak shift

Tensile strain effects on superconducting transition temperature (T_c) was observed in graphene-MgB_2 alloys to pursue high T_c in multi-gap superconductors. The enhancement of energy gap for π-band indicates the weak rescale of density of state on Fermi surface. The E_{2g} mode split into two parts: one dominant soften mode responding to tensile strain and another harden mode responding to carbon substitution effects.

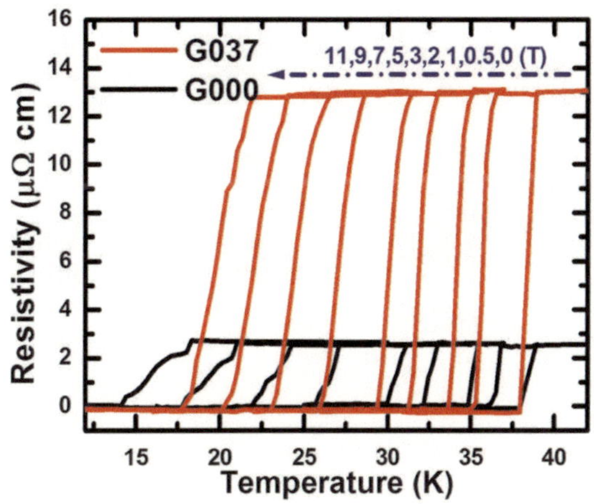

Fig. 8. The temperature dependence of the resistivity (ρ) measured in different fields for doped and undoped samples.

The existence of soften E_{2g} mode in bulk samples suggests that modified graphene-MgB_2 alloys are the potential candidates for the high performance superconducting devices.

To confirm the effect of tensile strain on EPC, Raman scattering was employed for measurement of phonon properties by a confocal laser Raman spectrometer (Renishaw inVia plus) with a 100× microscope. The 514.5 nm line of an Ar^+ laser was used for excitation and several spots were selected on the same sample to collect the Raman signals to make sure that the results were credible. Fig. 9(a) shows the typical spectrum of pure MgB_2 consisting of three broad peaks. The most prominent phonon peak located at lower frequency (ω_2: centered at ~600 cm^{-1}) is assigned to the E_{2g} mode. The other two Raman bands (ω_1: centered at 400 cm^{-1} and ω_4: centered at 730 cm^{-1}) have also been observed earlier in MgB_2 and attributed to phonon density of states (PDOS) due to disorder. The EPC strength in MgB_2 depends greatly on the characteristic of E_{2g} mode, both frequency and FWHM, while the other two modes, especially the ω_4 mode, are responsible for the T_c depression in chemically doped MgB_2 (Kunc, K. et al, 2001). The graphene addition in MgB_2 induces splitting of E_{2g} mode: one soften mode (ω_2) and another harden mode (ω_3), as shown in Fig. 9. ω_2 shifts to low frequency quickly with the graphene addition because of the strong tensile strain. The softness of E_{2g} mode was observed only in MgB_2–SiC thin films due to tensile strain-induced bond-stretching, which resulted in a T_c as high as 41.8 K. Although ω_2 modes are dominant in low graphene content samples, T_c drops slightly. This is in agreement with the energy gap behaviors because of the carbon substitution induced band filling and interband scattering. ω_2 is marginal in G10 and vanishes in G20. ω_3 shifts to high frequency slowly in low graphene content samples because the tensile strain has confined the lattice shrinkage. However, the tensile strain can not counteract the intensive carbon substitution effects when the graphene content is higher than 10 wt% and ω_3 takes the place of ω_2. It should be noted that ω_3 is not as dominant as ω_2 in pure MgB_2 and ω_4 is the strongest peak as in the other carbonaceous chemical doped MgB_2 due to lattice distortion. Furthermore, another peak ω_5 has to be considered in G10 and G20 to fit the spectra reasonably. The Raman spectrum of G20 was separated from the mixed spectra of MgB_2 and MgB_2C_2 based on their different scattering shapes: MgB_2 shows broaden and dispersed waves, while MgB_2C_2 shows sharp peaks (Li, W. X. et al., 2008).

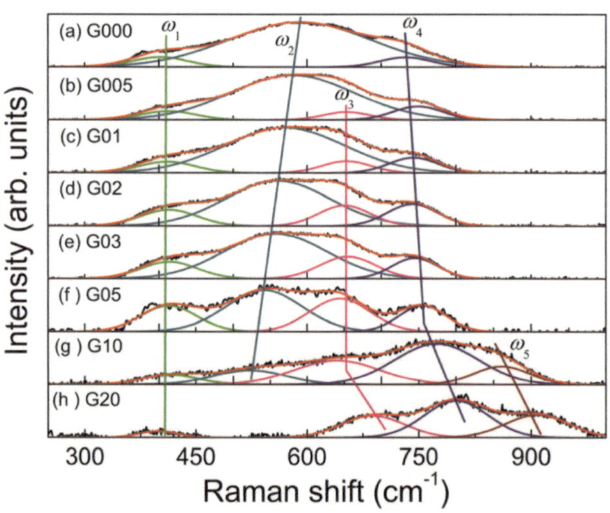

Fig. 9. The the typical spectrum of MgB_2 consisting of three broad peaks

The tensile strain was unambiguously detected in graphene-MgB_2 alloys made by diffusion process and the π energy gap was broadening with the graphene addition. The bond-stretching E_{2g} phonon mode splits into one soften mode due to the tensile strain and another harden mode due to the carbon substitution on boron sites. Although E_{2g} mode splitting have been observed in C doped MgB_2, both the two peaks shift to higher frequency and this is the first time to observe the coexistence of two modes shifting to opposite directions. The T_c value does not show enhancement because of impurity scattering effects and carbon substitution. However, higher T_c values are expected in graphene-MgB_2 alloys processed by proper techniques or made of stabilized graphene.

2.3.4 Upper critical field and irreversibility field

Figure 10 shows the upper critical field, H_{c2}, and the irreversibility field, H_{irr}, versus the normalised T_c for all the samples. It is noted that both H_{c2} and H_{irr} are increased by graphene doping. The mechanism for enhancement of J_c, H_{irr}, and H_{c2} by carbon containing dopants has been well studied. The C can enter the MgB_2 structure by substituting into B sites, and thus J_c and H_{c2} are significantly enhanced due to the increased impurity scattering in the two-band MgB_2 (Gurevich, A.,2003). Above all, C substitution induces highly localised fluctuations in the structure and T_c, which have also been seen to be responsible for the enhancements in J_c, H_{irr}, and H_{c2} by SiC doping.

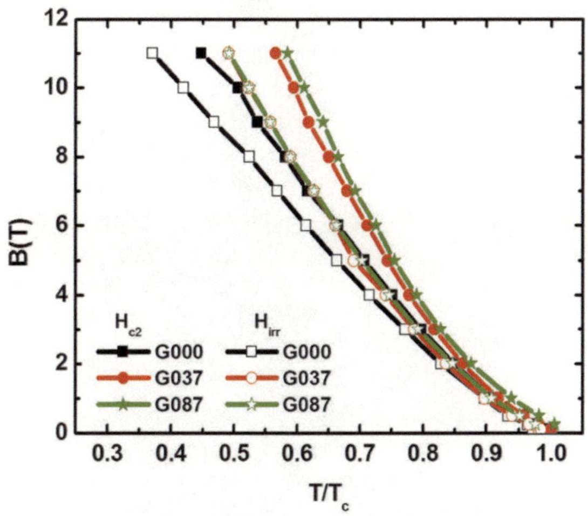

Fig. 10. Upper critical field, H_{c2}, and irreversibility field, H_{irr}, versus normalised transition temperature, T_c, for all graphenedoped and undoped MgB_2 samples (Xu, X. et al., 2010).

Furthermore, residual thermal strain in the MgB_2-dopant composites can also contribute to the improvement in flux pinning (Zeng, R. et al. 2009). In the present work, the C substitution for B (up to 3.7 at.%) graphene doping is lower, from the table 1, the change of the *a*-parameter is smaller, according to Avdeev et al result (Avdeev, M. et al., 2003), the level of C substitution, x in the formula $Mg(B_{1-x}C_x)$, can be estimated as $x=7.5 \times \Delta(c/a)$, where $\Delta(c/a)$ is the change in c/a compared to a pure sample. As both the *a*-axis and the *c*-axis lattice parameters determined from the XRD data showed little change within this doping range the level of carbon substitution is low at this doping level. This is in good agreement with the small reduction in T_c over this doping regime. At 8.7 at% doping, there is a noticeable drop in the *a*-axis parameter, suggesting C substitution for B, which is also consistent with the reduction in T_c. The source of C could be the edges of the graphene sheets, although the graphene is very stable at the sintering temperature (850°C), as there have been reports of graphene formation on substrates at temperatures ranging from 870-1320°C (Coraux, J. et al., 2009). The significant enhancement in J_c and H_{irr} for G037 can not be explained by C substitution only.

2.3.5 Microstructure by TEM

The microstructure revealed by high resolution transmission electron microscope (TEM) observations show that G037 sample has grain size of 100-200 nm which is consistent with value of the calculated grain size in table 1. The graphene doped samples have relatively higher density of defects compared with the undoped sample as shown in the TEM images of figure 11(a) and (c). The density of such defects is estimated to be 1/3 areas of TEM images, indicating high density in the doped samples. In figures 11(b) it should be noted that the order of fringes varies from grain to grain, indicates that the defect is due to highly anisotropic of the interface.

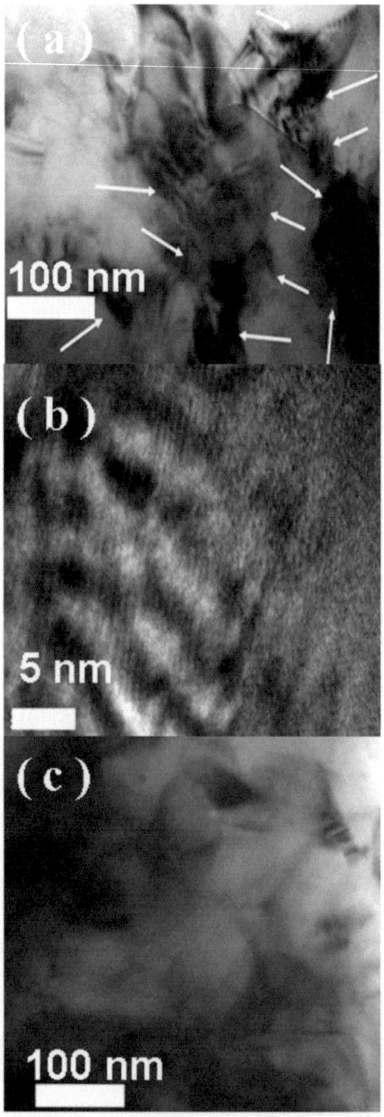

Fig. 11. (a) TEM image showing the defects with grains of the G037 sample with order of fringes varies between grains. Defects and fringes are indicated by arrow, and (b) HRTEM image of fringes. TEM images show large amount of defects and fringes can be observed in the graphene doped sample G037. (c) TEM image of the undoped sample for reference (Xu, X. et al., 2010).

Similar fringes have been reported in the MgB_2 (Zeng, R. et al. 2009), where these fringes were induced by tensile stress with dislocations and distortions which were commonly observed in the areas. As the graphene doped samples were sintered at 850°C for 10 hrs, the samples are expected to be relatively crystalline and contain few defects. Furthermore, as already shown above the C substitution level is low in graphene doped samples. Thus, the large amount of defects and amorphous phases on the nanoscale can be attributed to the residual thermal strain between the graphene and the MgB_2 after cooling because the thermal expansion coefficient of graphene is very small while that for MgB_2 is very large and highly anisotropic. The large thermal strain can create a large stress field, and hence structure defects and lattice distortion. These defects and distortions on the order of the coherence length, ξ, can play a role as effective pinning centres that are responsible for the enhanced flux pinning and J_c in the graphene doped MgB_2. The thermal strain-induced enhancement of flux pinning has also been observed in the SiC-MgB_2 composite as there is s noticeable difference in thermal expansion coefficient between MgB_2 and SiC (Coraux, J. et al., 2009).

3. Conclusion

In conclusion, the effects of graphene doping on the lattice parameters, T_c, J_c, and flux pinning in MgB_2 were investigated over a range of doping levels. By controlling the processing parameters, an optimised $J_c(B)$ performance is achieved at a doping level of 3.7 at.%. Under these conditions, J_c was enhanced by an order of magnitude at 8 T and 5 K while T_c was only slightly decreased. The strong enhancement in the flux pinning is argued to be attributable to a combination of C substitution for B and thermal strain-induced defects. Also, the evidence from collective pinning model suggests the δT_c pinning mechanism rather than the δl pinning for the graphene doped MgB_2, contrary to most doped MgB_2. The strong enhancement of J_c, H_{c2}, and H_{irr} with low levels of graphene doping is promising for large-scale MgB_2 wire applications.

Tensile strain effects on superconducting transition temperature (T_c) was observed in graphene-MgB_2 alloys to pursue high T_c in multi-gap superconductors. The enhancement of energy gap for π-band indicates the weak rescale of density of state on Fermi surface. The E_{2g} mode split into two parts: one dominant soften mode responding to tensile strain and another harden mode responding to carbon substitution effects. The existence of soften E_{2g} mode in bulk samples suggests that modified graphene-MgB_2 alloys are the potential candidates for the high performance superconducting devices.

The effects of graphene doping in MgB_2/Fe wires were also investigated. At 4.2K and 10T, the transport J_c was estimated to be for the wire sintered at 800°C for 30 minutes, the doped sample is almost improved as one order, compared with the best un-doped wire sample. The strong enchantment of the temperature dependence of the upper critical field (H_{c2}) and the irreversibility field (H_{irr}) is found from the resistance (R) – temperature (T). But the calculated active cross-sectional area fraction (A_F) represents the connectivity factor between adjacent grains is lower, which is the main factor to improve transport J_c in limitation. It should mention that in recently research activity, two groups can improve the mass density and the grain connectivity very well. One is the internal Mg diffusion processed (IMD) multi-filamentary wire, which is developed by Togano (Hur, J. M. et al., 2008). The other one is the cold high pressure densification (CHPD) in-situ MgB_2 wire by Flukiger [18]. If can combine these methods with the graphene doping, the strong enhancement of J_c, H_{c2}, and H_{irr} with low levels of graphene doping is promising for large-scale MgB_2 wire in industrial applications.

4. Acknowledgment

We acknowledge support from the ARC (Australia Research Council) Project (DP0770205, LP100100440). The author would like to thank Dr. T. Silver for her helpful discussions. This work was supported by Hyper Tech Research Inc., OH, USA, and the University of Wollongong.

5. References

Novoselov, K. S. et al. (2004). Electric field effect in atomically thin carbon films. *Science*, Vol.306, pp. 666–669, ISSN 0036-8075

Stankovich, S. et al. (2006). Graphene-based composite materials. *Nature*, Vol. 442, pp. 282–286, ISSN 0028-0836

Dou, S. X. et al. (2007). Mechanism of enhancement of electromagnetic properties of MgB_2 by nano-SiC doping. *Phys. Rev. Lett.*, Vol 98, pp. 097002-1, ISSN 0031-9007

Ma, Y. *et al.* (2006). Significantly enhanced critical current densities in MgB_2 tapes made by a scalable nanocarbon addition route. *Appl. Phys. Lett.*, Vol 88, pp. 072502-5, ISSN 0003-6951

Senlowocz, B. J. et al. (2005). Improved upper critical field in bulk-form magnesium diboride by mechanical alloying with carbon. *Appl. Phys. Lett.*, Vol 86, pp. 202502-5, ISSN 0003-6951

Kumakura, H. et al. (2004). Upper critical fields of powder-in-tube-processed MgB_2/Fe tape conductors. *Appl. Phys. Lett.*, Vol 84, pp. 3669-71, ISSN 0003-6951

Sumption, M. et al. (2005). Large upper critical field and irreversibility field in MgB_2 wires with SiC additions. *Appl. Phys. Lett.*, Vol 86, pp. 092507-10 ISSN 0003-6951

Dou, S. X. et al. (2003). Effect of carbon nanotube doping on critical current density of MgB_2 superconductor. *Appl. Phys. Lett.*, Vol 83, pp. 4996-9 ISSN 0003-6951

Kim, J. H. et al. (2006). Carbohydrate doping to enhance electromagnetic properties of MgB_2 superconductors. *Appl. Phys. Lett.*, Vol. 89, pp. 142505 -8 ISSN 0003-6951

Wilke, R. H. T. et al. (2008). Systematic effects of carbon doping on the superconducting properties of $Mg(B_{1-x}C_x)_2$. *Phys. Rev. Lett.*, Vol 92, pp. 062001 ISSN 0003-6951

Tung, V. C. et al. (2009). High-throughput solution processing of large-scale graphene. *Nature Nanotech.*, Vol 4, pp. 25-29, ISSN 1748-3387

Kim, K. S. et al. (2009). Large-scale pattern growth of graphene films for stretchable transparent electrodes. *Nature* , Vol 457, pp. 706-710, ISSN 0028-0836

Hernandez, Y. et al. (2008). High-yield production of graphene by liquid-phase exfoliation of graphite. *Nature Nanotech.* ,Vol 3, pp. 563–568, ISSN 1748-3387

Li, D. et al. (2008). Processable aqueous dispersions of graphene nanosheets. *Nature Nanotech.* ,Vol 3, pp. 101–105, ISSN 1748-3387

Li, X. et al. (2008). Chemically derived, ultrasmooth graphene nanoribbon semiconductors. *Science* Vol 319, pp. 1229–1232 ISSN 0036-8075

Choucair, M. Thordarson, P. and Stride, J. A. Gram-scale. (2009). Production of graphene based on solvothermal synthesis and sonication. *Nature Nanotech.* , Vol 4, pp. 30-33, ISSN 1748-3387

Nagamatsu, J. et al. (2001). Superconductivity at 39 K in magnesium diboride. *Nature*, Vol 410, No.6824, pp. 63-64, ISSN 0028-0836

Kunc, K. et al. (2001). MgB_2 under pressure: phonon calculations, Raman spectroscopy, and optical reflectance. *Journal of Physics -Condensed Matter*, Vol 13, No.44, pp. 9945-9962, ISSN 0953-8984

Soltanian, S. et al. (2005). High transport critical current density and large H_{c2} and H_{irr} in nanoscale SiC doped MgB_2 wires sintered at low temperature. *Superconductor Science & Technology*, Vol 18, No.5, pp. 658-666, ISSN 0953-2048

Yamamoto, A. et al. (2005). Universal relationship between crystallinity and irreversibility field of MgB_2. *Appl. Phys. Lett.*, Vol **86**, No.21 pp. 212502-5 ISSN 0003-6951

Dou, S. X. et al. (2002). Enhancement of the critical current density and flux pinning of MgB_2 superconductor by nanoparticle SiC doping. *Appl. Phys. Lett.*, Vol 81, No. 18, pp. 3419-3421, ISSN 0953-2048

Yeoh, W. K. et al. (2006). Control of nano carbon substitution for enhancing the critical current density in MgB_2. *Superconductor Science & Technology*, Vol 19, No.6, pp. 596-599, ISSN 0953-2048

Kim, J. H. et al. (2008). Correlation between doping induced disorder and superconducting properties in carbohydrate doped MgB_2. *Journal of Applied Physics*, Vol 104, No. 6, pp. 063911-6, ISSN 0021-8979

Dikin, D. A. et al. (2007). Preparation and characterization of graphene oxide paper. *Nature* Vol 448, pp. 457–460, ISSN 0028-0836

Xu, X. et al. (2010). Graphene doping to enhance the flux pinning and supercurrent carrying ability of a magnesium diboride auperconductor. *Superconductor Science & Technology*, Vol 23, pp. 085003-7, ISSN 0953-2048

Williamson, G. K., and Hall, W. H. (1953). X-ray line broadening from filed Aluminium and Wolfram. *Acta Metall.* Vol 1, pp. 22-31, ISSN 0001-6160

Horvat, J. et al. (2008). Transport and magnetic critical current in superconducting MgB_2 wires. *Superconductor Science & Technology*, Vol 21, pp. 065003-8, ISSN 0953-2048

Qin, M. J. (2002). Evidence for vortex pinning induced by fluctuations in the transition temperature of MgB_2 superconductors. *Phys. Rev. B.*, Vol 65, pp. 132508(4), ISSN 0163-1829

Wang, J. L. et al. (2008). Effects of C substitution on the pinning mechanism of MgB_2. *Phys Rev. B* Vol 77, pp. 174501(7), ISSN 0163-1829

Li, W. X. et al. (2008). Raman study of element doping effects on the superconductivity of MgB_2. *Phys. Rev. B.*, Vol 77, pp. 094517(9), ISSN 0163-1829

Gurevich, A. (2003). Enhancement of the upper critical field by nonmagnetic impurities in dirty two-gap superconductors. *Phys. Rev. B.*, Vol 67, pp.184515(7), ISSN 0163-1829

Zeng, R. et al. (2009). Thermal strain-induced enhancement of electromagnetic properties in SiC-MgB_2 composites. *Appl. Phys. Lett.*, Vol 94, 042510-3, ISSN 0003-6951

Avdeev, M. et al. (2003). Crystal chemistry of carbon-substituted MgB_2. *Physica C*, Vol 387 No.3-4, pp. 301-306. ISSN 0921-4534

Coraux, J. et al. (2009). Growth of graphene on Ir(111). *New J. Phys.*, Vol 11, pp. 023006-22. ISSN 1367-2630

Hur, J. M. et al. (2008). Fabrication of high-performance MgB_2 wires by an internal Mg diffusion process. *Superconductor Science & Technology*, Vol 21, pp. 032001-4, ISSN 0953-2048

Flukiger, R. Hossain, M. S. A. and Senatore, C. (2009) Strong enhancement of J_c and B_{irr} in binary *in situ* MgB_2 wires after cold high pressure densification. *Superconductor Science & Technology*, Vol 22, pp. 085002-8, ISSN 0953-2048

Rowell, J. M. (2003). The widely variable resistivity of MgB_2 samples. *Superconductor Science & Technology*, Vol 16, pp. R17-R27, ISSN 0953-2048

11

Preparation of Existing and Novel Superconductors using a Spatial Composition Spread Approach

Kevin C. Hewitt, Robert J. Sanderson and Mehran Saadat
Dalhousie University, Department of Physics and Atmospheric Science, Halifax NS Canada

1. Introduction

We describe in this chapter a promising system and method to search for novel superconductors by investigating appropriately chosen antiferromagnets and creating carriers by doping using a high throughput spatial composition spread (combinatorial) approach. The method has been applied to the cuprate superconductors and has the potential to enhance our understanding of these materials and push the boundaries of the field by quickly exploring novel ones.

Finding novel superconducting materials, which superconduct at much higher temperatures, now seems to be a realistic goal because of three recent developments: the 2008 discovery of superconductivity in iron arsenide based materials; the observation that a number of superconductors are doped antiferromagnets; and the tremendous progress researchers have made over the past 20 years in understanding the physical properties of existing superconductors. These developments suggest a path to novel superconductors - explore the electrical transport properties of doped antiferromagnets. While a number of applications of existing superconductors have been realized, their widespread use depends on raising the transition temperatures substantially above the current world record T_c of 138 K.

Expanding the number of known systems which exhibit superconductivity also allows researchers to identify its essential elements. These observations help reduce the number of models which purport to explain the mechanism of pair formation, and allow researchers to ignore irrelevant peculiarities of a particular system. In general, the preparation of novel phases of matter increases the likelihood of the discovery of novel material properties. The combinatorial approach to materials discovery (Xiang et al, 1995) allows one to realize these goals at an unprecedented rate.

Discoveries by groups in Japan and China over the past few years have added to the class of systems for which antiferromagnetic order exists in close proximity to superconductivity, and in some cases may even coexist. Iron arsenide-based, cuprate, fulleride and heavy fermion superconductors [detailed references below] populate this class, and the diversity of hosts highlight how useful it is to search for new examples of superconductors in aid of an empirical identification of the important parameters on which to build a correct theory. As the number of examples has grown, the importance of spin fluctuations has emerged. This chapter describes a route to discover novel superconductors in doped antiferromagnets to enhance our understanding of superconductivity. To accomplish these goals we use a

combinatorial approach to materials discovery, which we have recently demonstrated allows us to map the superconducting properties of the La214 cuprate superconductor and search for new superconductors by quickly and efficiently exploring phase space in a chosen system. The single layer K_2CuF_4 and double layer $K_3Cu_2F_7$ perovskites will be shown to be likely candidates to exhibit superconductivity on the border of antiferromagnetic or ferromagnetic phases. Rapid characterization using high throughput resistivity apparatus, such as the one (Hewitt et al, 2005) developed in our lab, allows one to identify superconducting phases at an accelerated pace.
The spatial composition spread approach is described in section 2, its application to cuprate superconductors in section 3 and section 4 proposes likely superconductors on the boundary of antiferromagnetic and ferromagnetic phases.

2. Spatial composition spread approach

Combinatorial materials science (CMS) methods represent a powerful technique to produce a large number of compositions on a spatially addressable substrate. In particular, one may produce a continuous variation in composition across a substrate using physical vapor deposition techniques. For example, one can locate three evaporation targets at the vertices of a triangle to prepare a ternary phase diagram (Kennedy et al, 1965); or four targets at the corners of a square to prepare a quaternary phase diagram (van Dover et al, 1998) by sputter deposition. The drawback of this approach is that a non-linear variation in composition is produced as a function of position on the substrate. In order to obtain a linear relationship which would allow for a much easier interpretation of the data, we altered the flux produced by sputtering targets through design of masks which intercept the flux to create a linear or constant deposition as a function of position. Although CMS methods have been used to show that particular superconducting phases can be prepared (Xiang et al, 1995), to the author's knowledge the composition spread approach has not been used for this purpose until recently by our group (Sanderson and Hewitt, 2005, 2007) and subsequently by a Brookhaven group (Logvenov et al, 2007) using molecular beam epitaxy.

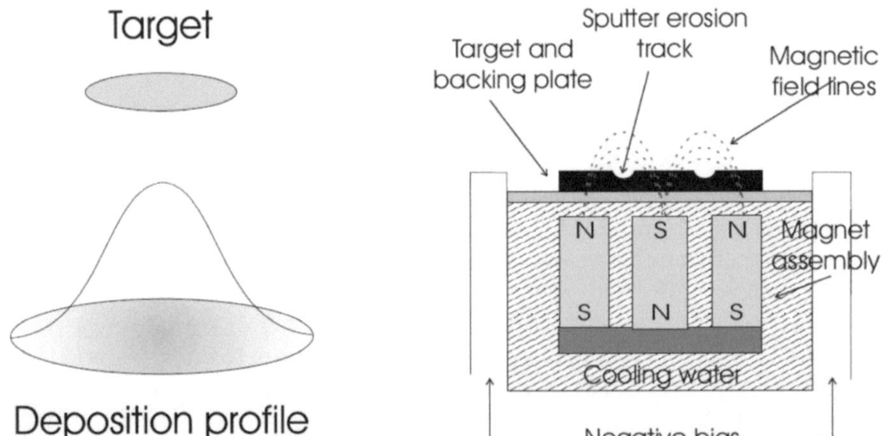

Fig. 1. Schematic showing the sputter flux generated by sputtering a circular target with a circular magnetron.

To implement a linear composition spread approach, the target deposition profile needs to be altered to provide the appropriate linear (or constant) variation across the substrate. One way to produce this variation is to interrupt the flux with a physical mask placed over the target. The process of determining the target mask profiles, for the Corona Vacuum Coaters V-3T sputtering system used in our lab is described in detail here, as outlined elsewhere (Dahn et al, 2002).

The sputtering flux from a circular magnetron can be determined by sputtering a target in front of a stationary substrate and measuring the mass as a function of position. The result for an aluminum target sputtered in an argon atmosphere at 5.6 mTorr is shown in Figure 2. A Gaussian profile of the form $J(q) = Ae^{-(q/w)^2}$ fits the data well, with the full width at half-max, $\omega = 10$ cm.

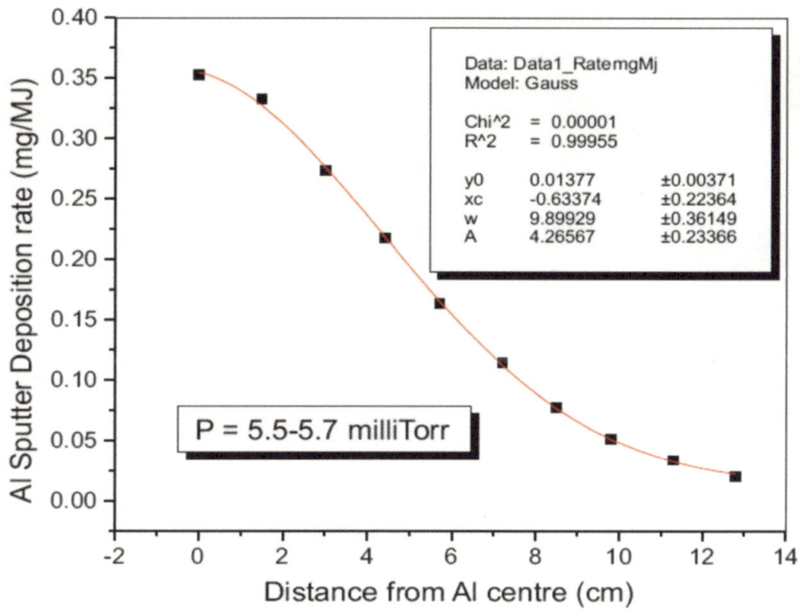

Fig. 2. Mass deposited per unit energy for an aluminum target sputtered in front of a stationary substrate table in a chamber with 5.6 mTorr Argon.

To intimately mix the elements the table must be rotated at a high rate, so one must calculate the deposition a rotating substrate. To obtain a specific deposition profile (linear out, linear in, constant) one must be able to calculate the amount of material deposited upon a rotating substrate, through a target mask. Figure 3 illustrates the geometry of the problem in which vectors $\mathbf{r}_T$ (table centre to target centre location on the table), $\mathbf{r}_1$ (from table centre to start of the deposited film) and $\mathbf{r}_2$ (distance from the table centre to the film end) are drawn. If $\mathbf{r}_1$ and $\mathbf{r}_2$ are equidistant from r_T , then the deposition would be the same at $\mathbf{r}_1$ and $\mathbf{r}_2$ when the substrate table is stationary. However, when the table is rotating, a point at $\mathbf{r}_1$ has a smaller tangential velocity than a point at $\mathbf{r}_2$, so more flux needs to reach $\mathbf{r}_2$ to produce the same deposition. For the Corona Vacuum Coaters V-3T sputtering system $\mathbf{r}_T$ = 13.33 cm, $\mathbf{r}_1$ = 9.5 cm and $\mathbf{r}_2$ = 17.1 cm.

To calculate the mask profile needed consider the geometry shown in Fig. 3B. **R** is a vector from the centre of the substrate table to a point below the centre of the target on the table, while **q** is a vector from the centre of the target to the point on the substrate where the flux is being calculated. By integrating the flux along the heavy arc shown in Figure 3B the deposition D(s) can be found.

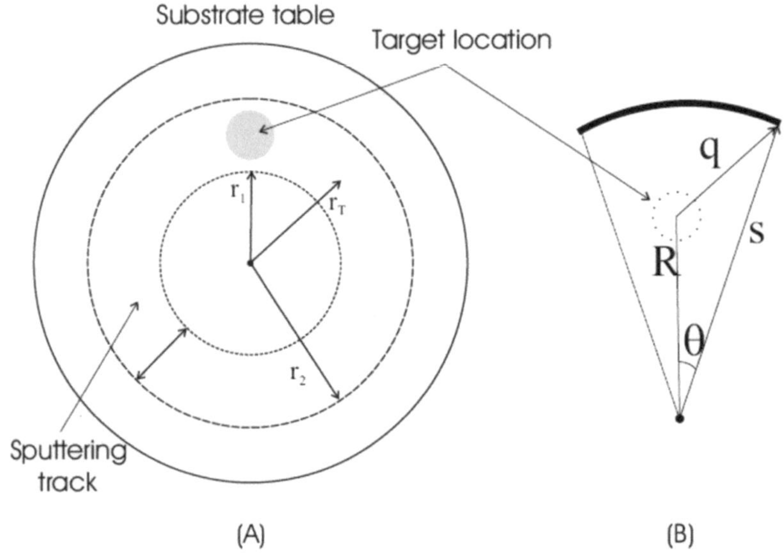

Fig. 3. Vectors defining important locations on the substrate table (A) and those needed to define the target mask (B).

It can be shown easily that,

$$q = \sqrt{R^2 + s^2 - 2sR\cos\theta} \tag{1}$$

and the deposition as a function of radial position (s) is,

$$D(s) = \int_0^{\theta_{max}} e^{-(\frac{q}{\omega})^2} s d\theta \tag{2}$$

The expression of D(s) can be used to determine the desired deposition profile, whether linearly increasing or decreasing with s, or simply a constant independent of s. Finally, the mask shape can be numerically determined by solving Equations 1 and 2 for θ_{max} and s. The mask shape is found by converting the θ_{max} and s values to x and y coordinates. Masks have been designed for a constant deposition, as well as depositions that vary linearly, increasing inwards and outwards. With these three masks a wide variation of film compositions can be made. Figure 4 shows an image of each mask as well as masses of the film deposited onto a rotating substrate through the corresponding mask.
The results presented in the next section usually employ the set-up depicted in Figure 5. Using targets of composition A (placed behind the linear-in mask, blue in Fig. 5) and B (placed behind the linear out mask, red in Fig. 5), results in compositions varying linearly as $A_{1-x}B_x$ with radial position (s).

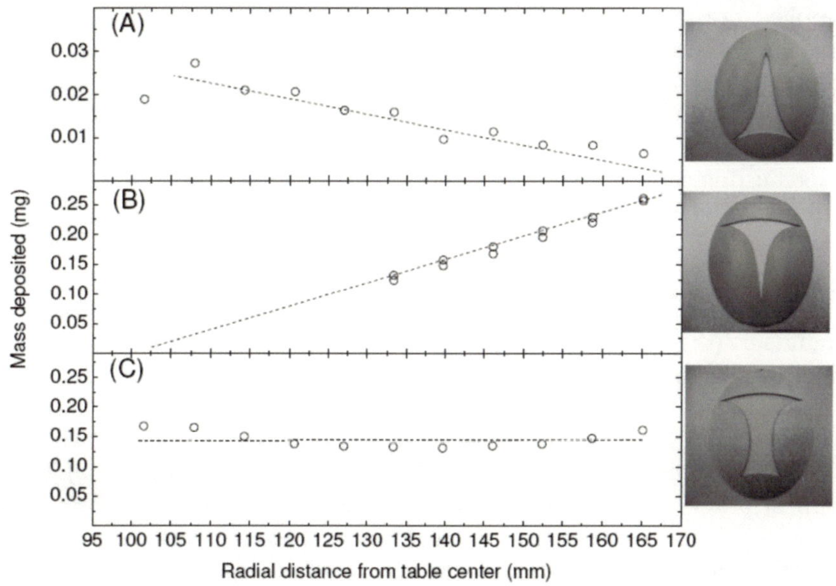

Fig. 4. Mass deposited through the linear in, linear out and constant masks. The dashed lines are a guide to the eye.

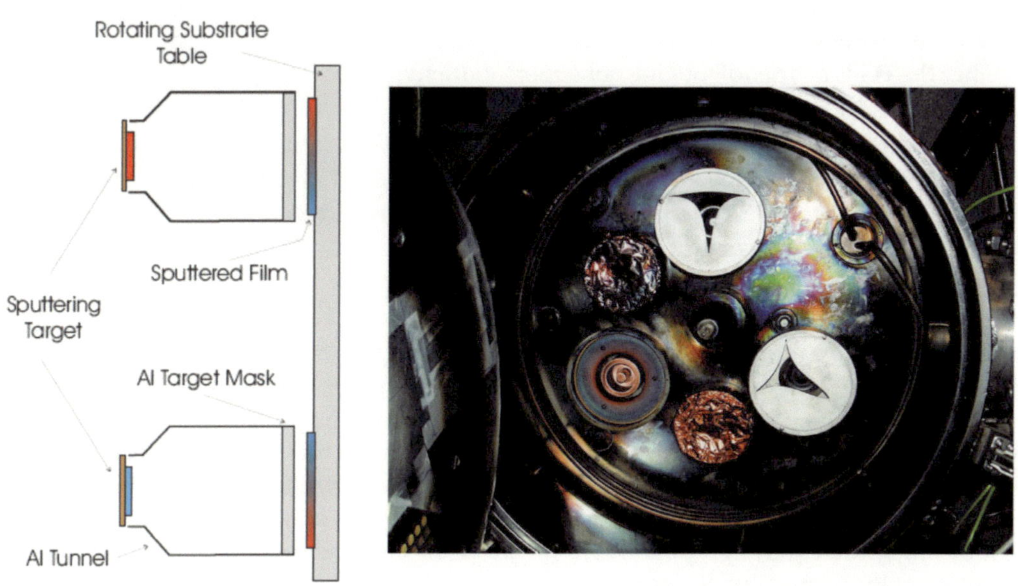

Fig. 5. Schematic side view (left) and actual (right) front-view image of the sputtering machine set-up with linear-in (at 4:00) and linear-out (at 12:00) masks placed in front of each target (color on-line).

3. Densely mapping the phase diagram of the cuprate superconductors

We first applied the linear composition spread approach to the cuprate superconductor Bi2212 (Sanderson and Hewitt, 2005, 2007), and then La214 (Saadat and Hewitt, 2010). The results for La214 are described here.

The cuprates are doped antiferromagnetic Mott-Hubbard insulators, becoming Fermi liquid metals at large hole concentrations. (They can also be electron doped, although the T_c's are not as high, reflecting some type of electron-hole asymmetry.) The limiting regimes are well described by the two-dimensional Hubbard model (Hubbard, 1963). When the onsite coulomb repulsion (U) dominates the kinetic energy of hopping (t) (U>>t), the solutions to the Hubbard Hamiltonian produce highly localized electron wavefunctions which gives rise to an insulator, and super-exchange interaction produces the Néel state. When the hopping term (t) dominates (t>>U), the solution is a modified Bloch wave function which gives rise to the Fermi liquid state. At intermediate hole concentrations where t and U are comparable, the transition from localized to itinerant electrons produces interesting physics and remains enigmatic. It is within this region of the phase diagram where superconductivity occurs. Thus it is important to understand the normal state properties of metal-insulator transitions (Mott, 1968), in order to decipher the nature of the superconducting state. Also, in this region of phase space there exists a partial suppression of low energy excitations, a Pseudogap (PG) (Timusk and Statt, 1999), which appears below a temperature $T^* > T_c$. Its exact description is a subject of intense study because it may hold the key to understanding the transition to the superconducting state. Theories of the PG can be divided into two broad categories: ones which identify it as a precursor to the onset of superconductivity, or others which classify it as a competing phase. It is thought that a dense map of the doping dependence of T* is the key to deciphering the nature of the PG (Norman et al, 2005). Whether T* merges with T_c in the overdoped regime or ends at optimal doping determines whether the PG is a "friend" or "foe" of superconductivity.

We have recently obtained data showing that one can densely map the temperature-hole concentration phase diagram of $La_{2-x}Sr_xCuO_{4+\delta}$ ($0 \le x \le 0.18$) using the spatial composition spread approach, as described in section 2 of this chapter, and presented in our recent article (Saadat et al, 2010). First applied to the cuprate superconductor Bi2212 (Sanderson and Hewitt, 2005, 2007), it was used to synthesize a $La_{2-x}Sr_xCuO_{4+\delta}$ ($0 \le x \le 0.18$) library. In this approach, targets of La_2CuO_4 and $La_{1.82}Sr_{0.18}CuO_4$ were co-sputtered with specially designed target masks, which ultimately produce a linear composition gradient varying from x = 0 (at 0 mm) to x = 0.18 (at 75 mm) on a set of eight single crystal substrates. The libraries' structures are characterized by X-ray diffraction (Fig. 6) and cation composition by Energy and Wavelength Dispersive Spectroscopy (EDS/WDS) (Fig. 7).

While we have shown it is possible to prepare single phase films in this manner, they are polycrystalline. Epitaxial films are required to separate contributions to T* from c-axis and ab-plane transport. Polycrystalline films are sufficient, however, to measure T_c. Therefore, a high-throughput resistivity apparatus (Hewitt et al, 2005) was used to measure the DC resistivity of the 52 member library.

T_c and T* were determined and plotted versus Sr content as shown in Fig. 9. We found that T_c is suppressed near 1/8 (x = 0.125) doping, consistent with the formation of a stripe phase (Tranquada et al, 1995). The lowest Sr content (x) at which superconductivity appears is 0.03, not at the expected value of 0.05 (Ando et al, 2004). Independent measurements of the hole

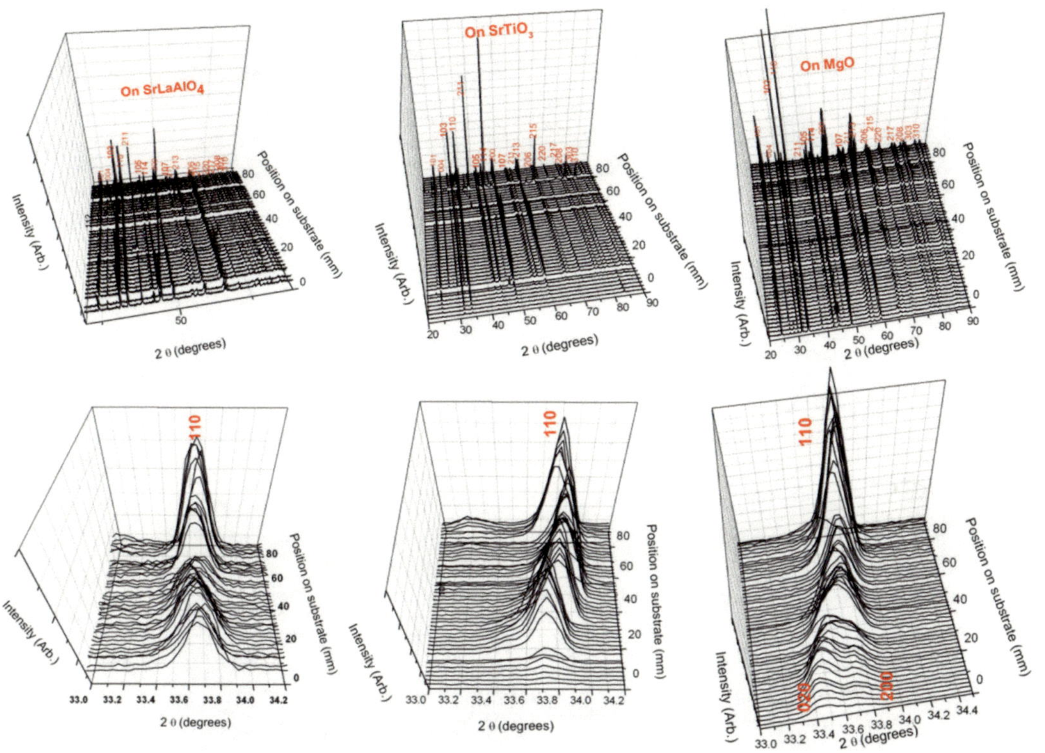

Fig. 6. X-ray diffraction of film library deposited onto three substrates ($SrLaAlO_4$, $SrTiO_3$ and MgO). The results (top panel of three) show the peaks can all be indexed to the La214 compound, and in the bottom panel of three the region where peaks sensitive to the tetragonal to orthorhombic phase transition ([110] – [020/200]) are found.

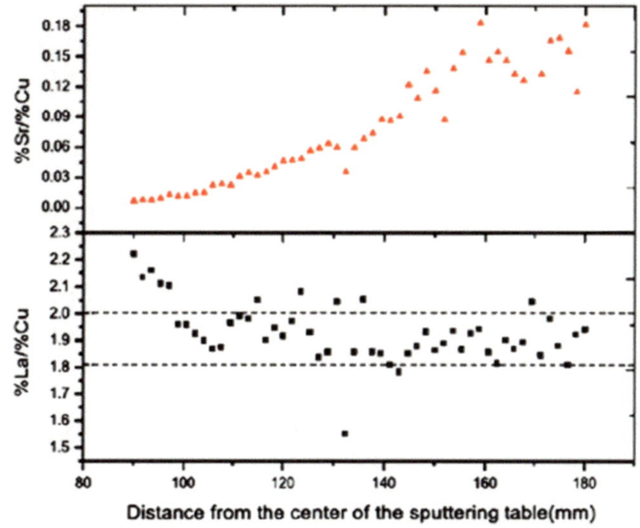

Fig. 7. Elemental composition of the $La_{2-x}Sr_xCuO_4$ ($0<x<0.18$) library on MgO (100).

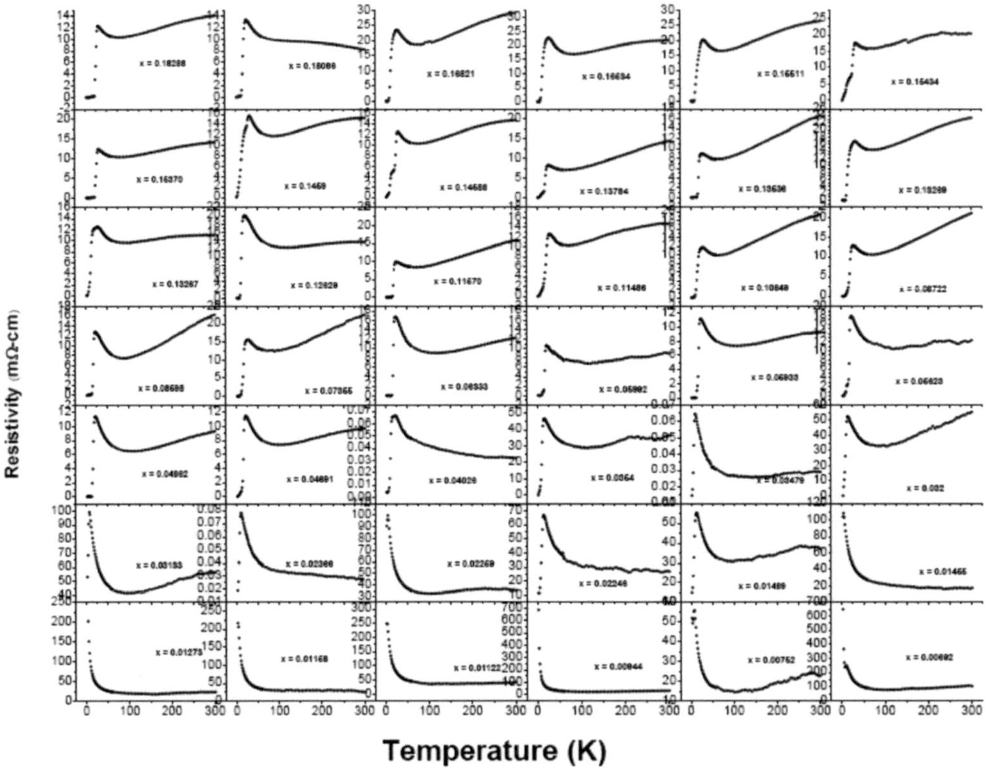

Fig. 8. The DC resistivity of the $La_{2-x}Sr_xCuO_4$ (0<x<0.18) library deposited onto $SrLaAlO_4$ substrates.

concentration can be obtained by measuring the intensity of a feature located 2 eV below the O Kedge (530 eV) by X-ray absorption spectroscopy (XAS) (Kuiper et al, 1988). These measurements would not only allow us to evaluate whether this unexpected result is a truly novel feature and not simply a consequence of oxygen non-stoichiometry and/or film strain, it would also test a theory that predicts the existence of charge 2e bosons in the PG state (Leigh et al, 2007; Choy et al, 2008).

4. Prospective 2D antiferromagnetic and ferromagnetic superconductors

In this chapter we also propose promising hosts for superconductivity and describe the use of the combinatorial approach to rapidly and efficiently scan the horribly large phase space of possible dopings. In the field of superconductivity there are very few "right" substitutions which produce superconductivity, and then only over a rather limited range of dopings.

Until recently, the cuprates and heavy fermion systems have presented a rather unique example of superconductivity in doped antiferromagnets. The phase diagram of the recently discovered FeAs-based (e.g. RE(O,F)FeAs, RE = La, Sm, or Ce (Luetkens et al, 2009; Drew et al, 2009; Zhao et al, 2008) and $(Ba,K)Fe_2As_2$ (Chen et al, 2009) superconductors share many features with the cuprates: a) they are doped antiferromagnets, b) superconductivity occurs

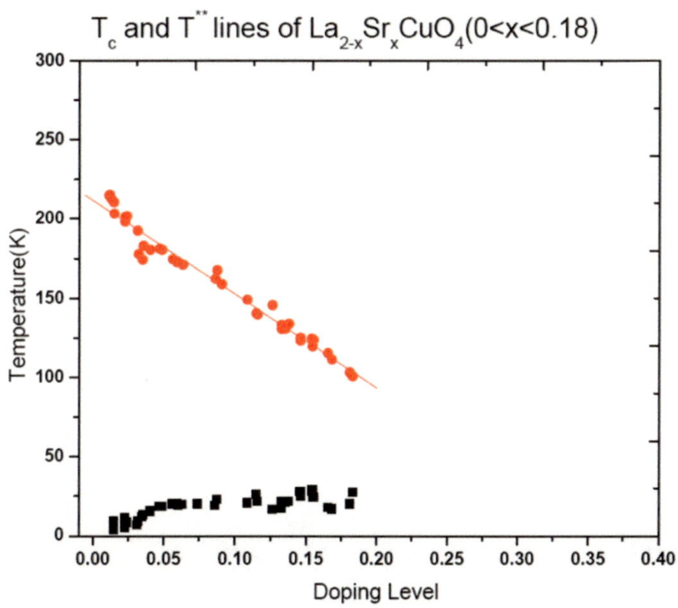

Fig. 9. T_c and T* derived from the resistivity data of Fig. 8, for a $La_{2-x}Sr_xCuO_4$ (0<x<0.18) library deposited onto $SrLaAlO_4$ substrates using the spatial composition spread approach.

in 2D planes, though corrugated in FeAs-based materials, c) the planes are doped by adjacent charge reservoir layers, and d) the maximum superconducting transition temperature is a similar fraction (~1/3) of the maximum Néel temperature. Recent re-examination of the fulleride (A_3C_{60}; A = K, Cs, and Rb) superconductors has revealed that they are also doped antiferromagnetic insulators (Arvanitidis , 2007; Takabayashi, 2009). The phase diagram of cobaltate ($Na_xCoO_2.yH_2O$) (Foo et al, 2004; Takada et al, 2003) and heavy fermion (e.g. $CeCoIn_5$) (Petrovic et al, 2001) materials have this property as well.

A consistent picture is emerging (Uemura, 2009) that unconventional superconductivity is intimately related to antiferromagnetism (Fig. 10). Imai and coworkers have shown recently (Imai, 2009) that the strength of antiferromagnetic spin fluctuations in FeSe are correlated with the pressure-induced increase in T_c, suggesting a link between spin fluctuations and the mechanism of superconductivity, and have prompted some to suggest that it is a common thread linking organic, heavy-fermion, actinide, cuprate and Fe superconductors (Scalapino, 2009; Uemura, 2009). Spin fluctuation mediated pairing has always been a strong candidate for the mechanism of superconductivity in the cuprates, among the more than twenty candidates (Cho, 2006).

Monthoux and Lonzarich have proposed (Monthoux & Lonzarich, 1999; Monthoux & Lonzarich, 2001; Monthoux et al, 2007) a spin fluctuation based mechanism for superconductivity in systems close to a ferromagnetic or antiferromagnetic instability, making the convincing argument that on the border of long-range magnetic order the dominant interaction channel must be of magnetic origin and depend on the relative spin orientations of the interacting quasiparticles. Superconductivity is predicted to be more robust in doped antiferromagnets vs ferromagnets, and the amplitude of the oscillations in the interaction is enhanced by low dimensionality. For example, the range of temperature and pressure over which superconductivity is observed was increased by about one order of

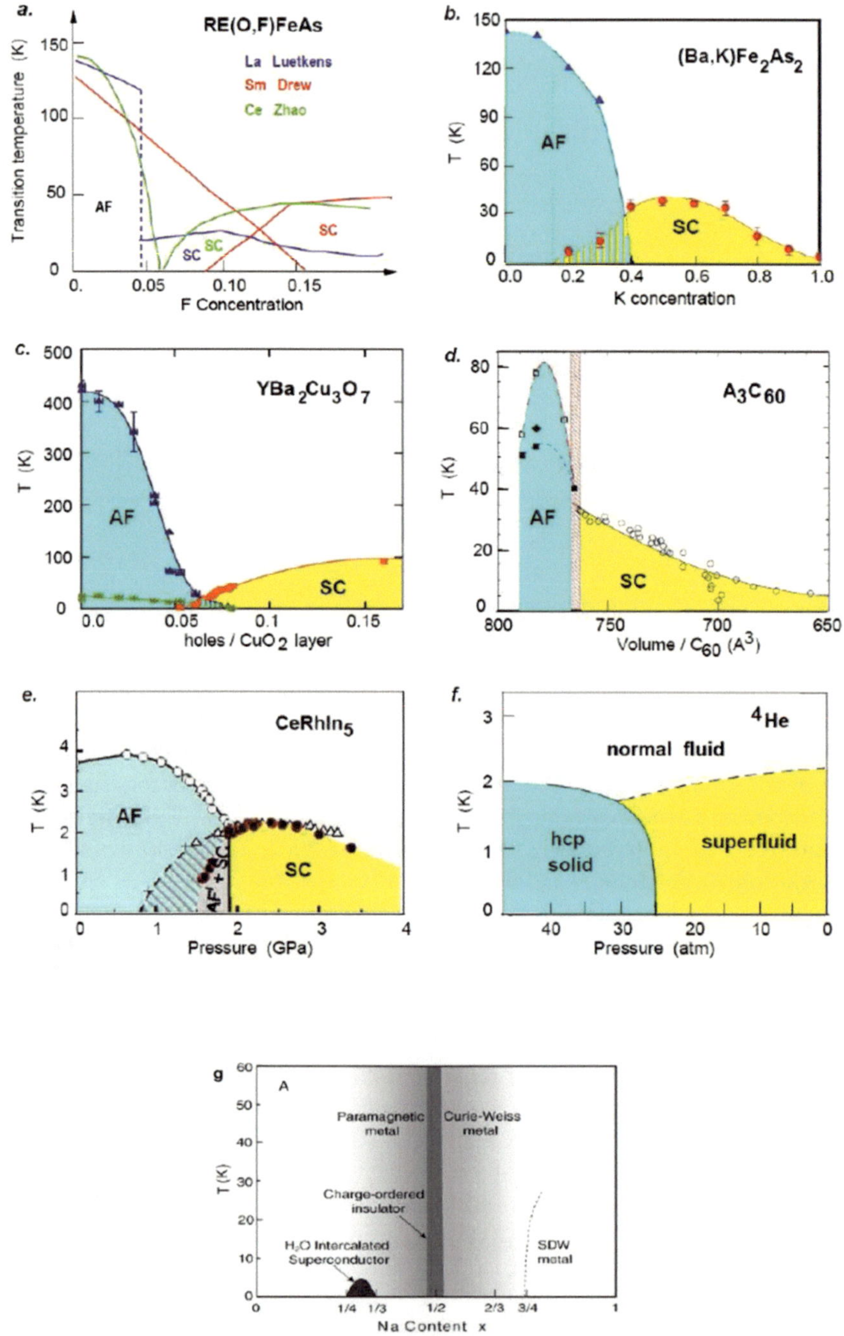

Fig. 10. Phase diagram of (a) RE(O,F)FeAs, (b) (Ba,K)Fe_2As_2, (c) $YBa_2Cu_3O_7$, (d) A_3C_{60}, (e) $CeRhIn_5$, (f) ^4He and (g) $Na_xCoO_2.yH_2O$ demonstrating the proximity of superconducting and AF phases (adapted from Uemura, 2009).

magnitude in going from cubic $CeIn_3$ to its tetragonal analogues $CeMIn_5$ (M = Rh, Ir or Co) as anticipated by the magnetic interaction model (Monthoux & Lonzarich, 2002). Thus in the search for higher temperature superconductors one should explore the border of antiferromagnetism in a quasi two-dimensional tetragonal system with high characteristic spin fluctuation frequencies.

The conditions favourable for magnetic pairing include: (i) strong quasi two-dimensional antiferromagnetic correlations (large J) for spin singlet pairing and for large amplitude oscillations of the spin-spin interaction (gives small correlation length ξ which is inversely proportional to T_c), (ii) a single band of relatively high characteristic energy scale, and (iii) a crystal structure that enables the repulsive regions of the pairing potential to be optimally neutralized. Favourable T_c's can be achieved in layered d-electron systems of moderate electron densities (n) and bandwidths (t) and can be controlled by chemical doping or hydrostatic pressure (Monthoux et al, 2007).

One system which satisfies most of these requirements is the perovskite-type single layer compounds of composition A_2MX_4 and double layer compounds of composition $A_3M_2X_7$, where A^{1+} = K, Rb, Cs, M^{2+} = Mg, Mn, Fe, Co, Cu, Cd and X = F, Cl , or Br (see Geick, 2001 for a review). In these perovskite-type layer structures the dominant magnetic interaction is the nearest-neighbour Heisenberg exchange within the layers which causes their 2D character. These compounds have a metal ion (M) surrounded by 6 halides (X) in an octahedral arrangement. The magnetic properties depend on the intralayer superexchange interaction (J) mediated by the halide (X) between two M ions. Theory predicts an exponential dependence of J on the nearest neighbour distance (a_{MXM}) and experiments find a power law dependence $J(a_{nn}) = J(a_{nn,0})(a_{nn}/a_{nn,0})^{-12}$ for small a_{nn}.

A classic perovskite layer compound is La_2CuO_4 ($X = O^{2-}$, $M = Cu^{2+}$, and $A = La^{2+}$) which when appropriately doped (e.g. hole doped by replacing Sr^{2+} for La^{3+} or electron doped by replacing La^{3+} with $Nd^{3+}$and doping with $Ce^{3+,4+}$) forms a high temperature superconductor. It must be noted that *a priori* one could not have predicted these dopings would produce superconductivity. La_2CuO_4 has AFM order in-plane and out of plane and it is thought that superconductivity above liquid helium temperatures are possible because of (a) the large exchange interaction $J/k_B \sim 766$ K (Hayden et al, 1991); and (b) that the electrons in the Cu $d_{x^2-y^2}$ band possesses the correct symmetry to avoid Coulomb repulsion.

In looking for promising hosts, the single layer K_2CuF_4 and double layer $K_3Cu_2F_7$ compounds seem to have the right structure, and the d-band of Cu is the highest partially filled band. However, the intralayer interaction is small (J/k_B = 11K), produces ferromagnetic order (Feldkemper et al, 1995) and the Cu^{2+} ions exhibit alternating occupation of $z^2 - x^2$ and $z^2 - y^2$ hole states unlike the $x^2 - y^2$ ordering in La_2CuO_4 (Fig. 11). However, by inducing distortive changes at pressures larger than 9.5 GPa in the basal plane of the CuF_6 octahedra (Ishizuka et al, 1996; Ishizuka et al, 1998) was able to obtain (Fig. 12) antiferromagnetic order in K_2CuF_4 with $x^2 - y^2$ hole orbital overlap, exactly as found in the prototype cuprate superconductor La_2CuO_4.

SCAN PHASE SPACE: Since the high pressure phase of K_2CuF_4 is so similar to La_2CuO_4 in its orbital ordering, structure and magnetic properties, it satisfies the conditions set out by Monthoux and Lonzarich and should become a superconductor when appropriately doped. To obtain the high pressure phase of K_2CuF_4 one may attempt (a) pseudomorphic growth of

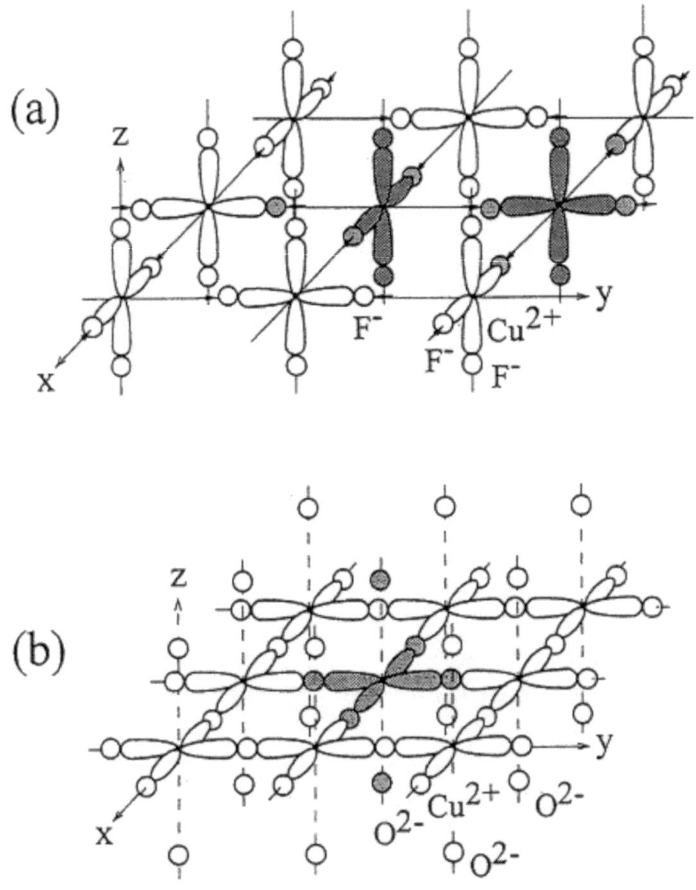

Fig. 11. The two kinds or orbital ordering in the basal plane of a K_2NiF_4-type compound. (a) Antiferrodistortive orbital ordering of $d_{x^2-z^2}$ and $d_{y^2-z^2}$ in K_2CuF_4 and (b) Ferrodistortive orbital ordering of $d_{x^2-y^2}$ in La_2CuO_4. In (a) the CuO_6 octahedra elongate alternately along a- and b-axis whereas it elongates along the c-axis only in (b). (from Ishizuka et al, 1996).

films onto substrates which produce compressive strain. The M-X-M distance in K_2CuF_4 is ~ 4.124 Å. Therefore, $SrLaAlO_4$ (3.756 Å, -10%), SrTiO3 (3.905 Å, -6%), $LaAlO_3$ (3.821 Å, -8%) and $SrLaGaO_4$ (3.843 Å, -8%) substrates should all produce compressive strain, while MgO (a = 4.212 Å, +1%) should produce tensile strain in epitaxial films. It must be noted that film stresses of more than 10 GPa have been achieved in pseudomorphic Fe layers (Sander, 1999). Epitaxial films of the cuprate superconductors sometimes show enhanced T_c perhaps because of the increase in J. One may also (b) dope smaller cations (Na^+, Li^+) to create pressure by cation substitution. Substrate or cation-induced decreases in the nearest neighbour M-X-M distances should also exponentially enhance the intra-layer interaction. In addition, we plan to replace the fluorine anion with other halides (Cl , Br , I) to enhance our understanding of the effect of the exchange interaction on the appearance of superconductivity in the films.

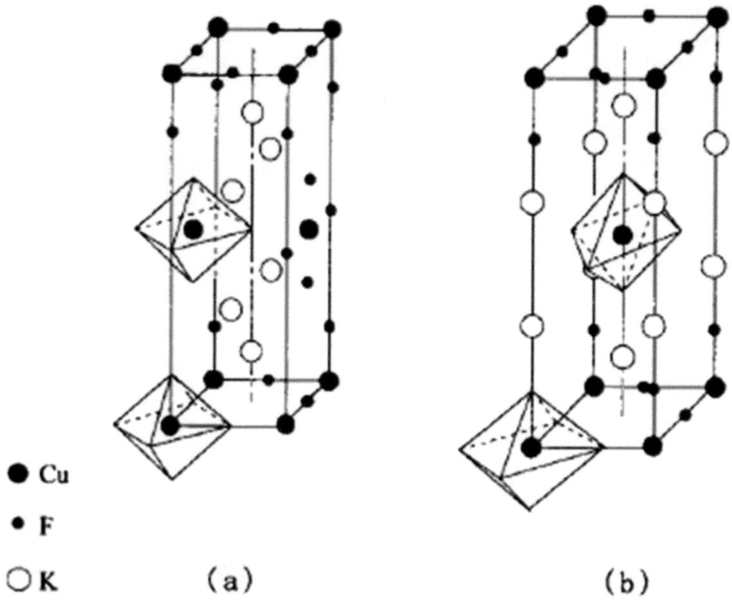

Fig. 12. The structures of K_2CuF_4 in (a) the high pressure phase (P>8GPa), and (b) at ambient pressure (from Ishizuka et al, 1998).

As pointed out earlier, it is difficult to predict *a priori* which doping would produce superconductivity, even when you've selected the right host. This is where the use of combinatorial methods to explore phase space rapidly and efficiently becomes a great asset. One should be able to replace K with higher valent cations C = Mg, Ca, Sr, Ba or Y, La to introduce carriers and produce the phases $K_{2-x}C_xCuF_4$ ($0<x<2$). Our 52-sample mask produces 52 unique compositions to be tested. In addition, K may be replaced with other alkali elements (A = Na, Li, Rb, Cs) at the same time to yield $(K_{1-y}A_y)_{2-x}C_xCuF_4$ phases ($0<y<1$, $0<x<2$), which ultimately produces 52 x 52 = 2,704 unique compositions in one experiment. Every phase can then be tested for superconductivity using a high throughput resistivity apparatus. The full composition range of a pair of substituents (A, C) can be deposited in one sputtering run. Where superconductivity is found one can then explore phase space in the interesting region at higher density, followed by conventional solid state reaction techniques to produce the bulk phases. For every pair of elements A (5 choices) and C (6 choices) 2,704 unique compositions are created. With 30 different dopant pairs A-C we therefore produce 81,120 unique phases. If we have chosen the right host the probability of finding a superconductor should be nonzero. Assuming a very conservative 0.1% probability of finding superconductivity one should discover 81 superconducting phases. Each dopant pair requires at least 3 months to investigate fully, so 7.5 years are required to cover 30 pairs. Paul Canfield (Ames Lab and Iowa State U.) said (Canfield, 2008), "In deference to the term 'fishing trip', a real fisherman goes where the fish are known to congregate and reaps an abundant harvest." By casting our net wide, in the right host, it is **very likely** that the exploration described here will discover novel superconducting phases.

5. Conclusions

New materials form the basis of new products which drive economic development. Superconducting materials have held great promise for some time because they pass a current without resistance and expel magnetic fields. These properties make them the most sensitive magnetic sensors, best source of large magnetic fields (e.g. for use in medical imaging - MRI), most efficient transmisson lines; and are a leading candidate for high speed quantum computers (B. G. Levi, 2009). However, they have not found widespread application mainly because the materials require cooling to at least -136 degrees C. Finding materials that superconduct at much higher temperatures is now thought to be a realistic goal with the recent discovery of superconductivityin iron arsenide based materials, the observation that a number of superconductors are doped antiferromagnets, and the tremendous progress researchers have made in understanding the physical properties of existing superconductors. These developments have re-ignited the field by offering a path to novel superconductors - explore the transport properties of doped antiferromagnets.

To explore the properties of a large number of samples, a spatial composition spread approach has been developed at Dalhousie to more quickly and efficiently prepare new materials. In a single experiment hundreds of compositions can be studied, whereas a serial preparation approach would take several years. We have described in this chapter the feasibility of the approach to densely map the physical properties of an existing superconductor, $La_{2-x}Sr_xCuO_4$. To identify novel superconductors, we have proposed that layered fluoride perovskite compounds be screened using the high-throughput resistivity apparatus developed in our labs. To enhance our understanding of existing superconductors we have also shown that the phase diagram can be mapped at very high density to deduce the doping dependence of a feature, the pseudogap onset temperature, which helps determine the class of theories that apply to the cuprate superconductors.

6. Acknowledgements

We acknowledge the financial support of the Natural Science and Enginnering Research council of Canada, and use of the facilities of the Institute for Research in Materials. We also acknowledge recent fruitful discussion with Paul Canfield during a recent visit to Dalhousie.

7. References

Ando, Y., Komiya, S., Segawa, K., Ono, S., & Kurita, Y. (2004). Electronic phase diagram of high-T_c superconductors from mapping of the in-plane resistivity curvature. *Phys. Rev. Lett.*, Vol. 93, No. 26, (December 2004), pp. 267001-1-4, ISSN 0031-9007

Arvanitidis, J., Papagelis, K., Takabayashi, Y., Takenobu, T., Iwasa, Y., Rosseinsky, M. J., & Prassides, K. (2007). Magnetic ordering in the ammoniated alkali fullerides $(NH_3)K_{3-x}$ Rb_xC_{60} (x=2,3). *J. Phys.: Condens. Matter*, Vol. 19, No. 38, (September 2007), pp. 386235-1-13, ISSN 0953-8984

Canfield, P. C. (2008) Fishing the Fermi sea. *Nature Physics,* Vol. 4, No. 3, (March 2008), pp. 167-169, ISSN 1745-2473

Chen, H., Ren, Y., Qiu, Y., Bao, W., Liu, R. H., Wu, G., Wu, T., Xie, Y. L., Wang, X. F., Huang, Q. & Chen X. H. (2009). Coexistence of of the spin-density wave and superconductivity in $Ba_{1-x}K_xFe_2As_2$. *Europhys. Lett.*, Vol. 85, No. 17006, (January 2009), pp. 17006-p1-p5, ISSN 0295-5075

Cho, A. (2006). High-T_c: The mystery that defies solution. *Science,* Vol. 314, No. 5802, pp. 1072-1075, ISSN 0036-8075

Choy, T.-P., Leigh, R. G., & Phillips, P. (2008). Hidden charge-2e boson: Experimental consequences for doped Mott insulators. *Phys. Rev. B,* Vol. 77, No. 10, (March 2008), pp. 104524-1-9, ISSN 1098-0121

Dahn, J. R., Trussler, S., Hatchard, T. D., Bonakdarpour, A., Mueller-Neuhaus, J. R., Hewitt, K. C., & Fleischauer, M. (2002). Economical Sputtering system to produce large-size composition-spread libraries having linear and orthogonal stoichiometry variations. *Chem. Mater.,* Vol. 14, No. 8, (July 2002), pp. 3519-3523, ISSN 0897-4756

Drew, A. J., Niedermeyer, C., Baker, P. J., Pratt, F. L., Blundell, S. J., Lancaster, T., Liu, R. H., Wu, G., Chen, X. H., Watanabe, I., Malik, V. K., Dubroka, A., Rossie, M., Kim, K. W., Baines, C., & Bernhard, C. (2009). Coexistence of static magnetism and superconductivity in $SmFeAsO_{1-x}F_x$ as revealed by muon spin rotation. *Nature Materials,* Vol. 8, No. 4, (April 2009), pp. 310-314, ISSN 1476-1122

Feldkemper, S., Weber, W., Schulenburg, J., Richter, J. (1995). Ferromagnetic coupling in non-metallic Cu^{2+} compounds. *Phys. Rev. B,* Vol. 52, No. 1, (July 1995), pp. 313-323, ISSN 0163-1829

Foo, M. L., Wang, Y. Y., Watauchi, S., Zandbergen, H. W., He, T., Cava, R. J., Ong, N. P. (2004). Charge ordering, commensurability, and metallicity in the phase diagram of the layered Na_xCoO_2. *Phys. Rev. Lett.*, Vol. 92, No. 24, (June 2004), pp. 247001-1-4, ISSN 0031-9007

Geick, R. (2001). Halide Perovskite-type Layer Structures (Subvolume J3), In:. *Group III: Condensed Matter – Volume 27, Magnetic Properties of Non-Metallic Inorganic compounds based on transition elements (Landolt-Börnstein: Numerical Data and Functional Relationships in Science and Technology),* H. P. J. Wijn, pp. 159-161, 188-190, 254, Springer-Verlag, ISBN 1616-9549, Berlin, Germany

Hayden, S. M., Aeppli, G., Osborn, R., Taylor, A. D., Perring, T. G., Cheong, S. W., Fisk, Z. (1991). High energy spin waves in La_2CuO_4. *Phys. Rev. Lett.*, Vol. 67, No. 25, (December 1991), pp. 3622-3625, ISSN 0031-9007

Hewitt, K. C., Casey, P. A., Sanderson, R. J., White, M. A. & Sun, R (2005). High-throughput resistivity apparatus for thin-film combinatorial libraries. *Rev. Sci. Instrum.,* Vol. 76, No. 9, (September 2005), pp. 093906-1-9, ISSN 0034-6748

Hubbard, J. (1963). Electron correlations in narrow energy bands. *Proc. Roy. Soc. Lon. A,* Vol. 276, No. 1365, (November 1963), pp. 257, ISSN

Imai, T., Ahilan, K., Ning, F. L., McQueen, T. M., Cava, R. J. (2009). Why does undoped FeSe become a high-T_c superconductor under pressure? *Phys. Rev. Lett.,* Vol. 102, No. 17, (May 2009), pp. 177005-1-4, ISSN 0031-9007

Ishizuka, M., Yamada, I., Amaya, K., & Endo, S. (1996). Change of magnetism in the two-dimensional Heisenberg ferromagnet K_2CuF_4 observed at high pressure. *J. Phys. Soc. Jpn.*, Vol. 65, No. 7, (July 1996), pp. 1927-1929, ISSN 0031-9015

Ishizuka, M. Terai, M., Endo, S., Hidaka, M., Yamada, I., & Shimomura, O. (1998). Pressure induced phase transition in the two-dimensional Heisenberg ferromagnet K_2CuF_4. J. *Magn. Mag. Mater.*, Vol. 177, No. 1, (January 1998), pp. 725-726, ISSN 0304-8853

Kennedy, K., Stefansky, T. Davy, G., Zakay, V. F. & Parker, E. R. (1965). Rapid Method for determining ternary-alloy phase diagrams. *J. Appl. Phys.*, Vol. 36, No. 12, (December 1965), pp. 3808-3810, ISSN 0021-8979

Kuiper, P., Kruizinga, G., Ghijsen, J., Grioni, M., Weijs, P. J. W., de Groot, F. M. F., Sawatzky, G. A., Verweij, H., Feiner, L. F., & Petersen, H. (1988). X-ray absorption study of the O 2p hole concentration dependence on O stoichiometry in $YBa_2Cu_3O_x$. *Phys. Rev. B*, Vol. 38, No. 10, (October 1988), pp. 6483-6489, ISSN 0163-1829

Leigh, R. G., Phillips, P., & Choy, T. P. (2007). Hidden charge 2e boson in doped Mott insulators. *Phys. Rev. Lett.*, Vol. 99, No. 4, (July 2007), pp. 046404-1-4, ISSN 0031-9007

Levi, B. G. (2009). Superconducting qubit systems come of age. *Physics Today*, Vol. 62, No. 7, (July 2009), pp. 14-16, ISSN 0031-9228

Logvenov, G., Sveklo, I., & Bozovic, I. (2007). Combinatorial molecular beam epitaxy of $La_{2-x}Sr_xCuO_{4+\delta}$. *Physica C*, Vol. 460, (September 2007), pp. 416-419, ISSN 0921-4534

Luetkens, H., Klaus, H. H., Kraken, M., Litterst, F. J., Dellmann, T., Klingeler, R., Hess, C., Khasanov, R., Amato, A., Baines, C., Kosmala, M., Schumann, O. J., Braden, M., Hamann-Borrero, J., Leps, N., Kondrat, A., Behr, G., Werner, J., & Buchner, B. (2009). The electronic phase diagram of the $LaO_{1-x}F_xFeAs$ superconductor. *Nature Materials*, Vol. 8, No. 4, (April 2009), pp. 305-309, ISSN 1476-1122

Monthoux, P. & Lonzarich, G. G. (1999) p-wave and d-wave superconductivity in quasi-two-dimensional metals. *Phys. Rev. B*, Vol. 59, No. 22, (June 1999), pp. 14598-14605, ISSN 0163-1829

Monthoux, P., & Lonzarich, G. G. (2001). Magnetically mediated superconductivity in quasi-two and three dimensions. *Phys. Rev. B*, Vol. 63, No. 5, (February 2001), pp. 054529-1-10, ISSN 0163-1829

Monthoux, P, Pines, D., Lonzarich, G. G. (2007). Superconductivity without phonons. *Nature*, Vol. 450, No. 7173, (December 2007), pp. 1177-1183, ISSN 0028-0836

Monthoux, P, & Lonzarich, G. G. (2002). Magnetically mediated superconductivity: Crossover from cubic to tetragonal lattice. *Phys. Rev. B*, Vol. 66, No. 22, (December 2002), pp. 224504-1-8, ISSN 1098-0121

Mott, N. (1968) Metal-Insulator Transitions. *Rev. Mod. Phys.*, Vol. 40, No. 4, (October 1968), pp. 677-683

Norman, M. R., Pines, D., & Kallin, C. (2005). The Pseudogap: friend or foe of high T_c? *Advances in Physics*, Vol. 54, No. 8, (December 2005), pp. 715-733, ISSN 0001-8732

Petrovic, C., Pagliuso, P. G., Hundley, M. F., Movshovich, R., Sarrao, J. L., Thompson, J. D., Fisk, Z., & Monthoux, P. (2001). Heavy-fermion superconductivity in $CeCoIn_5$ at 2.3

K. *J. Phys.:Condens. Mat.*, Vol. 13, No. 17, (April 2001), pp. L337-L342, ISSN 0953-8984

Saadat, M., George, A. E., & Hewitt, K. C. (2010). Densely mapping the phase diagram of cuprate superconductors using a spatial composition spread approach. *Physica C*, Vol. 470, No. Sp. Iss. SI Suppl. 1, (December 2010), pp. S59-S61, ISSN 0921-4534

Sander, D. (1999). The correlation between mechanical stress and magnetic anisotropy in ultrathin films. *Rep. Prog. Phys.*, Vol. 62, No. 5, (May 1999), pp. 809-858, ISSN 0034-4885

Sanderson, R. J. & Hewitt, K. C. (2007). Magnetron sputter deposition of a 48-member cuprate superconductor library: $Bi_2Sr_2Y_xCa_{1-x}Cu_2O_{8+\delta}$ ($0.5 \leq x \leq 1$) linearly varying in steps of $\Delta x = 0.01$. *Appl. Surf. Sci.*, Vol. 254, No. 3, (November 2007), pp. 760-764, ISSN 0169-4332

Sanderson, R. J. & Hewitt, K. C. (2005) Stoichiometry control of magnetron sputtered $Bi_2Sr_2Ca_{1-x}Y_xCu_2O_y$ ($0 \leq x \leq 0.5$), thin film, composition spread libraries: Substrate bias and gas density factors. *Physica C*, Vol. 425, No. 1-2, (September 2005), pp. 52-61, ISSN 0921-4534

Scalapino, D. J. 9th International Conference on Materials and Mechanism of Superconductivity. Tokyo, Japan, Sept. 10 (2009)

Takabayashi, Y., Ganin, A. Y., Jeglic, P., Arcon, D., Takano, T., Iwasa, Y., Ohishi, Y., Takata, M., Takeshite, N., Prassides, K., & Rosseinsky, M. J. (2009). The Disorder-free non-BCS superconductor Cs_3C_{60} emerges from an antiferromagnetic insulator parent state. *Science*, Vol. 323, No. 5921, (March 2009), pp. 1585-1590, ISSN 0036-8075

Takada, K., Sakurai, H., Takayama-Muromachi, E., Izumi, F., Dilanian, R. A., & Sasaki, T. (2003). Superconductivity in two-dimensional CoO_2 layers. *Nature*, Vol. 422, No. 6927, (March 2003), pp. 53-55, ISSN 0028-0836

Timusk, T. & Statt, B. (1999). The Pseudogap in high-temperature superconductors: An experimental survey. *Rep. Prog. Phys.*, Vol. 62, No. 1, (January 1999), pp. 61-122, ISSN 0034-4885

Tranquada, J. M., Sternlieb, B. J., Axe, J. D., Nakamura, Y., & Uchida, S. (1995). Evidence for stripe correlations of spins and holes in copper-oxide superconductors. *Nature*, Vol. 375, No. 6532, (June 1995), pp. 561-563, ISSN 0028-0836

Uemura, Y. J. (2009). Commonalities in phase and mode. *Nature Materials*, Vol. 8, No. 4, (April 2009), pp. 253-255, ISSN 1476-1122

Van Dover, R. B., Schneemeyer, L. F., & Fleming, R. M. (1998). Discovery of a useful thin-film dielectric using a composition-spread approach. *Nature*, Vol. 392, No. 6672, (March 1998), pp. 162-164, ISSN 0028-0836

Xiang, X. D., Sun, X.-D., Briceno, G., Lou, Y. L., Wang, K. A., Chang, H. Y., Wallacefreedman, W. G., Chen, S. W., & Schultz, P. G. (1995). A Combinatorial approach to materials discovery. *Science*, Vol. 268, No. 5218, (June 1995), pp. 1738-1740, ISSN 0036-8075

Xiang, X-D. (1999). Combinatorial materials synthesis and screening: An integrated materials chip approach to discovery and optimization of functional materials. *Annu. Rev. Mater. Sci.*, Vol. 29, pp. 149-171, ISSN 0084-6600

Zhao, J., Huang, Q., de la Cruz, C., Li, S. L., Lynn, J. W., Chen, Y., Green, M. A., Chen, G. F., Li, G., Li, Z., Luo, J. L., Wang, N. L., & Dai, P. C. (2008). Structural and Magnetic phase diagram of $CeFeAsO_{1-x}F_x$ and its relation to high-temperature superconductivity. *Nature Materials,* Vol. 7, No. 12, (December 2008), pp. 953-959, ISSN 1476-1122.

12

Superhard Superconductive Composite Materials Obtained by High-Pressure-High-Temperature Sintering

Sergei Buga, Gennadii Dubitsky, Nadezhda Serebryanaya, Vladimir Kulbachinskii and Vladimir Blank
Technological Institute for Superhard and Novel Carbon Materials, Ministry of Education and Science of the Russian Federation Russian Federation

1. Introduction

Superhard superconducting materials are of considerable interest for the creation of high pressure devices for investigating electrical and superconducting properties of various materials. The superconducting composites consisting of superconductors and superhard materials that are in thermal and electrical contacts may satisfy very conflicting requirements imposed on superconducting materials for special research cryogenic technique, wear-resistive parts of superconductor devices, superconducting micro-electro-mechanical systems (MEMS), etc. The design of materials combining such properties as superconductivity, superhardness, and high strength is an interesting task for both scientific and applied reasons. Superconducting composites may be used for the production of large superconducting magnetic systems (Gurevich et al., 1987).

The discovery of superconductivity in heavily boron-doped diamonds (Ekimov et al., 2004; Sidorov et al., 2005) has attracted much attention. Superconducting diamonds are the hardest known superconductors. The potential applications of superconducting diamonds are broad, ranging from anvils in research high-pressure apparatus to supecronducting MEMS. However, the highest value of the superconductivity onset temperature in boron-doped diamonds was found just about 7 K in thin CVD-grown films (Takano et al., 2004) and at about 4 K in bulk diamonds grown at high-pressure and high-temperature (Ekimov et al., 2004; Sidorov et al., 2005). In these pioneering works bulk polycrystalline diamonds with micron grainsize have been synthesized from graphite and B_4C composition (Ekimov et al., 2004) and graphite with 4 wt% amorphous boron (Sidorov et al., 2005). The synthesis have been carried out at 8-9 GPa pressure and 2500-2800 K temperature in both cases. Later Dubrovinskaya et al., 2006, carried out synthesis of graphite with B_4C composition at much higher pressure value 20 GPa but the same temperature of 2700K and found the superconducting state transition at lower temperature 2.4 - 1.4 K in the obtained doped polycrystalline diamonds. Due to the sharpening of the temperature interval of the superconductivity transition in magnetic field they suggested that superconductivity could arise from filaments of zero-resistant material. An alternative method for the creation of composite diamond superconductors was suggested by one of the authors of the present article, G. Dubitsky, who used sintering of diamond powders with molybdenum to fabricate special research high-pressure anvils with $T_C = 10$ K (Narozhnyi et al., 1988). Such a unique high-strength superconducting anvils for research high-pressure apparatus were employed for investigations of the pressure effect up to 22 GPa on the superconductor transition temperatures in the metallic high-pressure phase of GaP. Modern technologies for large-scale industrial powder diamonds and cubic boron nitride manufacturing provide an easy opportunity to produce a wide range of superhard sintered superconductors with various mechanical and electronic properties.

By sintering diamond micropowders with metal powders (Nb, Mo) and using metal-coated diamond micropowders at high static pressure and temperature we obtained superhard superconductors with T_C substantially higher than in boron-doped diamonds (Dubitsky et al., 2005, 2006). Interacting with diamond, Nb and Mo metals form carbides bonding diamond crystallites into a united compact material having relatively high critical temperatures of the transition to the superconducting state.

The alternative route is the sintering of superconductor powders with superhard fullerites - new carbon materials produced from C_{60} and C_{70} fullerenes (Blank et al., 1998, 2006). Under high pressure and temperature treatment soft C_{60} and C_{70} powders transform into fullerene polymers and other carbon structures with various hardness including superhard and even superior to diamond. There are known many alkali metal-fullerene superconductors with relatively high T_C up to about 30K (Holczer & Whetten, 1993, Kulbachinskii, 2004, Kulbachinskii et al., 2008). However alkali metal-fullerenes react with oxygen when exposed to air. Sintering with inert superhard materials may protect such compounds from oxidation and provide superconducting properties of such superhard composites.

The highest critical temperature of superconductor transition among known "regular" superconductors has magnesium diboride MgB_2 with T_C = 39K. The superconductor composites based on MgB_2 and superhard materials are promising materials as well (Kulbachinskii et al., 2010).

Using high-pressure-high-temperature sintering method we manufactured the following composite superhard superconducting materials: diamond-Nb, diamond-Mo, diamond-MgB_2, cubic boron nitride-MgB_2, fullerite C_{60}- MgB_2, diamond-$Ti_{34}Nb_{66}$, diamond-Nb_3Sn, what will be described in this chapter.

2. Experimental section

Experimental samples of the target materials were obtained by treatment at high static pressures and temperatures. The experiments were carried out using modified "anvils with cavity"-type high-pressure apparatus (Blank et al., 2007). Pressure value was calibrated by electrical resistance jumps in reference metals Ba (5.5 GPa), Bi (2.5, 2.7, 7.7 GPa), Pb (13 GPa) and ZnSe (13.7 GPa) at known phase transitions. The temperature graduation of the chambers was performed using Pt/Pt-10%Rh and W/Re thermocouples. The initial components were placed into a tantalum-foil shell of 0.1 mm thickness. Samples were heated by ac current through a graphite heater with a tantalum shell as a part of the sample system. The materials have been obtained at pressures in the range of 7.7 - 12.5 GPa and temperatures of 1373 - 2173 K. The heating time was 60 – 90 s. The samples were quenched under high pressure with a rate of 200 K per second. After pressure release the samples were extracted from the high-pressure cell. Small cylinder-shaped samples with a diameter of 4.5 mm and a height of 3.5 mm were obtained. The parallelepiped samples 3.9×2.51×1.54

mm^3 size were made by laser cutting and polishing. The examples of optical images of polished surfaces for 2 different samples are presented in Fig. 1.

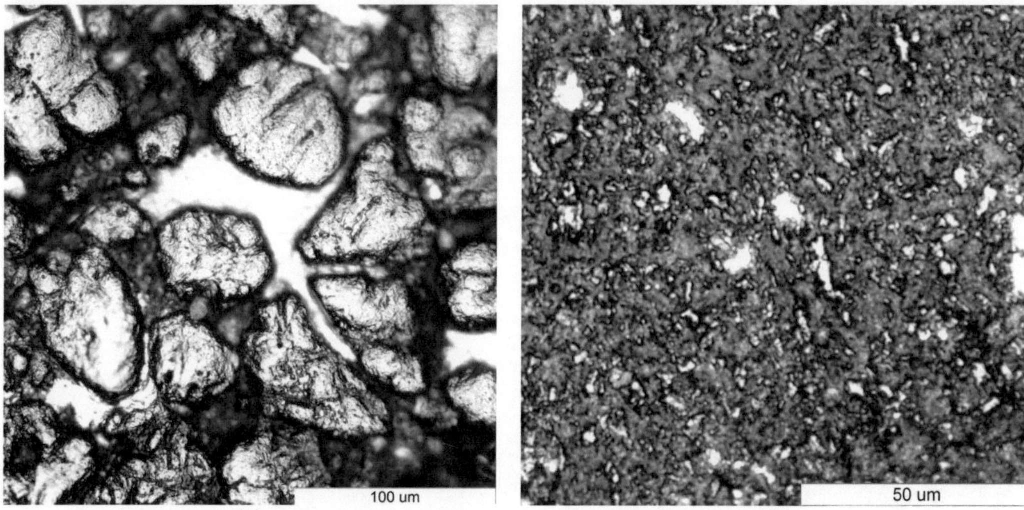

(a) (b)

Fig. 1. The examples of images of the polished surfaces of diamond-Nb sample (a) and micro-nanodiamod-NbTi sample (b).

The pictures show grains of superhard compound like diamonds and superconducting compounds like metal alloys in between the grains. The X-ray diffraction analysis revealed formation of metal carbides on the boundaries of diamond micro- and nanocrystals and nanocarbon phases originated from C_{60} fullerene in the samples synthesized with C_{60} powder. The carbide phases provide strong chemical bonding of superconductor matrix with superhard carbon grains, thus the target composites possess very high strength.

We investigated the phase content of the samples by the powder X-ray diffraction method using an ARL X'TRA diffractometer with Si-Li semiconductor detector and Cu Kα radiation (λ = 0.1541 nm) source. The electrical resistance of the resulting samples was measured by the conventional four probe method. Electrical contacts were made using conducting silver paste. The temperature dependence of the resistance was measured in the interval $1.7 < T < 300$ K. We applied the magnetic field up to 5T to determine the effect of magnetic field on the superconductivity transition temperature.

The Vickers microhardness was measured using PMT-3 device at an indenter load of 5.5 N. Sound velocities in 3 samples were measured using a wide-field pulse scanning acoustical microscope (WFPAM) in reflection mode (time-of-flight method) at a driving frequency of 50 MHz. The mean diameter of the acoustic beam in the specimen was about 0.1 mm. This experimental procedure was described in detail in (Prokhorov et al., 1999).

3. Superhard material-superconductive metal-systems

3.1 Diamond-niobium system

Synthetic diamond powder with 80 – 100 μm crystallites covered with a niobium film by vacuum sputtering was used as the initial material. The total amount of niobium in the initial material was 24 wt %. The experiments were carried out at a pressure of 7.7 GPa and a temperature of 1973 K for 60 s. The diffraction patterns exhibited peaks associated with diamond and NbC monocarbide (Fig. 2). A small fraction of NbO_2, practically traces, was also found. The NbC monocarbide synthesized at the boundaries of the crystallites had a face-centered cubic lattice with the cubic parameter a_0 = 0.447 nm. This value is consistent with the data obtained for NbC by another method (Toth, 1971).

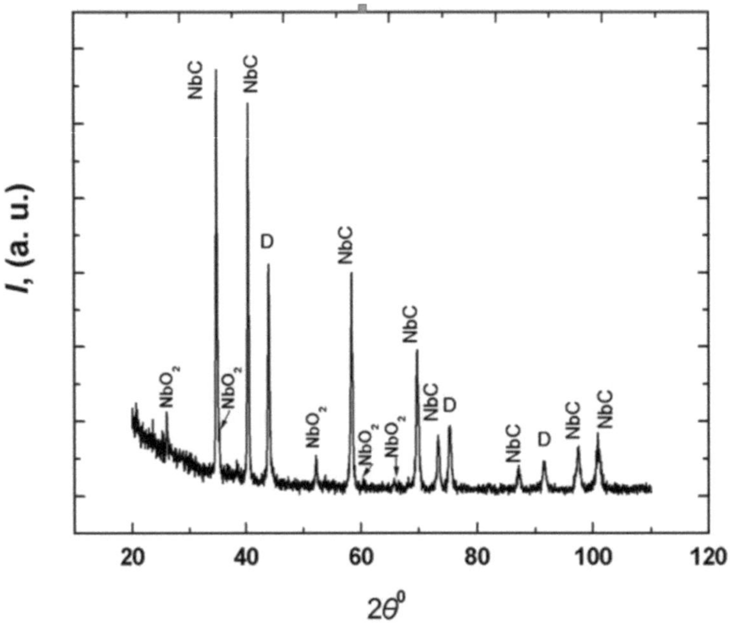

Fig. 2. X-ray diffraction pattern of sintered diamond-niobium sample. Diffraction reflections of NbC, diamond and NbO_2 are denoted.

The critical temperature of the transition to the superconducting state in all the measurements was fixed at the onset of the transition. According to an analysis of the temperature dependence of the resistance, the critical temperature of the transition of the synthesized samples to the superconducting state was equal to $T_C \approx 12.6$ K (Fig. 3a), which is characteristic of a NbC compound with almost full carbon positions (Karimov & Utkina, 1990; Shabanova et al., 1996; Krasnosvobodtsev et al., 1995). Carbon deficient NbC_{1-x} having lower T_C or superconductivity was entirely absent. The characteristic feature of the synthesized samples is a quite narrow superconducting transition, $\Delta T \approx 1.5$ K. The measured dependence of the resistance of a sample on the external magnetic field (Fig. 3b) is characteristic of NbC carbide. The value of the second critical field H_{C2} = 1.25 T (at T = 4.2 K) is consistent with H_{C2} for NbC films obtained by the laser-evaporation method (Shabanova et al., 1996). Thus, one may conclude that crystallites NbC with an almost perfect crystal structure were formed on the surface of the sintered diamond and a low concentration of defects has place in NbC which reduces the temperature of the superconducting transition (Pickett et al., 1986). An insignificant admixture of NbO_2 had no effect on the critical temperature of samples.

The microhardness values varied in the range of 35 –95 GPa depending on the actual place of indentation (Dubitski et al., 2005, 2006). According to (Toth, 1971), the Vickers microhardness of NbC is approximately equal to 17 GPa. This value is significantly lower than the results obtained in our work for microdomains enriched in NbC The hardness of

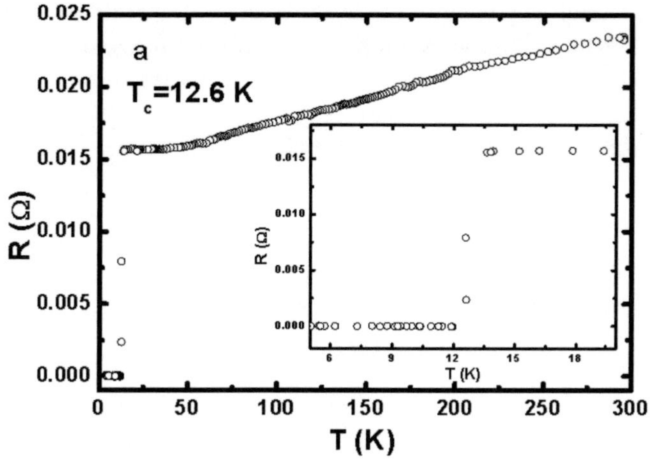

(a)

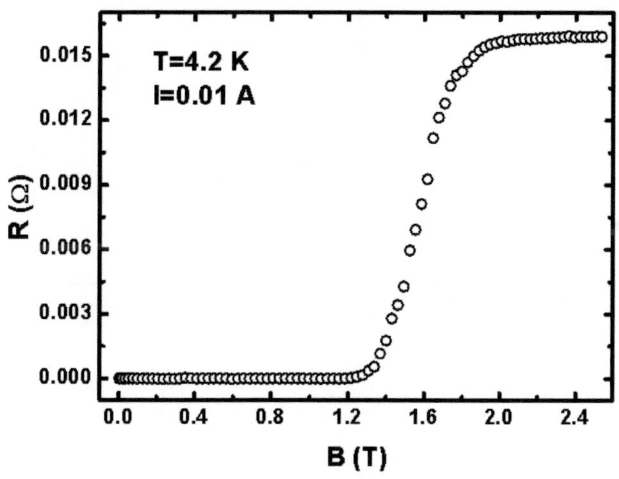

(b)

Fig. 3. Temperature (a) and magnetic field (b) dependences of the resistance for the sample obtained in the diamond-niobium system.

	ρ g cm^{-3}	V_L km s^{-1}	V_T km s^{-1}	E GPa	B GPa	G GPa	σ
Diamond-Nb	4.1± 0.1	12.1± 0.5	6.9± 0.25	490± 95	340± 30	195± 14	0.26± 0.04
Polycrystalline diamond "carbonado"	3.74± 0.05	16± 0.5	9.6± 0.3	850± 120	490± 30	340± 24	0.22± 0.04

Table 1. Diamond-Nb density ρ, longitudinal and transverse sound velocities V_L, V_T, Young modulus E, bulk and shear elastic modules B, G and the Poisson ratio σ. Data for polycrystalline sintered diamond "carbonado" type are given for reference.

such microdomains is high apparently due to the effect of the of diamond crystallites, which have much higher hardness (100 – 150 GPa along different faces, depending on the quality of the crystals). The velocities of sound and the elastic modules are 30 – 40% less than in pure bulk polycrystalline diamond (Table 1). Nevertheless they are comparable with the values for the next very strong superhard material: cubic boron nitride (c-BN).

3.2 C_{60}-diamond-niobium system

High-pressure - high-temperature treatment of C_{60} and C_{70} fullerenes leads to polymerization and transformation into new metastable carbon structures (Blank et al., 1998, 2006). Among various polymeric forms, the 3D-polymeric ones are the hardest

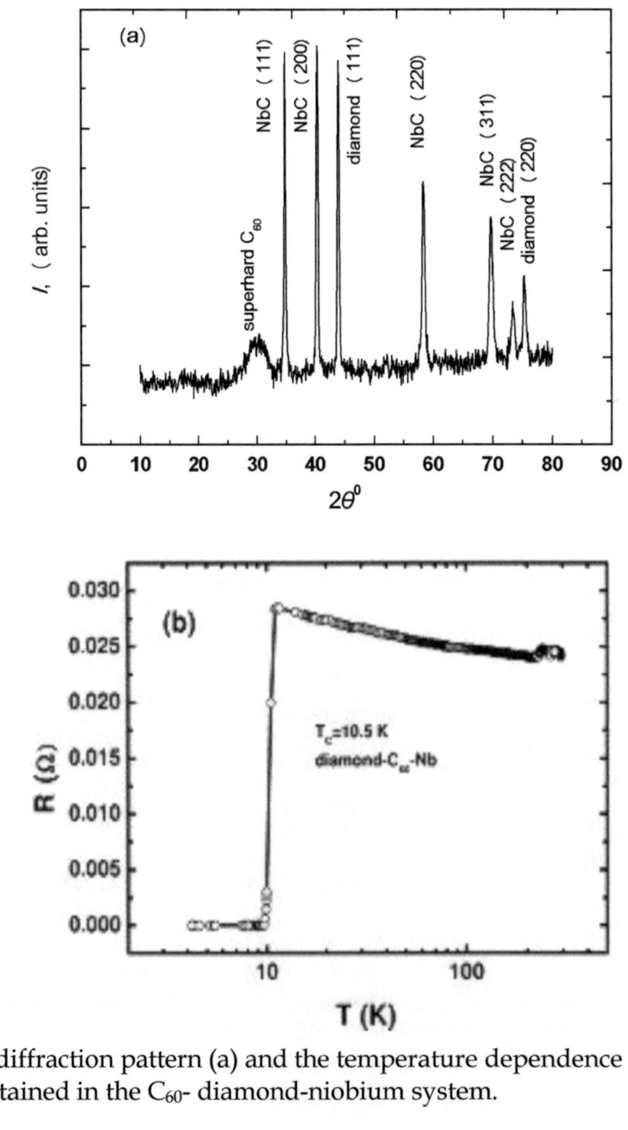

Fig. 4. The X-ray diffraction pattern (a) and the temperature dependence of the resistance (b) for the sample obtained in the C_{60}- diamond-niobium system.

(Blank et al., 1998). The density of superhard (H_v > 50 GPa) and ultrahard (H_v > 120 GPa) forms ranges from 2.5 to 3.3 g cm^{-3}, intermediate between the densities of graphite and diamond. Mechanical, electrical and other properties of such materials strongly depend on their particular crystalline or disordered structure. A set of metal-C_{60} compounds are superconducting with T_C = 2 - 34 K (Holczer & Whetten, 1993). However the conventional fullerene superconductors are not hard and chemically reactive; they oxidize in air. We investigated the synthesis of composite superconductors on the basis of a 50%-50% mixture of C_{60} powder with niobium coated diamond powder (Dubitsky et al., 2006) as described in the part above. Sintering was carried out at a pressure of 12.5 GPa and temperature 1650 K. The XRD analysis (Fig. 4a) showed formation of a disordered superhard component on the basis of C_{60} (broad diffraction peak) and niobium monocarbide with diamond crystals. The microhardness values of this composite material was 45 - 95 GPa. It is superconducting below 10.5 K (Fig. 4b).

3.3 Diamond-molybdenum system

In this system, a synthetic-diamond powder with a granularity of 40 - 100 μm and a molybdenum powder with a particle size of 1 - 5 μm were used as the initial materials consisting of 60 wt % of diamond and 40 wt % of molybdenum. A compact material was obtained by holding at a pressure of 7.7 GPa and a temperature of 2173 K for 90 s. The phase content of the samples was determined by the same method as for the diamond-Nb system. The following phases were identified in the samples: the diamond, the α-MoC phase with a cubic lattice with the parameter a_0 = 0.427 nm (B1 type), the η-MoC hexagonal phase with the lattice parameters a_0 = 0.300 and c_0 = 1.452 nm, and traces of the hexagonal γ-MoC phase (WC type) with the parameters a_0 = 0.290 and c_0 = 0.282 nm. The lattice parameters that were determined for molybdenum-carbon compounds are consistent with the data published by (Toth, 1971). The Vickers microhardness of the samples varies in the range of 27 - 83 GPa.

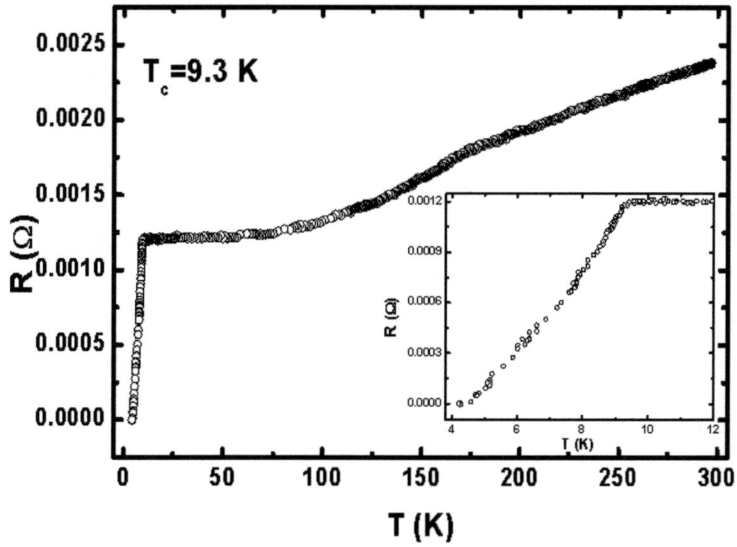

Fig. 5. Temperature dependence of the resistance for the sample obtained in the diamond-molybdenum system.

The composites obtained in the reaction of diamonds with molybdenum are superconductors with characteristic features. First, the onset of the transition to the superconducting state is T_C = 9.3 K, which is slightly lower than the T_C values for molybdenum carbide obtained by sintering powders of molybdenum and graphite (Willens et al., 1967). Second, the transition width $\Delta T \approx 5$ K is larger than that for the diamond-niobium system. Fig. 5 shows the temperature dependence of the resistance of these composites.

4. Superhard material-MgB_2-systems

4.1 Diamond-MgB_2 and cubic boron nitride-MgB_2-systems

Magnesium diboride, whose superconductivity was discovered 10 years ago (Nagamatsu et al., 2001; Zenitani & Akimitsu, 2003) has much higher critical temperature T_C = 39 K than niobium carbide and molybdenum carbide. Many investigations were devoted to the effect of the conditions of MgB_2 producing and treatement at high pressures and temperatures on the superconducting properties with various dopants (Jung et al., 2001; Prikhna et al., 2002; Pachla et al., 2003; Tampieri et al., 2004; Toulemonde et al., 2003; Zhao et al., 2003; Dou et al., 2003). It is of interest to obtain a superconducting composite material in which diamond or cubic boron nitride are used as the superhard components and MgB_2 is used as the superconducting component.

As the initial material, we used industrial MgB_2 powders in which the content of the basic product was equal to 98.5%. The particle size was reduced to 5 – 10 μm by additional powdering. The prepared mixtures consisted of 80 wt % of the superhard component and 20 wt % of MgB_2. The granularity of the diamond and cubic boron nitride powders was equal to 40 – 100 and 28 – 40 μm, respectively. The assembly of the high-pressure cells and the experimental procedure were the same as those used for the diamond-molybdenum system. In one of the experiments, a niobium-coated diamond powder was used. The samples were obtained by sintering at a pressure of 7.7 GPa and a temperature of 1373 K for 60 s.

Using XRD analysis, diamond, MgB_2 and MgO were identified in the system diamond-MgB_2 after the synthesis.

The temperature dependence of the resistance of the samples shows that the temperature of the transition to the superconducting state is $T_C \approx 37$ K (Fig. 6), which is close to the value known for MgB_2 (Nagamatsu et al., 2001). This closeness indicates that MgB_2 has a key role in the superconductivity of these composite materials, and the matrix consisting of cubic boron nitride or diamond changes Tc insignificantly, while the hardness of such superconducting material is much higher than the one of compacted MgB_2.

The microhardness of the samples indicates that the composite matrix (consisting of cubic boron nitride or diamond) occupying the major part of the body of the samples has a

	ρ, g cm^{-3}	V_L, km/s	V_T, km/s	E, GPa	B, GPa	G, GPa	σ	H_v,
Diamond-MgB_2	3.4±0.1	5.6±0.3	3,3±0.15	90±20	57±4	35±3	0.23±0.04	25÷78
cBN - MgB_2	3.3±0.1	6.0±0.3	3.7±0.15	110±25	60±7	45±4	0.19±0.04	24÷57

Table 2. Diamond- MgB_2 and cBN-MgB_2 composites density ρ, longitudinal and transverse sound velocities V_L, V_T, Young modulus E, bulk and shear elastic modules B, G and Poisson ratio σ and the Vickers microhardnes H_v.

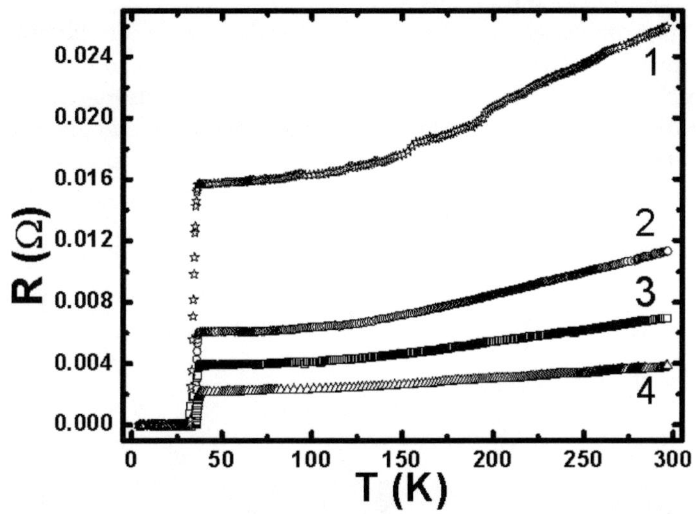

(a)

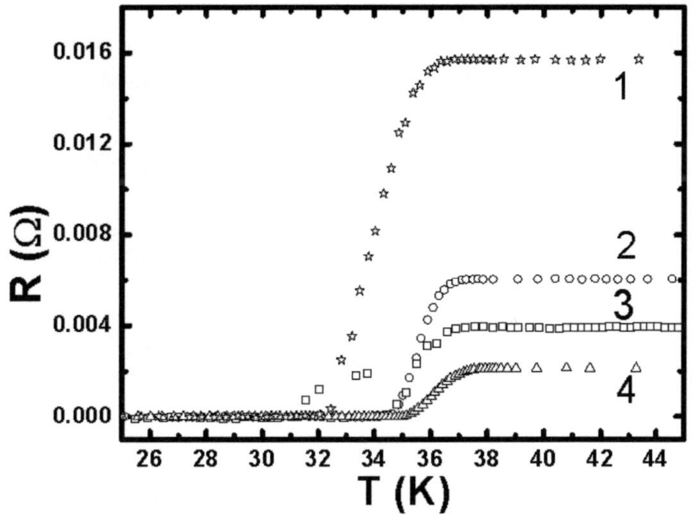

(b)

Fig. 6. Temperature dependence of the resistance (a) and transition range (b) for the composite samples obtained in the systems (1) cubic boron nitride-MgB_2, T_C = 36.1 K, (2) diamond-MgB_2, T_C= 36.9 K, (3) MgB_2, T_C = 37 K, and (4) diamond-niobium-MgB_2, T_C = 37.5 K.

microhardness of 57 – 95 GPa. Such microhardness values are characteristic for superhard compact polycrystalline materials based on cubic boron nitride and diamond that are used to produce various abrasive and cutting tools (Shul'zhenko et al., 1987). The specific gravity and velocities of longitudinal and transverse sound waves were measured and elastic modules evaluated (Table 2). Though the elastic modules are not very high, such materials have good potential for applications.

4.2 Polymerized fullerite C_{60}-MgB_2- system

We synthesized and investigated a set of composite materials obtained from MgB_2 superconductor and C_{60} fullerite mixed in various bulk ratios (Kulbachinskii et al., 2010). The same commercial MgB_2 powder with 98.5% of MgB_2 and commercial C_{60} powder with 99,8% C_{60} and the rest 0,2% of other carbon components were used as the initial compounds for the synthesis. The superconducting superhard composite possesses superconductivity due to MgB_2 fraction and superhardness due to polymerized C_{60} part (Blank et al., 1998).

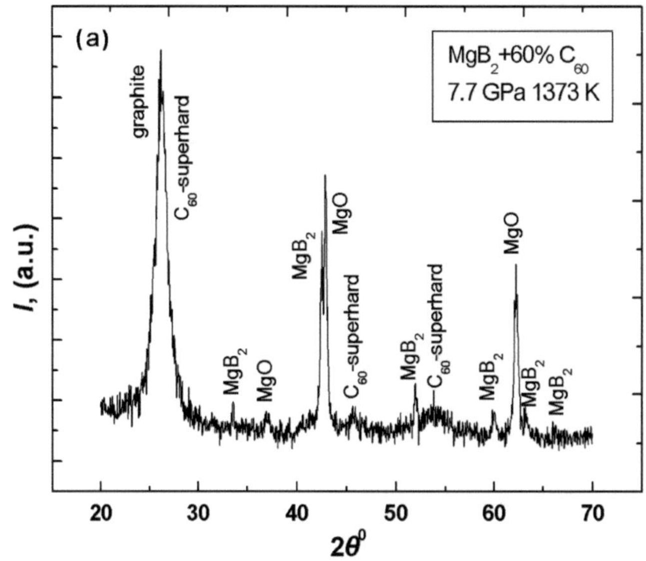

(a)

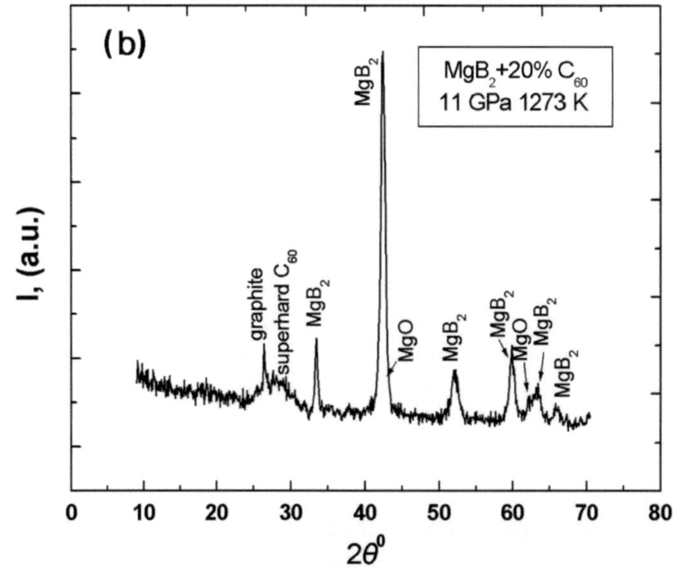

(b)

Fig. 7. Diffraction data composites MgB_2:C_{60}. Parameters of synthesis are shown at figures.

The particle size was reduced to 5–10 μm by additional powdering. We prepared polycrystalline composite MgB_2:C_{60} with different wt. content of C_{60} up to 60%. The samples of the materials were obtained at high static pressures 7.7 GPa and temperatures 1273 - 1373 K. The Vickers microhardness values of composites MgB_2-C_{60} were in the range of 18 - 59 GPa. The lowest value has been measured at the local point with dominating MgB_2 grains and the highest ones at the point with dominating superhard compound. According to (Prikhna et al., 2002), the Vickers microhardness H_V of MgB_2 with 2 wt. % of Ta synthesized at T_s=1073 K and P = 2 GPa was approximately equal to 12.8 GPa. In samples of composites

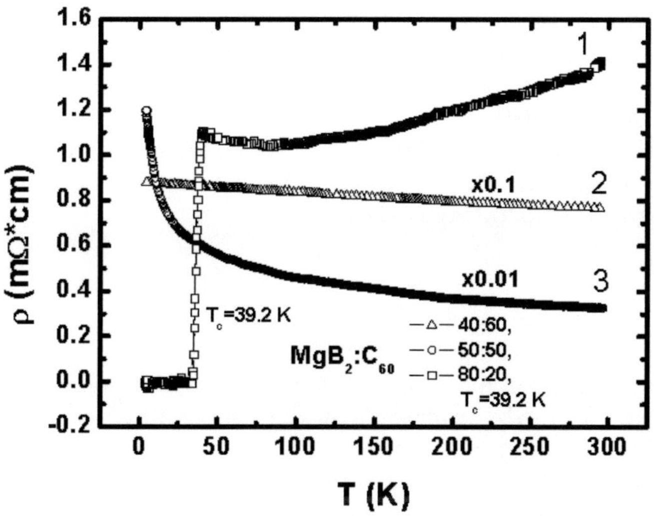

(a)

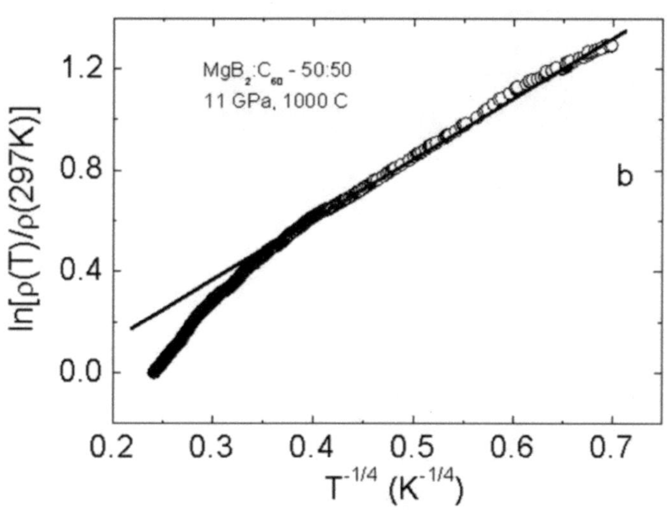

(b)

Fig. 8. a) Temperature dependence of resistivity for composites with different content MgB_2:C_{60}: 1 – 80:20; 2 – 40:60; 3 – 50:50 wt%; b) logarithm of the relative change of the resistivity versus $T^{-1/4}$.

$MgB_2:C_{60}$ we identified MgB_2, MgO and superhard 3D-polymerized C_{60}. MgO was revealed in the initial magnesium diboride, its content increased with the increasing of sintering temperature. In Fig. 7 diffraction data are shown for two samples with different $MgB_2:C_{60}$ ratio.

It was found that up to 20 wt. % C_{60} the composite $MgB_2:C_{60}$ is a superconductor. The superconducting transition temperature of composite $MgB_2:C_{60}$ – 80%:20% is T_C=39,2 K, that is the same as for host MgB_2 ($T_C \approx$ 39 K (Nagamatsu et al., 2001)). In Fig. 8 we plotted temperature dependences of resistivity for composites with different C_{60} content.

The resistivity of the superconducting composite $MgB_2:C_{60}$-80%:20% increases near the transition to the superconducting state (39<T<80 K). It is not typical for the host MgB_2. We suppose that in this temperature range C_{60} clusters play a significant role in the temperature dependence of the resistivity. For clusters of C_{60} the resistivity increases when temperature decreases (Buga et al., 2000, 2005).

When the content of C_{60} increases above 20 wt. %, the superconductivity disappears and the temperature dependence of the resistivity changes. Mott variable range hopping conductivity was observed, following the law:

$$\sigma = \sigma_0 \exp\left\{-(T_0 / T)^{1/4}\right\}$$

Fig. 8b. shows the relative change of resistivity in logarithmic scale on $T^{-1/4}$. The linear dependence is clearly visible in the temperature range 4.2- 60 K. When the content of C_{60} is >60 wt. % the resistivity slowly increases with the temperature decrease as it is shown in Fig. 8 for sample $MgB_2:C_{60}$-40%:60%. This takes place because the main superhard carbon component which appeared after C_{60} transformation has graphite-like cross-linked layered structure with semimetallic type of conductivity (Buga et al., 2000) .

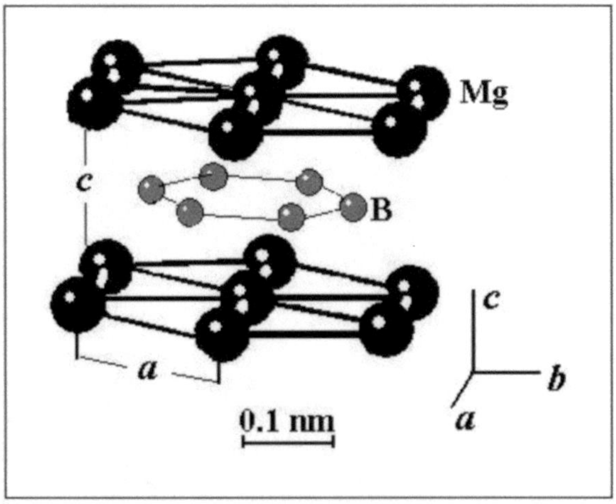

Fig. 9. The crystal structure of MgB_2. Axes of hexagonal unit cell *a* and *c* are designated.

MgB_2 has hexagonal crystal structure ((Nagamatsu et al., 2001). This structure (Fig. 9) consists from a sequence of hexagonal closed packed layers of Mg and graphite-like layers of B. In such structure atoms of Mg have ionic bonding with atoms of B. Inside the boron layer B atoms have covalent 2D bonding.

The critical field value B_c^c for magnetic field parallel to *c* axis is about 3–4 T down to T = 0 K (Lyard et al., 2002). The value of in plane critical magnetic field B_c^{ab} in the *ab* plane is significantly higher and reaches 15–20 T at T=0 K. The anisotropy coefficient is equal B_c^{ab}/B_c^c = 4–5. In the superconducting composite MgB_2:C_{60}-80%:20% we measured superconducting transition at different magnetic fields. (Fig. 10a). Using these data the temperature dependence of the critical magnetic field B_c was plotted in Fig. 10b. The temperature dependence of B_c is close to the same dependence in the initial MgB_2 (Lyard et al., 2002; Buzea & Yamashita, 2001). Thus composite with 20 wt. % of C_{60} has the same superconducting parameters as host MgB_2.

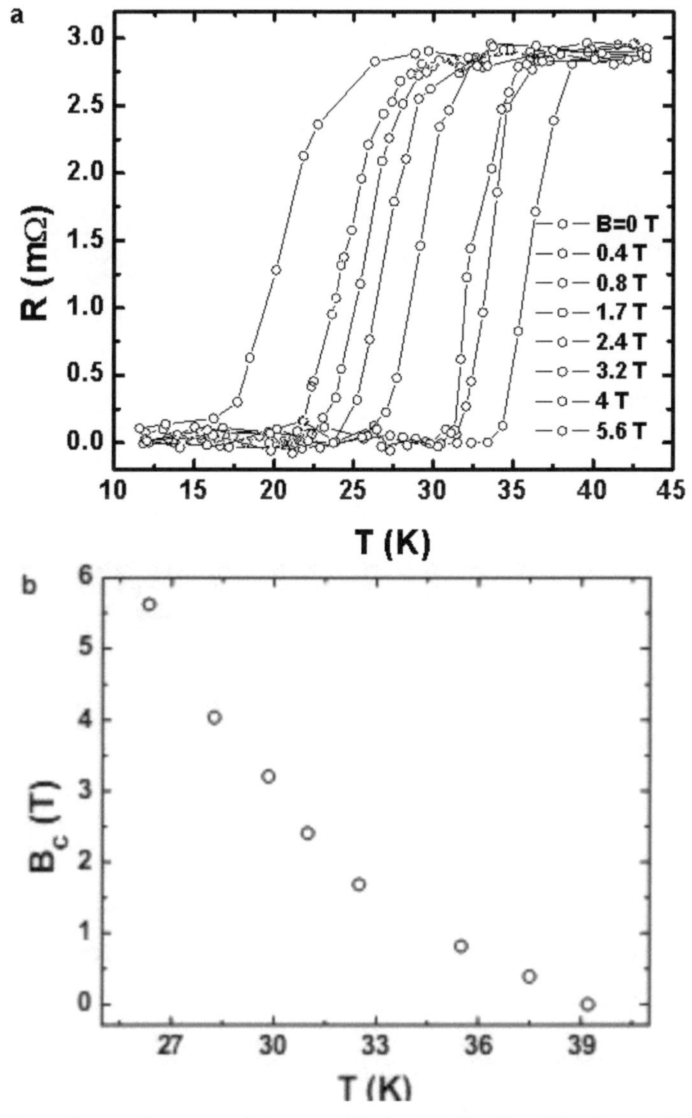

Fig. 10. Temperature dependence resistance of R for MgB_2:C_{60} - 80:20 wt. % at different magnetic field (a), and temperature dependence of critical magnetic field B_c (b) that is similar to that of the initial MgB_2.

4.3 Superconductivity of heterofullerides

Recently a method has been developed for synthesizing superconducting heterofullerides of the Fe and Cu groups with the composition K_2MC_{60} (Bulychev et al., 2004). For example, Fig. 11a shows the temperature dependence of the magnetic susceptibility χ for fullerides with M=Fe, Ni, Cu. Also shown in Fig. 11a is the $\chi(T)$ curve for the well-known superconductor K_3C_{60} for comparison. Superconducting heterofullerides of composition K_2MC_{60} have lower superconducting transition temperatures T_c than K_3C_{60} , and the crystal lattice constant a of

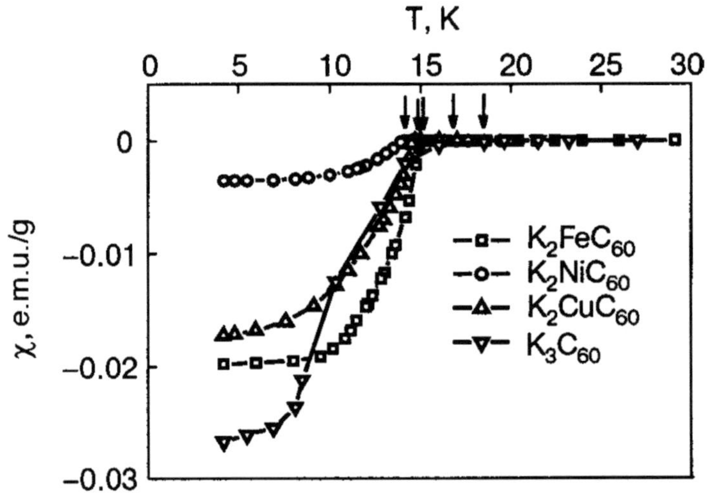

(a)

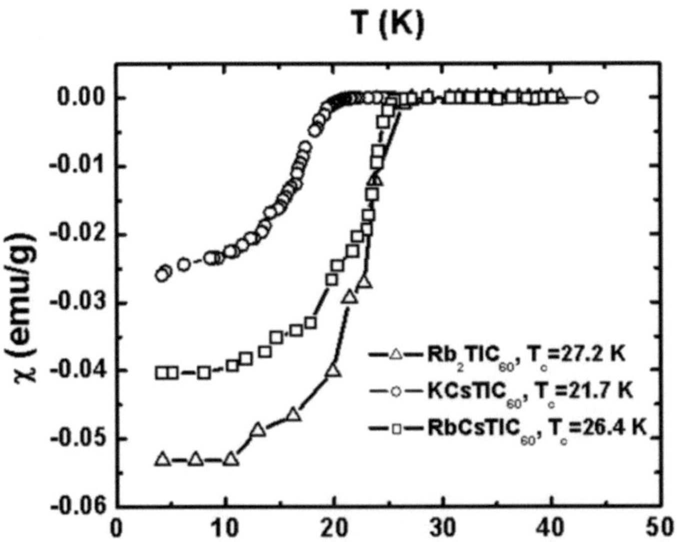

(b)

Fig. 11. Temperature dependence of the magnetic susceptibility x of fullerides of composition K_2MC_{60} (M = Fe, Ni, Cu) and K_3C_{60} for comparison (upper) and compounds $RbCsTlC_{60}$, $KCsTlC_{60}$, Rb_2TlC_{60} (down).

the heterofullerides studied is smaller than that of K_3C_{60} , apparently because substitution of the potassium ion by an ion of smaller diameter decreases the lattice constant. Thus there is a correlation between the superconducting transition temperature and the crystal lattice constant: a decrease of a leads to a decrease of T_c .

Fullerides $RbCsTlC_{60}$ and Rb_2TlC_{60} have superconducting transition temperature T_c = 26.4 K, and 27.2 K respectively. For $KCsTlC_{60}$, the value of T_c = 21.7 K is the highest for fullerides with potassium atoms. Temperature dependences of the magnetic susceptibility of these compounds are plotted in Figure 11b. The fulleride $KCsTlC_{60}$ possesses the highest T_c = 21.7 K among all synthesized in the present study of heterofullerides with potassium. This confirms the suggestion that the increase of the T_c value is due to the increase of the lattice constant caused by substitution of K and Rb by Cs. $KCsTlC_{60}$ has maximal value of a (a = 1.442 nm compared to K_3C_{60} a = 1.431 nm) among all superconducting heterofullerides with potassium. The same is for $RbCsTlC_{60}$ (a = 1.467 nm, Rb_2BeC_{60} a = 1.445 nm). Such appreciable increase of the parameter a of face-centred cubic lattice is quite naturally and also is one more proof of intercalation of atoms of bigger sizes in fulleride structure. There were no superconducting transitions in heterofullerides with more than one Cs atom per fullerene C_{60}. It is of substantial interest to investigate sintering of such compounds with superhard materials like diamond and cBN to learn if superhard compounds protect superconducting metal-fullerene compound against oxydation and provide high mechanical properties of the composite.

5. $Ti_{34}Nb_{66}$- and Nb_3Sn-diamond micropowder-systems

5.1 $Ti_{34}Nb_{66}$-diamond system

The NbTi alloy is a superconductor which has one of the best strength and high critical magnetic field. A set of new composite materials were synthesized from $Ti_{34}Nb_{66}$ and Nb_3Sn superconductors mixed with microcrystalline diamond and nanodiamond powders in various bulk ratios. The particle size of superconductors was reduced to 5–10 μm by powdering.

The initial $Ti_{34}Nb_{66}$ alloy (Fig. 12) has a body-centered cubic crystal structure with $I\,m\,\overline{3}\,m$ space group and the unit cell parameter a = 0.328 nm. The superconducting temperature T_C = 9.85 K is known for this alloy.

The crystal structure of $Nb_{34}Ti_{66}$ is a homogeneous solid solution with statistical distribution of elements in a unit cell. Figure 13 shows diffractograms of samples obtained after sintering of the mixture of diamond powder and superconductor under pressure of 7.7 GPa at temperatures 1373 K and 1623 K. The lower diffractogram No. 4 (Fig. 13a) was obtained from the initial alloy, the next one (No. 3) of the sample obtained after sintering at 1373 K. No reflections of TiC or NbC appeared at this synthesis temperature.

The next two samples were sintered at 1623 K. The diffractogram No. 2 corresponds to the sample $Nb_{34}Ti_{66}$ (50%) mixed with 45% diamond micropowder covered by Nb and 5% of nanodiamond is added in this composition. The diffractogram No. 1 corresponds to the sample $Nb_{34}Ti_{66}$ (50%) mixed with 50% diamond micropowder. We will discuss in detail the difference of diffraction patterns of these samples and how it may affect on the superconductive transition temperature.

At the dffractograms of samples No. 1 and 2 the diffraction peaks from Nb_{34} Ti_{66} alloy are easy visible because they have maximal amplitude. They are slightly shifted one to another and the shift is larger at wide angles ($2\theta \sim 90 \div 100^0$). A cubic unit cell parameter of $Ti_{34}Nb_{66}$

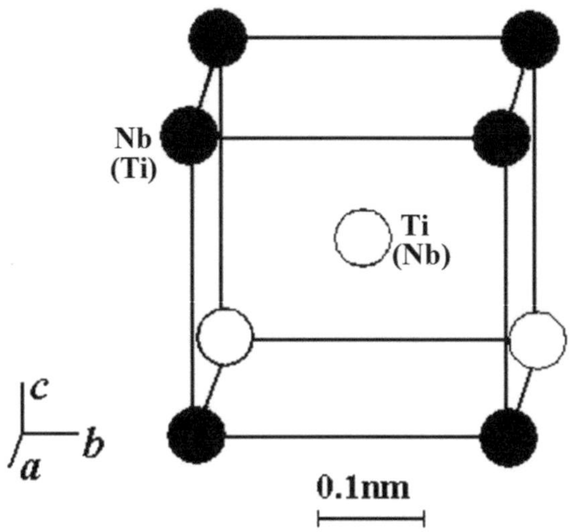

Fig. 12. The crystal structure of $Nb_{34}Ti_{66}$ alloy. Nb and Ti atoms have no determined positions.

alloy calculated with the wide angles (see fig. 13) a=0.329 nm of sample No. 1 is slightly less than a = 0.330 nm of sample No. 2. The intensities of the main peaks are different in these two samples. The intensities of TiC and NbC peaks are different in these samples as well. In sample No. 1 the intensity of TiC-reflections is higher than NbC-reflections while in sample No 2 the intensity of NbC-reflections is higher than TiC-reflections. As for example, at 2θ = $70 \div 75^0$ and $85 \div 90^0$ regions (Fig. 13b) NbC-reflections of diffractogram No. 1 disappear. The additional Nb covering diamond crystals in sample No. 2 leads to higher quantity of NbC in this sample. Apparently, 5% of nanodiamond in sample No. 2 increases a separation of Nb. Taking into account that the atomic radius of Ti (0.146 nm) is higher than the atomic radius of Nb (0.145 nm) and that the intensities of NbC - reflections increased, we suppose that the concentration of Nb in sample No. 1 is higher than in sample No. 2. That is why the cubic parameter is smaller. This leads to higher T_C in sample No 1 than in sample No. 2 (Fig. 14).

The appearance of NbC and TiC evidences partial decomposition of $Nb_{34}Ti_{66}$ alloy during the synthesis. Carbides were synthesized under high pressure 7.7 GPa without melting of metals though temperature of synthesis 1625 K is less than the necessary for direct synthesis of carbides. Their melting temperatures exceed 2000 K. The large part of carbon (50%) in samples promotes formation of carbides. However, the content of carbides is not sufficient to rise the T_C value up to 12.6 K known for NbC carbide.

Table 3 shows the Vickers microhardness of $Ti_{34}Nb_{66}$ –diamond composite samples No's 1 and 2. The value of microhardness has a wide range as well as in the other composites: the lowest value of hardness has been measured for the Ti-Nb-alloy located in superconductive channels (35 GPa in sample No. 1 and 42 GPa in sample No. 2) and the highest ones in superhard matrix (93-98 GPa). We suppose that the higher hardness of channels (42 GPa) in sample No. 2 has place due to nanodiamond fraction in this sample.

No. 1: 50% $Ti_{34}Nb_{66}$ + 50% diamond micropowder, T_C=8.9 K;

No. 2: 50% $Ti_{34}Nb_{66}$ + 45% diamond micropowder covered by metallic Nb + 5% nanodiamond, T_C=6 K.

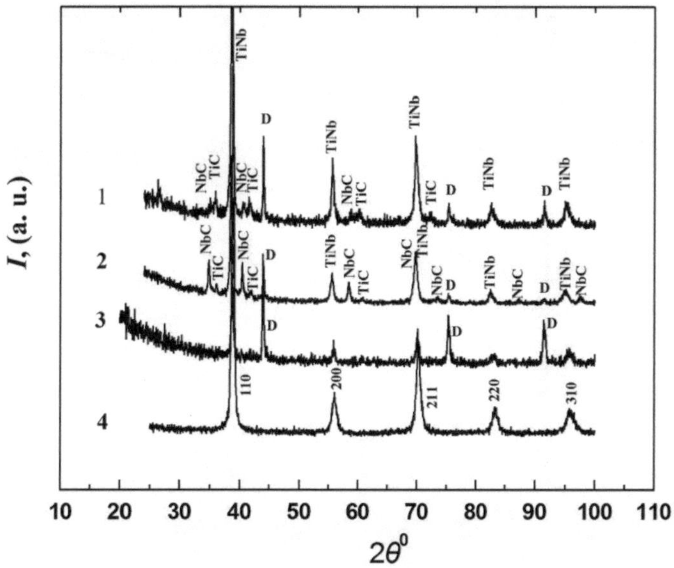

(a)

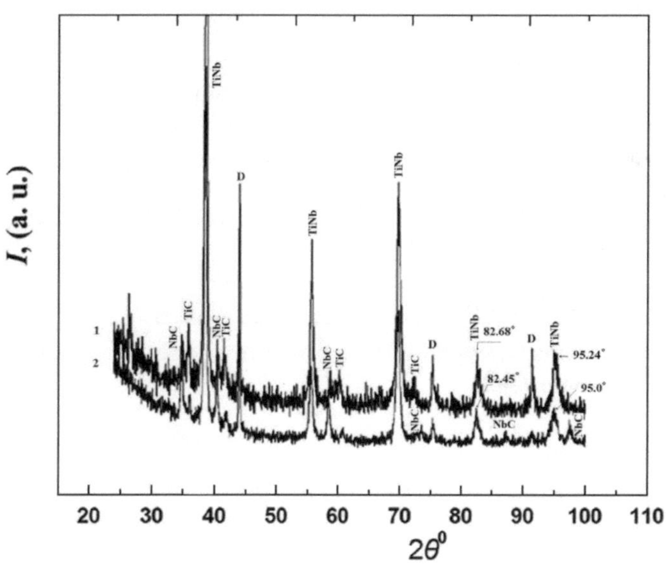

(b)

Fig. 13. (a) Diffractograms of ($Ti_{34}Nb_{66}$ + diamond micropowder) samples sintered at high pressure 7.7 GPa and high temperature 1373 K and 1623 K. The composition of samples: 1) 50% $Ti_{34}Nb_{66}$ + 50% diamond micropowder, T=1623 K; 2) 50% $Ti_{34}Nb_{66}$ + 45% diamond micropowder covered by Nb + 5% nanodiamonds, T=1623 K; 3) 50% $Ti_{34}Nb_{66}$ + 50% diamond micropowder, T = 1373 K; 4) initial $Ti_{34}Nb_{66}$ alloy.

(b) X-ray diffractograms of samples No's. 1 and 2.. D – diffractional reflections of diamond, NbC and TiC – diffractional reflections from Nb and Ti carbides. TiNb and the wide angles of $2\theta^0$ are denoted for $Ti_{34}Nb_{66}$ reflections.

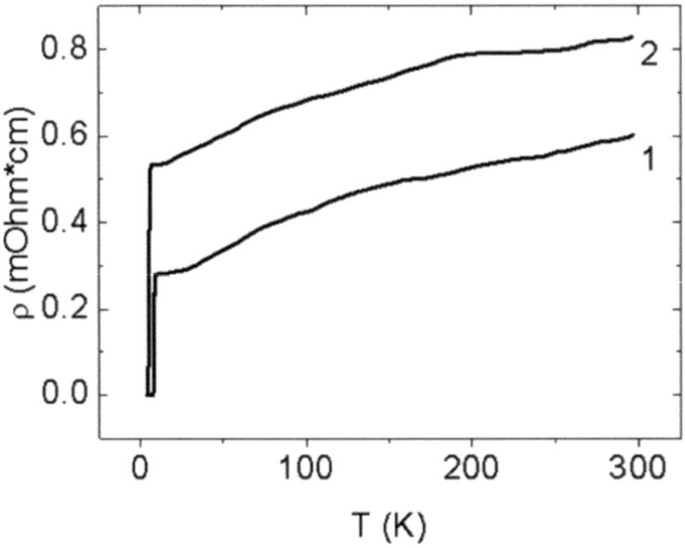

Fig. 14. Temperature dependence of resistance in two superhard composites sintered at T = 1623 K and P =7.7 GPa.

Sample No.	Composite	Component ratio, wt.%	Vickers microhardness, GPa	T_c, K
1	$Ti_{34}Nb_{66}$ -diamond micropowder	50:50	35-93	8.9
2	$Ti_{34}Nb_{66}$ -(diamond micropowder covered by Nb + 5% nanodiamond)	50:50 (45:5)	42-98	6

Table 3. The Vickers microhardness and the temperature of superconductive transition T_C of synthesized composite materials.

5.2 Nb_3Sn-diamod-system

An intermetallic Nb_3Sn compound crystallizes in cubic structure type A-15. Tin atoms are located in body-centered cubic positions, pairs of Nb atoms located on the cubic faces parallel to the coordinate axes (fig. 15). The unit cell contains 8 atoms: 2 Sn + 6Nb; the space group *Pm3n*, $a_{cub.}$ = 0.529 nm. Nb-atoms generate cross-cut chains (fig. 15a). The interatomic distance for Nb-atoms in one chain is appreciably less than the distance in the different chains. The chains of Nb-atoms respond for the generation of quasi one-dimensional electronic spectrum of *d*-state in this structure.

Nb_3Sn was mixed with micropowdered synthetic diamond and sintered at P = 7.7 GPa, T=1623 K. The diffraction pattern in Fig. 16 shows that under high temperature and pressure Nb_3Sn partially decomposes to atoms of Nb and Sn. Metallic Nb creates NbC, while Sn in sample is in metallic state. It is worth to note that the temperature of synthesis 1623 K is much less than the melting temperature of Nb_3Sn (2400 K). It means that niobium carbide was synthesized in solid state without melting of metal. The content of metallic Sn is very

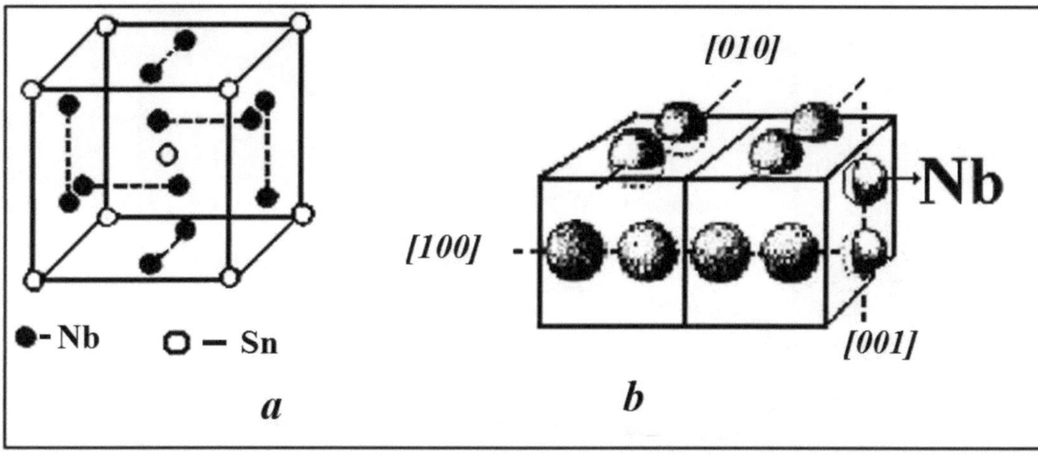

Fig. 15. The crystal structure of Nb_3Sn. *a* –Positions of atoms in the unit cell, *b* - chains of Nb-atoms. The coordinate axes are denoted in the *b*-part where two unit cell are painted.

small. Positions of diffractions peaks of Nb_3Sn correspond to the diffraction database (ICDD database PDF-2, card № 19-0875). Thus the decomposition of Nb_3Sn is insignificant in spite of high parameters of sintering. The composite Nb_3Sn with diamond powder is a superconductor with T_C about 15.5 K as it is shown in fig. 17. This value of T_C is close to the T_C of the initial Nb_3Sn.

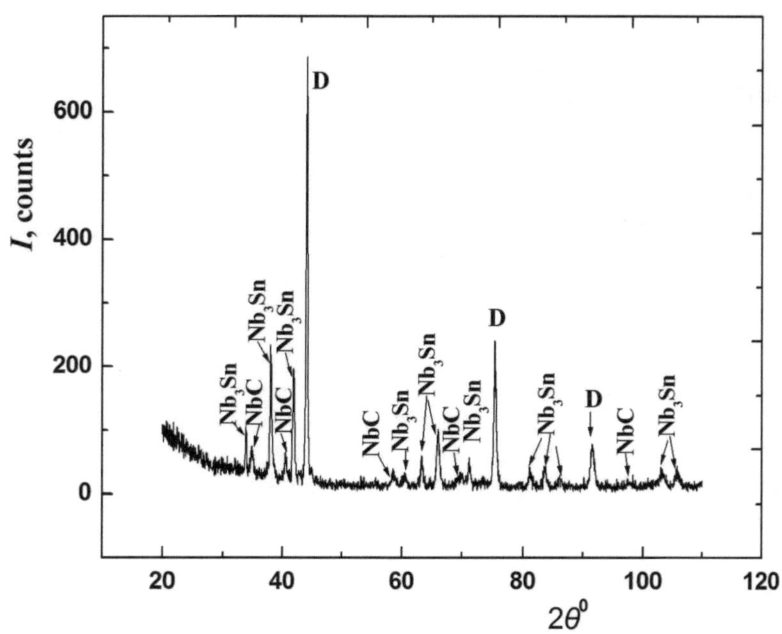

Fig. 16. X-ray diffraction pattern of (50%Nb_3Sn + 50% micropowder diamond) sample sintered at P = 7.7 GPa, T = 1625 K; D – reflections of diamond, NbC – reflections of niobium carbide.

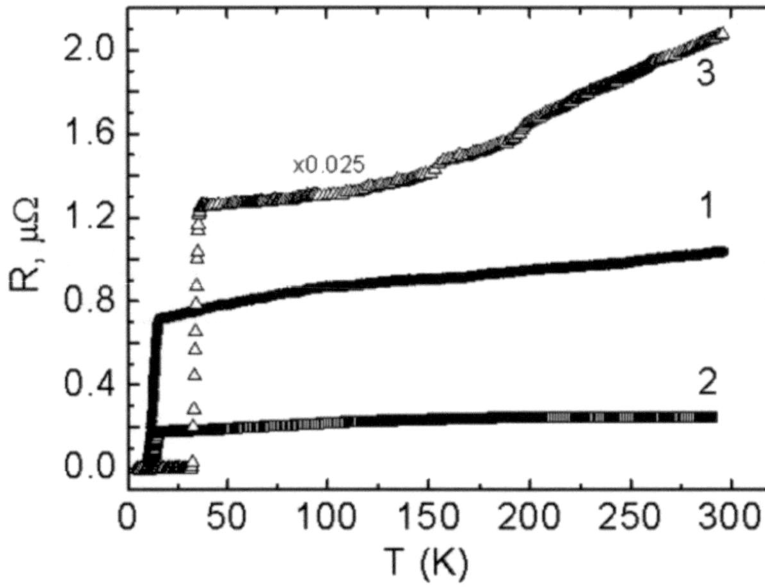

Fig. 17. Temperature dependence of resistance in different superhard composites: 1 – 50%Nb_3Sn + 50% micropowder diamond (T_C=15.6 K); 2 – 70%Nb_3Sn + 30% micropowder diamond (T_C=15.5 K) and 3 – MgB_2+cBN (T_C=36.5 K) for comparison.

6. Conclusion

The superhard superconducting composites are the new large family of materials for cryogenic electro-mechanical tools and devices. We employed high-pressure-high-temperature technique for synthesis of various superconducting composites on the basis of the hardest known materials: diamond, cubic boron nitride, C_{60}-fullerites. The best traditional superconductor alloys and relatively new, like MgB_2 have been used for synthesis to provide a superconductivity of the target materials. The structure and properties of the synthesized composites have been investigated. The highest values of microhardness up to 98 GPa and the highest elastic moduli have been found in diamond-based composites. Among them diomond-niobium composite is the hardest and it possesses 12.5K superconductor transition temperature. The highest T_C = 37.5 K has diamond-niobium-MgB_2 composite. The composites of superhard materials with conventional superconductor alloys like $Ti_{34}Nb_{66}$ and Nb_3Sn also possess superconductivity with the critical temperature 8.9 - 15.6 K. The optimal ratio of superconductor to superhad compounds in composites varies in the range from 20:80 to 50:50 wt%. The pressure and temperature parameters of synthesis are rather high: P = 7.7 - 12.5 GPa ; T_s = 1373 - 2173 K at the heating time τ = 60 – 90 s. However it may be supposed that with the increase of τ the pressure and temperature of synthesis may be reduced substantially. The X-ray diffraction analysis revealed formation of metal carbides on the boundaries of diamond micro- and nanocrystals and nanocarbon phases originated from C_{60} fullerene. The carbide phases provide strong chemical bonding of superconductor matrix with superhard carbon grains, thus the target composites possess very high strength. The obtained new composite materials can be successfully used in cryogenic electro-mechanical systems and in cryogenic research devices. The unique high-strength superconducting anvils for research pressure-induced apparatus were made and employed for investigations of the pressure effect up to 22 GPa on the superconductor transition temperatures in the metallic high-pressure phase of GaP.

7. References

Blank, V.; Buga, S.; Dubitsky, G.; Serebryanaya, N.; Popov, M. & Sundqvist, B. (1998). High-Pressure Polymerized Phases of C_{60}. *Carbon*, V. 36, No 4, (April 1998), pp. 319-343, ISSN 0008-6223

Blank, V.; Buga, S.; Dubitsky, G.; Serebryanaya, N.; Prokhorov, V.; Mavrin, B.; Denisov, V.; Chernozatonskii, L.; Berezina, S. & Levin, V. (2006). Synthesis of Superhard and Ultrahard Materials by 3D-polymerization of C_{60}, C_{70} Fullerenes under High Pressure (15 GPa) and Temperatures up to 1820 K. *Zeitschrift fur Naturforschung section B-A Journal of Chemical Sciences*, Vol. 61 b, No 12, (December 2006), pp. 1547-1554, ISSN 0932-0776

Blank, V., Buga, S., Dubitsky, G., Gogolinsky, K., Prokhorov, V., Serebryanaya, N.; Popov, V. (2007). High-Pressure Synthesis of Carbon Nanostructured Superhard Materials., In: *Molecular Building Blocks for Nanotechnology* , Mansoori, A., George, T., Assoufid, L., Zhang, G., pp. 393-418. Springer Science+Business Media LLC, ISBN-10: 0-387-39937-2, New York, U.S.A.

Buga, S.; Blank, V.; Dubitsky, G.; Edman, L.; Zhu, X.-M.; Nyeanchi, E. & Sundqvist, B. (2000). Semimetallic and Semiconductor Properties of Some Superhard and Ultrahard Fullerites in the Range 300–2 K. *Journal of Physics and Chemistry of Solids*, Vol. 61, No 7, (July 2000), pp. 1009-1015, ISSN 0022-3697

Buga, S.; Blank, V.; Serebryanaya, N.; Dzwilewski, A.; Makarova, T. & Sundqvist, B. (2005). Electrical Properties of 3D-Polymeric Crystalline and Disordered C_{60} and C_{70} fullerites. *Diamond and Related Materials*, Vol. 14, Issue: 3-7, Special Issue: Sp. Iss. SI, (Mart-July 2005), pp. 896-901, ISSN 0925-9635

Bulychev, B., Lunin, R., Krechetov, A., Kulbachinskii, V., Kytin, V., Poholok, K., Lips, K.; Rappich, J. (2004) *J. Phys. Chem. Solids* Vol. 65, (2004) pp. 337-343

Buzea, C. & Yamashita, T. (2001). Review of the Superconducting Properties of MgB_2. *Superconductor Science &Technology*, Vol. 14, No 11, (November 2001), pp. R115-R146, ISSN: 0953-2048

Dou, S.; Yeoh, W.; Horvat, J. & Ionescu, M. (2003). Effect of Carbon Nanotube Doping on Critical Current Density of MgB_2 Superconductor. *Applied Physics Letters*, Vol. 83, No 24, (15 December 2003), pp.4996-4998, ISSN 0003-6951

Dubitsky, G; Blank, V.; Buga, S.; Semenova, E.; Kul'bachinskii, V.; Krechetov, A. & Kytin, V. (2005). Superhard Superconducting Materials Based on Diamond and Cubic Boron Nitride. *JETP Letters*, Vol.. 81, No 6, (June 2005), pp. 323-326, ISSN 0021-3640

Dubitsky, G.; Blank, V.; Buga, S.; Semenova, E., Serebryanaya, N.; Aksenenkov, V.; Prokhorov, V.; Kul'bachinskii, V.; Krechetov, A. & Kytin, V. (2006). Superhard Superconductor Composites Obtained by Sintering of Diamond, c-BN and C_{60} Powders with Superconductors. *Zeitschrift fur Naturforschung section B-A Journal of*

Chemical Sciences, Vol. 61 b, No 12, (December 2006), pp. 1541-1546, ISSN 0932-0776

Dubrovinskaya, N., Eska, G., Sheshin, G., Braun, H.; (2006) Superconductivity in Polycrystalline Boron-doped Diamond Synthesized at 20 GPa and 2700K. *Journal of Applied Physics,* Vol. 99 (February 2006) pp. 0333903-1 -7, ISSN 0021-8979

Ekimov, E.; Sidorov, V.; Bauer, E.; Mel'nik N.; Curro, N.; Thompson, J. & Stishov S. (2004). Superconductivity in Diamond. *Nature,* Vol. 428, No 6982, (April 2004), pp. 542-545, ISSN 0028-0836

Gurevich, A.; Minc, R. & Rakhmanov, A. (1987). *Fizika kompositnyh sverhprovodnikov,* (Russian.), Nauka, Moscow, USSR

Holczer, K. & Whetten, R. (1993). Superconducting and Normal State Properties of the A_3C_{60} Compounds, In: *The Fullerenes,* Kroto, H.; Fischer, J. & Cox, D., pp. 123-138. Pergamon Press, ISBN 0-08-042152-0, Oxford, UK

Jung, C.; Park, M.-S.; Kang, W.; Kim, M.-S.; Kim, K.; Lee, S. & Lee, S.-I. (2001). Effect of Sintering Temperature under High Pressure on the Superconductivity of MgB_2. *Applied Physics Letters,* Vol. 78, No 26, (25 June 2001), pp. 4157-4159, ISSN 0003-6951

Karimov, Yu. & Utkina. (1990). Superconductivity of Nonstoichiometric Niobium Carbide. *Pis'ma v Zhurnal Eksperimental I Teoreticheskoi Fiziki,* Vol. 51, No. 9, (May 1990), 468-470. (*Journal Experimental and Theoretical Physics Letters,* Vol. 51, No. 9, (May 1990), pp. 528-531, ISSN 0021-3640

Krasnosvobodtsev, S.; Shabanova, N.; Ekimov, E.; Nozdrin, V. & Pechen', E. (1995). Critical Magnetic Field of NbC: New Data on Clean Superconductor Films. *Zhurnal Eksperimental i Teoreticheskoi Fiziki,* Vol. 108, No. 3(9), (September 1995), pp.970-976. (*Journal Experimental and Theoretical Physics,* Vol. 81, No. 3, (September 1995), pp. 534-537, ISSN 1063-7761

Kulbachinskii, V. (2004) . Electronic properties and superconductivity of low-dimensional carbon structures *Low Temperature Physics* Vol. 30 , No. 11, (November 2004), pp. 826-833.

Kulbachinskii, V, Bulychev, B., Kytin, V., Krechetov, A., Konstantinova, E.;. Lunin, R. (2008). Superconductivity, electron paramagnetic resonance, and Raman scattering studies of heterofullerides with Cs and Mg. *Advances in Condensed Matter Physics* Vol. 2008 (2008), Article ID 941372, 6 pages doi:10.1155/2008/941372

Kulbachinskii, V.; Buga, S.; Blank, V.; Dubitsky, G. & Serebryanaya N. (2010). Superconducting Superhard Composites Based on C_{60}, Diamond or Boron Nitride and MgB_2. *Journal of Nanostructured Polymers and Nanocomposites,* Vol. 6, No. 4, (2010), pp. 119-122, ISSN 1790-4439

Lyard, L.; Samuely, P.; Szabo, P.; Klein, T.; Marcenat, C.; Paulius, L.; Kim, K.; Jung, C.; Lee, H.; Kang, B.; Choi, S.; Lee, S.; Marcus, J.; Blanchard, S.; Jansen, A.; Welp, U.; Karapetrov, G. & Kwok, W. (2002). Anisotropy of the Upper Critical Field and Critical Current in Single Crystal MgB_2. *Physical Review B,* Vol. 66, No. 18, (November 2002), pp. 180502(4), ISSN 0163-1829

Nagamatsu, J.; Nakagawa, N.; Muranaka, T.; Zenitani, Y. & Akimitsu, J. (2001). *Nature,* Vol. 410, No. 6824, (March 2001), pp. 63-64, ISSN 0028-0836

Narozhnyi, V.; Stepanov, G.; Dubitsky, G.; Semenova, E. & Yakovlev, E. (1988). Kamera Vysokogo Davleniya iz sverhprovodyaschego Materiala (in Russian). *Fizika I Tehnika Vysokih Davlenii,* No. 27, (1988), pp. 88-91, ISSN 0203-4654

Pachla, W.; Kovac, P.; Diduszko, R.; Mazur, A.; Huek, I.; Morawski, A. & Presz, A. (2003). Supercondor Science and Technology, Vol. 16, No. 7, (January 2003), pp. 7-13, ISSN 0953-2048

Pickett, W.; Klein, B. & Zeller, R. (1986). Electronic structure of the carbon vacancy in NbC. *Physical Review B,* Vol. 34, No. 4 (August 1986), pp. 2517-2521, ISSN 1098-0121

Prikhna, T.; Gawalek, W.; Surzhenko, A.; Moshchil, V.; Sergienko, N.; Savchuk, Y.; Melnikov, V.; Nagorny, P.; Habisreuther, T.; Dub, S.; Wendt, M.; Litzkendorf, D.; Dellith, J.; Schmidt, C.; Krabbes, G. & Vlasenko, A.(2004) High-pressure Synthesis of MgB_2 with and without Tantalum Additions. *Physica C-Superconductivity and its Applications,* Vol. 372, (2004), pp. 1543-1545, ISSN: 0921-4534

Prokhorov, V.; Blank, V.; Buga, S. & Levin, V. (1999) Scanning Acoustic Microscopy Study of Superhard C_{60} Based Polimerized Fullerites. *Synthetic Metals,* Vol. 103, No. 1-3, pp. 2439-2442, ISSN 0379-6779

Shabanova, N.; Krasnosvobodtsev, S.; Nozdrin, V. & Golovashkin, A. (1996). Upper Critical Magnetic Field and Electron Characteristics of NbC, Nb_3Sn, $RBa_2Cu_3O_7$ Superconducting Compounds where (R=Y, Ho). *Fizika Tverdogo Tela,* Vol. 38, No. 7, (July 1996), pp. 1969-1985. (*Physics of the Solid State,* Vol.38, No. 7, pp.1085-, ISSN 1063-7834)

Shul'zhenko, A.; Ginsburg, B.; Hovah, N. & Pruss, A. (1987). New Superhard Materials abroad (reiew) (in Russian). *Sverhtverdye Materialy,* No. 1, pp. 23-28, ISSN 0203-3119.

Sidorov, V.; Ekimov, E.; Stishov, S.; Bauer, E. & Thompson, J. (2005). Superconducting and Normal-State Properties of Heavily Hole-Doped Diamond. *Physical Review B,* Vol. 71, No. 6, (June 2005), pp. 060502(4), Rapid Communications, ISSN 1098-0121

Tampieri, A.; Celotti, G.; Sprio, S.; Caciuffo, R. & Rinaldi, D. (2004). Study of the Sintering Behavior of MgB_2 Superconductor During Hot-Pressing. *Physica C,* Vol. 400, No. 3, (2004), pp. 97–104, ISSN 0921-4534

Toulemonde, P.; Musolino, N. & Flukiger, R. (2003). High-pressure Synthesis of Pure and Doped Superconducting MgB_2 compounds. *Superconductor Science & Technology,* Vol. 16, No. 2, (February 2003), pp. 231-236, ISSN 0953-2048

Toth, L. (1971) *Transition Metal Carbides and Nitrides.* Academic, New York, 1971. (Mir, Moscow, 1974)

Willens, R.; Boehler, E. & Matthias, B. (1967). Superconductivity of the Transition-Metal Carbides. *Physical Review,* Vol. 159, No. 2, (July 1967), pp. 327-330.

Zenitani, Y. & Akimitsu, J. (2003). Discovery of the New Superconductor MgB_2 and its Recent Development. *Association Asia Pacific Physical Societies Bulletin,* Vol. 13, No. 1, (February 2003), pp. 26-33, ISSN 0218-2203.

Zhao, Y.; Cheng, C.; Rui, X.; Zhang, H.; Munroe, P.; Zeng, H.; Koshizuka, N. & Murakami, M. (2003). Improved Irreversibility Behavior and Critical Current Density in MgB_2-Diamond Nanocomposites. *Applied Physics Letters,* Vol. 83, No. 14, (October 2003), pp. 2916-2918, ISSN 0003-6951.